THE JOURNAL OF EMBRYOLOGY
AND EXPERIMENTAL MORPHOLOGY

GROWTH AND THE DEVELOPMENT OF PATTERN

EDITED BY

R. M. GAZE, V. FRENCH, M. SNOW and D. SUMMERBELL

VOLUME 65 SUPPLEMENT

1981

Papers presented in relation to a discussion meeting held under the
auspices of the Company of Biologists Limited, at the
White House, Chelwood Gate, Sussex, U.K.,
from 26th to 29th June 1981.

CAMBRIDGE UNIVERSITY PRESS

CAMBRIDGE

LONDON   NEW YORK   NEW ROCHELLE

MELBOURNE SYDNEY

QL
971
.G84   Published by the Press Syndicate of the University of Cambridge
The Pitt Building, Trumpington Street, Cambridge CB2 1RP
32 East 57th Street, New York, NY 10022, USA
296 Beaconsfield Parade, Middle Park, Melbourne 3206, Australia

© Company of Biologists 1981

First published 1981

Printed in Great Britain at the
University Press, Cambridge

Library of Congress catalogue card number: 81-1545

British Library Cataloguing in Publication Data
Growth and the Development of Pattern. – (Journal of Embryology
and Experimental Morphology.
Supplement, ISSN 0022–0752; v. 65)
1. Embryology
I. Gaze, R. M.
591.3'3 QL955

ISBN 0 521 24557 5

Contents

J. Embryol. exp. Morph. Vol. 65 (Supplement), p. 1, 1981
Printed in Great Britain © Company of Biologists Limited 1981

FOREWORD

JEEM is owned by the Company of Biologists Ltd. This organization, the members of which are practising biologists, is registered as a charity and its aim is to promote knowledge and understanding of biology. One of the main activities of the Company of Biologists is the publication of three journals: this one, the *Journal of Experimental Biology* and the *Journal of Cell Science*.

This volume is the outcome of a decision by the Company of Biologists to support a discussion meeting on a topic to be chosen by the editors of *JEEM*. The meeting was held at the White House, Chelwood Gate, Sussex, from 25 to 30 May 1981. 'Growth and the Development of Pattern' comprises papers presented in relation to this discussion meeting. We hope that further such meetings may be arranged in due course.   R.M.G.

INTRODUCTION

Developmental biology is dominated by the reductionist approach. The idea that everything can ultimately be explained in terms of physics and chemistry, while valid in relation to the simpler manifestations of life, has been the cause of an unhelpful tendency to look for simplicity where it may not exist. This is particularly troublesome in relation to studies of the nervous system and of development; and it is relevant that in both cases the level of complexity is such that we are still arguing about the proper terminology to use to describe events. Embryological studies have now identified many of the main components of development, but we are in danger of losing something if we consider these in isolation. There has been a tendency to consider three-dimensional embryos as sets of independent axes, to make artificial dichotomies such as that between epimorphic and morphallactic regeneration, to use definitions as if they indicated understanding and to consider closely coupled phenomena as separate fields of study. The main theme of this meeting was specifically to examine the connexions between two aspects of development that have become somewhat dissociated over the years: growth and the development of pattern.

The discussions covered territory ranging from insects to man; from mitogens to abstract sets of rules; from anatomy to 19th century philosophy. Two general themes recurred. The first was that development is complicated and, while one can dissect it enough to identify some of the components, one should be wary of studying them in isolation. Growth and pattern are often inextricably linked and one can destroy all hope of understanding one by ignoring the other. The second point was that each of the different model systems investigated appeared to have a lot in common with most of the others. The same language (or at least, languages easily translatable one into the other) seems to be used in development throughout the animal kingdom.   V.F. M.S. D.S.

J. Embryol. exp. Morph. Vol. 65 (Supplement), pp. 3–18, 1981
Printed in Great Britain © Company of Biologists Limited 1981

Distal transformation in regenerating double anterior axolotl limbs

By DAVID L. STOCUM

From the Department of Genetics and Development, University of Illinois

SUMMARY

Previous investigations have shown that axolotl double anterior thighs regenerate only a tapered extension of the femur after simple amputation, but undergo intercalary regeneration of the femur and a symmetrical tibia when a normal wrist blastema is grafted to the thigh. There are two possibilities to account for these results: (1) distal transformation in both terminal and intercalary regeneration depends upon special properties of posterior cells (such as polarizing ability) that are lacking in the anterior half of the limb, but which would be provided by the posterior half of a normal wrist blastema graft, or (2) the pattern of cellular interactions required for distal transformation during terminal regeneration is different from that required for distal transformation during intercalary regeneration.

These alternatives were tested by grafting axolotl double anterior wrist blastemas to double anterior thighs. The host thighs regenerated a femur and symmetrical tibia that structurally were the same as the bones intercalated after grafting a normal wrist blastema to a double anterior thigh. In all cases, the graft developed as a symmetrical hand comprising one to three carpals and one to two digits. These results rule out the necessity of any special properties of posterior cells for distal transformation or polarization of the anterior–posterior axis and suggest instead that distal transformation during intercalary regeneration involves a different pattern of cellular interactions than in terminal regeneration.

INTRODUCTION

The rule of distal transformation is a formalization of the observation that cells of the amphibian limb regeneration blastema normally form only those structures distal to the level of amputation (Rose, 1962), even when regeneration takes place from a limb whose proximal–distal (PD) axis has been reversed (Butler, 1955). The cellular basis of this rule is that a continuous sequence of PD positional information (Wolpert, 1969, 1971) is recorded in the limb cells during ontogeny (Pescitelli & Stocum, 1981), and this cellular memory prevents blastema cells from forming morphological patterns proximal to their level of origin (Pescitelli & Stocum, 1980; Maden, 1980). This constraint ensures that the structural patterns formed during regeneration are always those which complete the limb distally.

Symmetrical morphological patterns are inimical in varying degrees to distal

---

[1] *Author's address:* Department of Genetics and Development, University of Illinois, 515 Morrill Hall, 505 South Goodwin Avenue, Urbana, Illinois 61801, U.S.A.

transformation (Bryant, 1976; Bryant & Baca, 1978; Stocum, 1978; Tank, 1978; Tank & Holder, 1978; Holder, Tank & Bryant, 1980). The most inhibitory pattern is a double anterior half (dA) upper arm or thigh. When these limbs are amputated 15–30 days after construction, they most often regenerate only a short cone of cartilage capped with soft tissue. In contrast, double posterior half (dP) thigh stumps amputated after the same healing interval are often able to regenerate proximodistally complete double posterior limbs which in the majority of cases are hypomorphic with regard to midline structures. Both dA and dP lower arms and shanks regenerate proximodistally complete limbs with loss of midline structures (Stocum, 1978; Krasner & Bryant, 1980).

The results of double half-limb regeneration have led to the formulation of a model in which distal transformation leading to restoration of the PD axis depends upon interactions between cellular positional values that specify pattern in the transverse plane (Bryant, 1978; Bryant & Baca, 1978; Holder, et al. 1980; Krasner & Bryant, 1980). In this model, which is a modification of the polar coordinate model of French, Bryant & Bryant (1976), the transverse limb pattern is represented by a series of numbers around the circumference of the limb and the PD axial pattern is represented by a series of letters along the radii of the circle formed by the circumference. Non-adjacent cells on the circumference (dermal connective tissue cells) are assumed to interact across short arcs to form a new complete circle by intercalary regeneration. The new circle, which lies inside the circumference, then adopts the next most distal PD (radial) value. Repetitions of this process restore all the missing PD values in a proximal to distal sequence.

The difference in the degree of distal transformation between dA and dP limbs is hypothesized to depend upon the extent of short-arc interactions between cells with the same or adjacent positional values on opposite sides of the midline. The pattern of short-arc interactions in dA limbs causes de-differentiated cells having the same or adjacent positional values to quickly confront each other across the midline, eliminating all but a few of the most anterior values, halting intercalation and, therefore, distal transformation. In contrast, short-arc interactions in dP limbs are postulated to take place in a pattern that does not lead to such midline confrontations, thereby allowing intercalation to continue until all the PD levels of the limb are restored (Holder et al. 1980; Krasner & Bryant, 1980).

Restoration of missing PD positional values also takes place in normal limbs when distal blastemas are grafted to a more proximal stump level (Iten & Bryant, 1975; Stocum, 1975). The intercalated structures are derived from blastema cells contributed by the host stump (Pescitelli & Stocum, 1980; Maden, 1980). Furthermore, dA thigh stumps will regenerate missing intermediate structures when normal wrist blastemas are grafted to them (Stocum, 1980a). The latter result suggests that the distal transformation taking place after this operation is not dependent upon interactions across short arcs in the transverse plane, but is the result of direct intercalation of missing intermediate PD positional values between the values represented by the amputation levels of graft and host. Intercalary regeneration between two PD levels and terminal regeneration after simple amputation may therefore depend on different patterns of cellular interactions.

However, there is another hypothesis which might account for the difference in degree of terminal regeneration between dA and dP limbs and for the fact that PD intercalary regeneration from dA thighs is evoked by normal wrist blastemas. Distal transformation may require some special activity that is the exclusive property of cells in the posterior half of the limb. Thus, amputated dP limbs and dA limb stumps provided with a normal wrist blastema would undergo distal transformation due to the presence of posterior cells. Other evidence suggests that a special zone of posterior cells (zone of polarizing activity) is involved in polarization of the anterior–posterior axis of the embryonic urodele and amniote limb, and a dorsal zone may also be involved in polarizing the DV axis (Swett, 1938; Slack, 1976, 1977, 1980a; Fallon & Crosby, 1977; Summerbell & Tickle, 1977).

The latter hypothesis can be tested by experiments in which dA wrist blastemas are grafted proximal to their level of origin to either normal or dA thigh stumps. The results of such experiments are the subject of this paper.

METHODS AND MATERIALS

White and dark axolotls (*Ambystoma mexicanum*), four months post-hatching and 60–70 mm in snout–tail tip length were used for all the experiments. In the first experiment, dA lower arms were surgically constructed on the right forelimbs of dark and white animals by replacing the posterior half of the lower arm with the anterior half of the left lower arm. The operations were done in petri dishes with the animals lying on paper towelling saturated with a 0·025 % (w/v) solution of Chlorotone (Eastman Chemical Co.) made up in 100 % Holtfreter solution, after first anaesthetizing the animals in a 0·075 % (w/v) solution of Chlorotone made up in 1 % Holtfreter solution. The donor and host limb halves were sufficiently adherent to one another after 2–4 h to transfer the animals to individual finger bowls containing 1 % Holtfreter solution, where they were fed every other day with freshly-hatched brine shrimp. The dA lower arms were allowed to heal for 10 days, and were then amputated through the doubled carpus or distal zeugopodium. The right (normal) hindlimb was amputated through the midthigh at the same time.

In the second experiment, dA lower arms were first constructed on the right forelimbs of dark and white animals. After a 10-day healing period, dA thighs were constructed on the right hindlimbs of the same animals by grafting the skin and muscle from the anterior half of the left thigh in place of the skin and muscle of the posterior half of the right thigh. At the end of a further 10-day

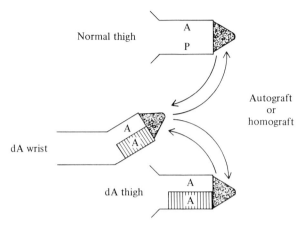

Fig. 1. Diagram illustrating the exchange between dA wrist blastemas and normal or dA thigh blastemas. Hatched areas represent the anterior (A) half of the left limb that was grafted in place of the posterior (P) half of the right limb.

healing period, the dA forelimbs were amputated through the wrist and the dA hindlimbs through the midthigh.

In both experiments, when medium bud to early redifferentiation blastemas appeared (staging according to Stocum, 1979), the dA wrist blastemas were exchanged as autografts or homografts with either the normal or dA thigh blastemas, as diagrammed in Fig. 1.

In a number of cases in both experiments, dA donor wrists did not receive a reciprocal thigh graft, but were allowed to regenerate a second time so that the structure of their regenerates could be compared to that formed by dA wrist blastemas grafted to the thigh.

All limbs were allowed to regenerate for 40–50 days, after which they were removed from the animal, fixed in Gregg's solution and stained for cartilage with methylene blue according to the Van Wijhe technique as modified by Gregg and Butler (Hamburger, 1960).

These experiments allowed an assessment of (1) the ability of normal and dA thigh blastemas to develop according to their origin on dA wrist stumps, and (2) the ability of dA wrist blastemas to evoke intercalary regeneration from normal or dA thigh stumps.

RESULTS

(1) *dA wrist re-regenerates*

Double anterior wrist stumps regenerating for the second time formed hands with one or two symmetrical digits and one to three symmetrically arranged carpals (Figs. 2, 3), and thus regenerated no differently from the grafted dA blastemas formed after the first amputation (see below).

(2) *Normal and dA thigh blastemas grafted to dA wrist stumps*

Three of five normal thigh blastemas developed according to origin when grafted to dA wrist stumps, forming a normal hindlimb distal to the wrist (Fig. 4). In one of the two remaining cases, the graft underwent resorption of prospective proximal structures, but redifferentiated a normal foot distal to the host zeugopodium. In the other case, the graft resorbed totally, and the host dA stump regenerated a short, tapered rod of cartilage.

The majority (7 of 12) of dA thigh blastemas grafted to dA wrist stumps also developed according to origin, redifferentiating a tapered rod of cartilage distal to the host wrist (Fig. 5). Of the remaining five cases, the graft failed to differentiate in two cases, resorbed in one case, followed by regeneration of a symmetrical single-digit hand from the host stump, and in two cases was pushed to one side as the host regenerated a symmetrical single-digit hand.

(3) *dA wrist blastemas grafted to normal or dA thigh stumps*

Table 1 summarizes the results of grafting dA wrist blastemas to normal thigh stumps. The grafts did not develop well. In half the cases, the graft resorbed and the host stump regenerated a normal hindlimb. In 20 % of the cases, the host regenerated a normal hindlimb while the graft developed separately on the anterodorsal side of the host foot. In 30 % of the cases, the graft formed the anterior digits of a host–donor chimaera (Fig. 6). There were no cases in which typical intercalary regeneration of the host stump took place so that normal hindlimb skeletal structures were intercalated up to the distal end of the tibia-fibula, with the limb terminating in a symmetrical one- to two-digit hand.

Table 2 summarizes the results of grafting dA wrist blastemas to dA thigh stumps. Only 2 of 13 cases (both homografts) failed to exhibit typical intercalary regeneration of stylopodial and zeugopodial skeletal elements. In one of these two cases (white donor, dark host), the graft formed a single digit-like spike of unpigmented tissue on the non-regenerating host. At the time of fixation, the spike exhibited haemostasis, indicating that the donor tissue was undergoing chronic immunorejection. In the other case, the stylopodium and a single zeugopodial bone regenerated, but the limb did not terminate in carpals or digits, suggesting that while intercalary regeneration was proceeding, graft resorption was also taking place.

The remaining 11 cases (3 autograft, 8 homograft) formed regenerates consisting of a distal stylopodium (femur), a single symmetrical zeugopodial bone (tibia) and a one- to two-digit hand identical in structure to the hands regenerated by donor wrist stumps. Figs. 7–9 illustrate the skeletal structure of the three autograft cases.

The homograft cases, in general, did not develop as well as the autograft cases. Of the eight homografts, four displayed pigmentation patterns in which the donor colour was restricted to the hand and host colour to the intercalary

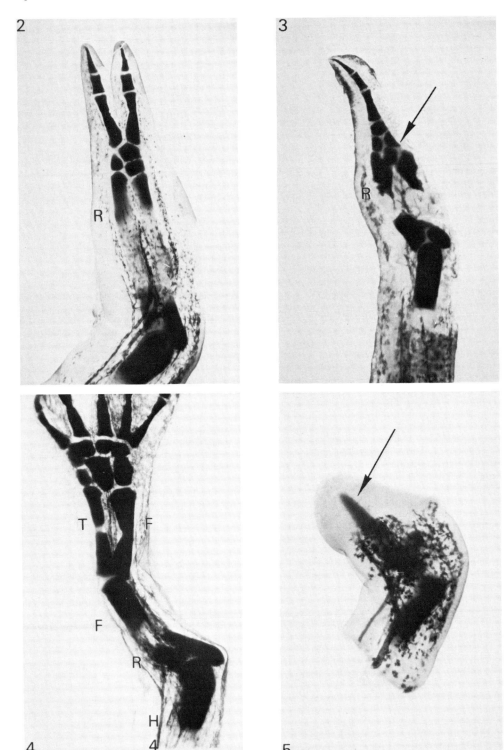

Table 1. *Results of grafting double anterior (dA) wrist blastemas to normal thigh stumps*

	No.	Normal* hindlimb	Normal hindlimb† + dA hand	Hindlimb-forelimb‡ chimera	dA hand
Autograft	6	5	0	1	—
Homograft	4	0	2	2	—
Total	10	5	2	3	—
Donor re- regenerate	6	—	—	—	6

* Grafted blastema resorbed, host stump regenerated.

† Grafted blastema developed according to origin, but host stump regenerated completely.

‡ Grafted blastema developed according to origin and formed the anterior part of a normal hindlimb along with the regenerating host stump.

regenerate (Figs. 10, 11, 12). This colour separation confirms the host origin of the intercalary regenerate. In two of these cases, the intercalated skeleton consists of a single bone which cannot be identified as either stylopodial or zeugopodial because it does not articulate with the femur (Figs. 11, 12).

The remaining four homograft cases (all dark donors and white hosts) exhibited donor pigmentation in the skin or around the blood vessels of the intercalary regenerate. The area occupied by donor melanocytes was usually restricted to either the distal end of the zeugopodium (one case) or to a thin band extending along a portion of the length of the zeugopodium (two cases, Fig. 13). In one case, however, pigment extended over the whole zeugopodium, the skeleton of which consisted of two radii fused in the longitudinal axis for most of their length. The pigmentation and skeletal patterns of the donor and

Fig. 2. Donor dA wrist stump regenerated for the second time. Digit 1 is regenerated twice in a symmetrical arrangement. Three carpals, probably representing the radiale and carpal D1-2 (the most anterior carpals), regenerated. Note the loss of midline structures (middle column of carpals and digit 2) distal to the doubled radius (R).

Fig. 3. Another re-regenerate of a donor dA wrist showing extensive pattern convergence distal to the doubled radius (R). There are four carpal elements (arrow) arranged in a symmetrical pattern, and a single digit, probably composed of the anterior halves of two digit 1's fused in the midline.

Fig. 4. Normal thigh blastema autografted to a dA distal zeugopodium. A normal five-toed hindlimb developed distal to the doubled radius (R) of the host. H, Host humerus; F, T, f, femur, tibia and fibula of donor.

Fig. 5. dA thigh blastema homografted to dA distal zeugopodial stump (white animal to dark animal). The skin has been removed from the dorsal side of the specimen for better viewing of the skeleton. The blastema developed according to origin, forming a tapered cone of cartilage (arrow). The pigmentation pattern of the limb is coincident with the different morphological patterns of donor and host.

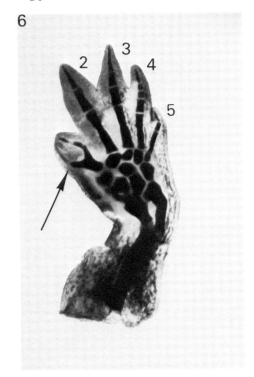

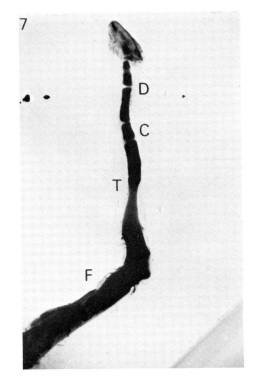

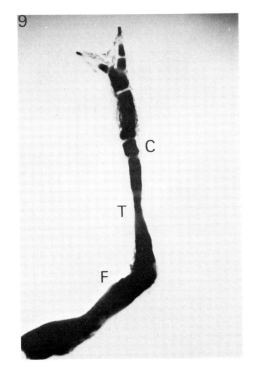

Table 2. *Results of grafting double anterior (dA) wrist blastemas*
to double anterior thigh stumps

	No.	dA hand, no intercalation	Intercalation, graft resorbed	dA hand, intercalation	dA hand
Autograft	3	0	0	3	—
Homograft	10	1	1	8	—
Total	13	1	1	11	—
Donor re-regenerate	3	—	—	—	3

host of this case are in accord, however, for the records show that the donor lower arm had been amputated through the doubled radius, at a more proximal level than other cases.

DISCUSSION

The fact that normal and dA thigh blastemas grafted to dA wrist stumps develop according to origin reinforces the conclusion, drawn from other work (Stocum, 1968, 1980b; Holder & Tank, 1979) that enough positional information to specify the pattern of the regenerate is inherited by the blastema from the pattern of the stump tissues during their dedifferentiation. The regenerate pattern is not primarily imposed upon or secondarily reinforced by a signal transmitted from the stump to the blastema after the latter has formed, as has recently been claimed by Slack (1980b).

Fig. 6. dA wrist blastema homografted to normal thigh stump (white animal to dark animal). The skin has been partially removed from the dorsal side of the specimen for better viewing of the skeleton. The host regenerated a femur, a short and somewhat abnormal tibia and fibula, a tarsal pattern exhibiting some abnormalities on the anterior side, and four toes numbered from anterior to posterior. The graft (arrow) formed two symmetrically arranged digit 1's fused in the metacarpal region; these digits were in their proper location on the anterior side of the host foot and completed the digital array.

Fig. 7. dA wrist blastema autografted to dA thigh stump. The host thigh intercalated the femur (F) and a single symmetrical tibia (T). The graft developed according to origin, forming a single carpal (C) and digit (D). The skin was removed from the specimen for better viewing of the skeleton.

Fig. 8. dA wrist blastema autografted to dA thigh stump. The thigh stump intercalated the femur (F), and a single symmetrical tibia (T). The graft formed one rod-shaped and one round carpal (arrows) and a single digit. The skin was removed from the specimen for better viewing of the skeleton.

Fig. 9. dA wrist blastema autografted to dA thigh stump. Again, the host intercalated femur (F) and tibia (T), while the graft developed a double anterior hand consisting of a single carpal (C) and two digit 1's fused along the longitudinal axis in the regions of the metacarpal and the first phalange. The skin was removed from the specimen for better viewing of the skeleton.

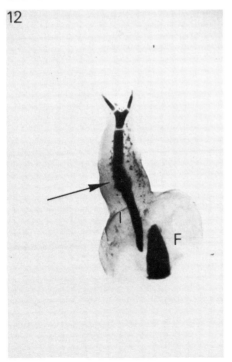

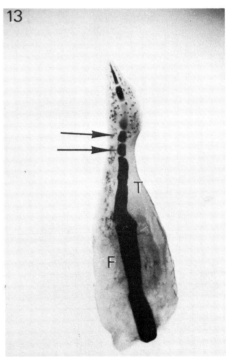

Limbs composed of dA wrist blastemas grafted to normal thighs did not undergo intercalary regeneration. Instead, the host seemed to ignore the presence of the graft, or else the graft resorbed, with the host thigh regenerating a complete PD array of skeletal structures. In those cases where the graft survived, it was carried distally as the host thigh regenerated. The graft ultimately wound up at its normal position on the anterior side of the host foot, either integrated with the host toes to form a hand-foot chimaera, or as a separate structure on the anterodorsal side of the host foot. The positioning of the graft structures with respect to the host foot structures agrees with the finding of Shuraleff & Thornton (1967) that a foot blastema grafted to the thigh of a regenerating limb seeks its level of origin in the final regenerate. This fact strongly suggests that the molecular basis of positional information in the regenerating limb resides in the plasmalemma or surface coat of the limb cells. The reason for the high rate of resorption of the dA wrist blastema grafts is unknown. It is possible that they have difficulty in becoming established because their development is subordinated to and overwhelmed by the greater regenerative potential which may be afforded by the complete transverse positional value map at the amputation surface of the host.

Fig. 10. dA wrist blastema homografted to dA thigh stump (white animal to dark animal). The graft did not develop well, forming a single digit (arrow), but the host intercalated the femur (F) and a short tibia (T). Note that the host stump and intercalated region are covered with skin of host colour, while the digit has donor colour.

Fig. 11. dA wrist blastema homografted to dA thigh stump (dark animal to white animal). The limb did not develop well. A single bone (arrow) was intercalated from the host stump (covered by skin of host colour), but whether this bone is femur or tibia cannot be determined with certainty. The graft developed a single digit with the donor pigmentation pattern, and exhibited intense signs of immunorejection (haemostasis) at the time of fixation.

Fig. 12. dA wrist blastema homografted to dA thigh stump (dark animal to white animal). The graft formed two digits fused along the longitudinal axis in the regions of the metacarpal and first phalange, and a single carpal element (arrow). The skin over these structures has the donor pigment pattern. A single bone (I) was intercalated from the host stump, but could not be identified as femur or tibia. The proximal two-thirds of this bone appeared to be embedded in the soft tissues of the host stump and its proximal one-third overlapped the distal end of the host femur stump (F). A small amount of fine pigment can be seen in the skin covering the distal one-third of the intercalated bone. The haemostasis of immunorejection was visible in the phalanges and distal half of the metacarpal. Blood vessels were dilated in the proximal half of the metacarpal region and in the carpal region, but blood flow was still excellent.

Fig. 13. Well-developed homograft of a dA wrist blastema to a dA thigh stump (dark animal to white animal). The graft formed two carpals in tandem (arrows) and a single digit. The host intercalated the femur (F) and a single symmetrical tibia (T). A thin band of pigmented skin lies on the dorsal side over the whole length of the zeugopodium. There was haemostasis in the phalangeal region, and the blood vessels in the metacarpal and carpal regions were heavily dilated, with sluggish blood flow. In the pigmented band of skin over the zeugopodium, the blood vessels appeared to be slightly dilated, but blood flow was normal.

In contrast to the above results, 85 % of the limbs consisting of a dA wrist blastema grafted to a dA thigh stump exhibited typical intercalary regeneration, resulting in limbs consisting of a femur and/or a single, symmetrical tibia intercalated from the host stump, and terminating in a hand comprising a few symmetrically arranged carpals and one to two symmetrical digits. The hand structure was the same as that produced by dA wrist stumps that underwent re-regeneration, and the symmetrical structure of the tibia is what would be expected from the fusion along the midline of two tibiae, each derived from one half of the double anterior-half thigh. The intercalary regenerates were exactly the same in structure as those observed after intercalary regeneration evoked by *normal* wrist blastemas grafted to dA thigh stumps (Stocum, 1980a).

Evidence that the graft redifferentiated according to its origin and for the host origin of the stylopodial and zeugopodial skeleton in these regenerates is provided by those homograft cases in which donor colour was restricted to the hand while the rest of the regenerate had host colour. In several cases, however, donor pigmentation was found over the distal end of the intercalary regenerate, or in a thin band extending along the longitudinal axis of the intercalary regenerate. The former pattern might reflect some mixing of graft and host cells in the most distal portion of the intercalary regenerate (see Pescitelli & Stocum, 1980). The latter pattern superficially might suggest that proximal transformation of graft cells took place. However, it is unlikely that this interpretation is valid. Studies of intercalary regeneration using ploidy markers have shown that the cells of the intercalated structures are derived solely from the host stump (Pescitelli & Stocum, 1980). The pigmented bands of skin can be explained by wound healing patterns in which fingers of donor epidermis bridge the graft–host junction during initial wound healing. This epidermis could provide factors favourable for migration of melanocytes along it (see Frost & Malacinski, 1980, for review).

Since intercalary regeneration of dA thighs is evoked by grafts containing only a double anterior half-map of positional values, it is clear that cells in the posterior half of the urodele limb do not act in any special way to promote distal transformation, nor are they organized as a polarizing zone which specifies the anterior–posterior axial structure of the regenerate. Special polarizing or distal transformation-promoting properties of posterior limb cells are also ruled out by the finding that dP upper arms which have undergone prolonged healing times (30 days or more) after their construction exhibit regenerative behaviour characteristic of dA upper arms; they regenerate short, tapered spikes (Bryant, 1976; Tank, 1978; Tank & Holder, 1978). Carlson (1975) was also led to conclude that no polarizing zone exists in regenerating urodele limbs because he could observe no preferential location for the emergence of supernumerary limbs after skin rotation and amputation. We may also conclude that the inhibition of terminal regeneration in dA limbs is not due to a subthreshold innervation pattern in double anterior-half limbs. The fact that little or no pattern convergence occurs in the dA intercalary regenerates of the present experiments suggests instead that distal transformation during intercalary regeneration is accomplished by interactions occurring directly between PD positional values of graft and host cells without requiring any interactions between transverse values. However, it is not ruled out that transverse interactions may be taking place along with proximal–distal interactions during PD intercalary regeneration, but in a pattern that avoids convergence.

Either way, the results are compatible with the hypothesis of Bryant & Baca (1978) and Holder *et al.* (1980) that short-arc interactions take place between transverse positional values during terminal regeneration of the normal limb, and lead to pattern convergence in dA upper arms and thighs, and in dP upper arms that have healed for an extensive length of time. According to this hypothesis, the extent of distal transformation is proportional to the degree to which confrontation between like or adjacent transverse positional values across the midline is avoided by the pattern of short-arc interactions. Double anterior limbs healed for 15 days or more, and dP limbs healed for 30 days or more regenerate after amputation as if the pattern of short-arc interactions progressively eliminates midline positional values, leading to confrontations of transverse values inhibitory to further intercalation and halting distal trans-formation early in the course of blastemal outgrowth. In contrast, dP limbs healed for less than 30 days often regenerate as if the pattern of short-arc interaction does not eliminate as many midline values, allowing intercalation between transverse values to continue until all missing PD levels have been restored.

Presumably, there is a temporally related difference in healing patterns which makes it possible for dP limbs with 'short' healing times to undergo complete distal transformation, while dP limbs with 'long' healing times and dA limbs cannot. An effect of healing time has also been observed in dA forelimbs by Tank & Holder (1978), but is not as pronounced as that observed in dP forelimbs. When healed for less than 15 days after construction, dA upper arm stumps often regenerate past the elbow, but with longer healing times, the stumps either fail to regenerate at all, or regenerate a short spike of cartilage. Double anterior thighs healed for only 10 days always regenerate only a short cartilage spike (Stocum, 1978). In contrast to dP upper arm stumps, however, dA upper arm stumps never regenerate digits, regardless of healing time. Thus, Holder *et al.* (1980) postulate factors in addition to the pattern of healing to explain the difference in regenerative ability between dA and dP limbs amputated 5–15 days after construction. If it is assumed that the posterior half of the limb carries more positional values than the anterior half, confrontation of like or ad-jacent positional values across the midline will occur somewhat earlier during blastemal outgrowth in dA limbs than in dP limbs. Histological evidence consistent with this notion is that the posterior half of the limb contains more cells, at the same cell density, than the anterior half (Tank & Holder, 1980).

Furthermore, Maden (1979) has found that posterior half-upper arms exhibit a greater regulative ability during terminal regeneration than anterior halves. At present, however, there is no histological or cytological evidence for the existence of interactions between cells across short arcs in the amputation plane. It would be of interest to look for such interactions using light microscopy, and scanning and transmission electron microscopy.

Although existing evidence is compatible with the idea that cellular interactions in the plane of amputation can lead to pattern convergence (deletion of midline structures) in double half-limbs, it does not justify the conclusion that convergent interactions actually lead to the cessation of distal transformation in these limbs; i.e., that the actual generation of PD positional values during regeneration is dependent upon interactions between transverse positional values. The convergent patterns in the regenerates of double half-limbs have been interpreted as being truly truncated, with the pattern ending in an abnormally proximal PD positional value. However, it is possible that what we have viewed as lack of distal transformation is an illusion created by convergence, the reality being that during terminal regeneration, positional values representing all the PD levels of the double half-limbs are reconstituted. However, due to progressive pattern convergence, only the most peripheral positional values are represented at the level of the zeugopodium and autopodium, giving a tapered, but actually proximodistally complete regenerate.

With this possibility in mind, we reasoned that if the latter notion has any validity, the distal tip of the blastema which forms on a dA thigh represents a PD positional value at least one and possibly two segments distal to the stylopodium. In a preliminary experiment, several dA thighs were constructed on large dark axolotls (80–90 mm snout–tail tip length). After a 10-day healing period, these limbs were amputated through the midthigh and a medium bud blastema allowed to form (approximately 12 days postamputation). The distal tip, comprizing about one-fourth the length of the blastema was then severed and grafted to the amputated normal thigh of a small white axolotl (25–30 mm snout–tail tip length). The exchange was made between large and small animals because when exchanges were made between animals of the same size, there was a wide disparity between cross-sectional areas of the graft and host, and the grafts always resorbed. Only one of three cases has survived to date, but an intercalary regenerate has formed from the host stump (white colour), distal to which is a pigmented conical structure typical of the distal end of a dA thigh regenerate. Although definite conclusions cannot be based on one case, this result lends support to the idea that all PD levels are represented in the blastema that forms after amputation of a dA thigh. Thus, the truncated appearance of double half-limbs cannot yet be offered as evidence that PD values are serially generated by interactions between transverse values in the amputation plane.

This research was supported by Grant HD-12659 from the National Institutes of Health.

REFERENCES

BRYANT, S. V. (1976). Regenerative failure of double half limbs in *Notophthalmus viridescens*. *Nature* **263**, 676–679.

BRYANT, S. V. (1978). Pattern regulation and cell commitment in amphibian limbs. In *The Clonal Basis of Development* (ed. S. Subtelny & I. M. Sussex), pp. 63–82. Academic Press.

BRYANT, S. V. & BACA, B. A. (1978). Regenerative ability of double half and half upper arms in the newt, *Notophthalmus viridescens*. *J. exp. Zool.* **204**, 307–324.

BUTLER, E. G. (1955). Regeneration of the urodele limb after reversal of its proximo-distal axis. *J. Morph.* **96**, 265–282.

CARLSON, B. M. (1975). The effects of rotation and positional change of stump tissues upon morphogenesis of the regenerating axolotl limb. *Devl Biol.* **47**, 269–291.

FALLON, J. F. & CROSBY, G. M. (1977). Polarizing zone activity in limb buds of amniotes. In *Vertebrate Limb and Somite Morphogenesis* (ed. D. A. Ede, J. R. Hinchliffe & M. Balls), pp. 55–69. Cambridge University Press.

FRENCH, V., BRYANT, P. J. & BRYANT, S. V. (1976). Pattern regulation in epimorphic fields. *Science* **193**, 969–981.

FROST, S. & MALACINSKI, G. M. (1980). The developmental genetics of pigment mutants in the Mexican axolotl. *Devl Gen.* **1**, 271–294.

HAMBURGER, V. (1960). *A Manual of Experimental Embryology*, p. 196. University of Chicago Press.

HOLDER, N. & TANK, P. W. (1979). Morphogenetic interactions occurring between blastemas and stumps after exchanging blastemas between normal and double-half forelimbs in the axolotl, *Ambystoma mexicanum*. *Devl Biol.* **68**, 271–279.

HOLDER, N., TANK, P. W. & BRYANT, S. V. (1980). Regeneration of symmetrical forelimbs in the axolotl, *Ambystoma mexicanum*. *Devl Biol.* **74**, 302–314.

ITEN, L. E. & BRYANT, S. V. (1975). The interaction between the blastema and stump in the establishment of the anterior–posterior and proximal-distal organization of the limb regenerate. *Devl Biol.* **44**, 119–147.

KRASNER, G. N. & BRYANT, S. V. (1980). Distal transformation from double-half forearms in the axolotl, *Ambystoma mexicanum*. *Devl Biol.* **74**, 315–325.

MADEN, M. (1979). Regulation and limb regeneration: The effect of partial irradiation. *J. Embryol. exp. Morph.* **52**, 183–192.

MADEN, M. (1980). Intercalary regeneration in the amphibian limb and the rule of distal transformation. *J. Embryol. exp. Morph.* **56**, 201–209.

PESCITELLI, M. J., JR. & STOCUM, D. L. (1980). The origin of skeletal structures during inter-calary regeneration of larval *Ambystoma* limbs. *Devl Biol.* **79**, 255–275.

PESCITELLI, M. J., JR. & STOCUM, D. L. (1981). Nonsegmental organization of positional information in regenerating *Ambystoma* limbs. *Devl Biol.* **82** (in the Press).

ROSE, S. M. (1962). Tissue arc control of regeneration in the amphibian limb. In *Regeneration* (ed. D. Rudnick), pp. 153–176. Ronald Press.

SHURALEFF, N. C. & THORNTON, C. S. (1967). An analysis of distal dominance in the regenerating limb of the axolotl. *Experientia* **23**, 747–752.

SLACK, J. M. W. (1976). Determination of polarity in the amphibian limb. *Nature (Lond.)* **261**, 44–46.

SLACK, J. M. W. (1977). Determination of anteroposterior polarity in the axolotl forelimb by an interaction between limb and flank rudiments. *J. Embryol. exp. Morph.* **39**, 151–168.

SLACK, J. M. W. (1980*a*). Regulation and potency in the forelimb rudiment of the axolotl embryo. *J. Embryol. exp. Morph.* **57**, 203–217.

SLACK, J. M. W. (1980*b*). A serial threshold theory of regeneration. *J. theor. Biol.* **82**, 105–140.

STOCUM, D. L. (1968). The urodele limb regeneration blastema: A self-organizing system II. Morphogenesis and differentiation of autografted whole and fractioned blastemas. *Devl Biol.* **18**, 457–480.

STOCUM, D. L. (1975). Regulation after proximal or distal transposition of limb regeneration blastemas and determination of the proximal boundary of the regenerate. *Devl Biol.* **45**, 112–136.

STOCUM, D. L. (1978). Regeneration of symmetrical hindlimbs in larval salamanders. *Science* **200**, 790–793.

STOCUM, D. L. (1979). Stages of forelimb regeneration in *Ambystoma maculatum*. *J. exp. Zool.* **209**, 395–416.

STOCUM, D. L. (1980a). Intercalary regeneration of symmetrical thighs in the axolotl, *Ambystoma mexicanum*. *Devl Biol.* **79**, 276–295.

STOICUM, D. L. (1980b). Autonomous development of reciprocally exchanged regeneration bastemas of normal forelimbs and symmetrical hindlimbs. *J. exp. Zool.* **212**, 361–371.

SUMMERBELL, D. & TICKLE, C. (1977). Pattern formation along the antero-posterior axis of the chick limb bud. In *Vertebrate Limb and Somite Morphogenesis* (ed. D. A. Ede, J. R. Hinchliffe & M. Balls), pp. 41–53. Cambridge University Press.

SWETT, F. H. (1938). Experiments upon the relationships of surrounding areas to polarization of the dorsoventral limb-axis in *Ambystoma punctatum* (Linn.). *J. exp. Zool.* **78**, 81–100.

TANK, P. W. (1978). The failure of double-half forelimbs to undergo distal transformation following amputation in the axolotl, *Ambystoma mexicanum. J. exp. Zool.* **204**, 325–336.

TANK, P. W. & HOLDER, N. (1978). The effect of healing time on the proximo distal organization of double-half forelimb regenerates in the axolotl, *Ambystoma mexicanum. Devl Biol.* **66**, 72–85.

TANK, P. W. & HOLDER, N. (1980). The distribution of cells in the upper forelimb of the axolotl, *Ambystoma mexicanum. J. exp. Zool.* **209**, 435–442.

WOLPERT, L. (1969). Positional information and spatial pattern of cellular differentiation. *J. theoret. Biol.* **25**, 1–47.

WOLPERT, L. (1971). Positional information and pattern formation. *Curr. Top. Devl Biol.* **6**, 183–224.

J. Embryol. exp. Morph. Vol. 65 (Supplement), pp. 19–36, 1981
Printed in Great Britain © Company of Biologists Limited 1981

Pattern formation and growth in the regenerating limbs of urodelean amphibians

By NIGEL HOLDER[1]

From the Anatomy Department, King's College London

SUMMARY

The results of numerous types of grafting experiments involving the amputation of symmetrical limbs are described. These experiments were designed to test the tenets of the polar coordinate model. The analysis of the results of these grafts coupled with a quantitative analysis of blastemal shape strongly indicates that pattern regulation during amphibian limb regeneration can be understood in terms of the model.

INTRODUCTION

The amphibian limb undergoes a specific series of morphologically defined events following amputation at any level of the proximal to distal axis (Iten & Bryant, 1973; Tank, Carlson & Connelly, 1976; Stocum, 1979). Following wound healing the mesodermal cells close to the wound surface undergo a process of dedifferentiation when cells with clear phenotypic characteristics, such as muscle and cartilage cells, become visually homogeneous and form a founding cell population which divides at a consistently high rate to produce a blastema. The blastema then increases in size and cell number until the new parts of the limb begin to reform in a proximal to distal sequence. Once the missing parts have all been replaced they grow to the correct degree and thus result in a complete limb which is the size and form of the part that was initially removed.

Growth is occurring at all stages in this regeneration process, but only part of the process is concerned with pattern formation. The object of this paper is to discuss which part of the growth process is linked to the spatial patterning mechanisms and how this link might be characterized. The term growth is used here in the general sense of increase in size with reference to the new tissues being formed during regeneration. There are three ways in which a tissue can increase in size: the blastemal cell population may increase in number by cell division or cell recruitment, the cells may increase in size (cell hypertrophy) and they can produce extracellular material which swells the tissue mass.

The initial stages of limb regeneration involve both cell division and cell recruitment and some growth is accounted for by swelling of the tissues as a

[1] *Author's address:* Anatomy Department, King's College London, Strand, London WC2R 2LS, U.K.

result of general oedema (Chalkley, 1954). By far the greatest component of growth in the first stages of blastemal outgrowth is an increase in cell number. Cell division occurs at a consistently high level once the blastemal population has been established (Maden, 1976; Smith & Crawley, 1977; Stocum, 1980a). This population of cells is derived from the stump tissues during the dedifferentiation process as a result of cell division and cell recruitment (Chalkley, 1954). The relationship between cell recruitment and cell division in the formation of the blastemal cell population is important to understand in terms of the relationship between pattern formation and growth. It is especially important because amphibian limb regeneration is thought of as an epimorphic process where cell division is crucial for the regeneration of the limb pattern; and it is equally important if we are to establish the relationship between the production of new positional values and cell division.

Pattern formation and growth: the polar coordinate model

The main part of this paper is given over to discussion of the evidence which supports the view that cell division and generation of new positional values are intimately linked. The experiments discussed clearly establish that growth of the regenerating limb can be dramatically affected as a result of surgical manipulations which alter the normal spatial relationship of cells giving rise to the blastemal population. The experiments described were performed in the attempt to test the tenets of a formal model which theoretically links the process of cell division to the generation of new positional values.

The polar coordinate model (French, Bryant & Bryant, 1976; Bryant, French & Bryant, 1981) postulates that the positional values existing in the limb are arranged in a circular and an angular sequence (Fig. 1a). When cells are confronted which have normally non-adjacent positional values, intercalation of the normally intervening values occurs as a result of cell division. With regard to the circular sequence of values this intercalation occurs via the shortest route of the sequence. It is this observation, incorporated into the model as the shortest route intercalation rule, which is of central importance to our understanding of the cellular interactions leading to regeneration of the pattern and to the relationship between pattern formation and growth. The actual number of intercalatory events, and thus the number of new cells with new positional values, will depend upon which cells (i.e. which positional values) contact during the regeneration process. This point is best illustrated by examining how the polar coordinate model accounts for normal limb regeneration.

Blastemal accumulation produces a mound of cells on the distal tip of the amputated limb. This conical mound is narrower at its tip than at its base and therefore, as outgrowth proceeds, cells from different circumferential positions will contact at the distal tip. As cells with normally non-adjacent positional values contact in a localized manner around the circumference intercalation occurs and new cells are produced (Fig. 1b). This process occurs all around the

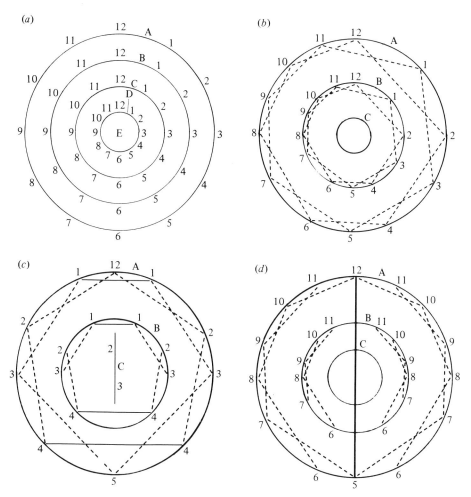

Fig. 1. (*a*) The polar coordinate model can be formally represented as a two-dimensional array of positional values. One dimension is angular and is represented by twelve numbers arranged around the circumference, the other is radial and is represented by letters A (proximal) through E (distal). (*b*) Short-arc intercalation following amputation of a normal limb. The dashed lines represent local cell contacts resulting in intercalation of new circumferential positional values which also have the next most distal radial positional value. Local contacts around the circle produce a complete new, more distal circular sequence. This process continues until the distal end of the pattern is reached. (*c*) Short-arc intercalation in a double anterior symmetrical upper arm stump from a long-healing-time group. Dashed lines represent cell contacts leading to intercalation of cells with more distal circumferential positional values. Solid straight lines represent cell contacts which will not stimulate such cell divisions. Due to loss of midline circumferential values and non-productive cell contacts, regeneration ceases at radial level C. (*d*) Short-arc intercalation in a double posterior symmetrical upper arm stump from a short-healing-time group. No cell contacts occur across the central solid line which represents the wound incurred by the initial surgery. This, coupled with the proposed greater number of circumferential positional values in the posterior limb region, allow more distal regeneration to occur than in Fig. 1*c*.

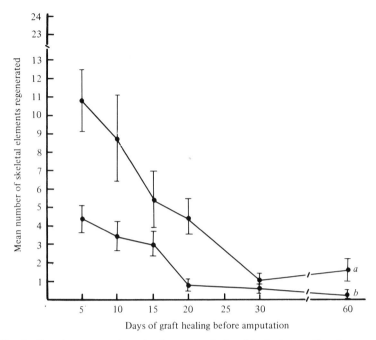

Fig. 2. Graph representing the inverse relationship between the degree of distal transformation and graft healing time prior to amputation in double posterior (*a*) and double anterior (*b*) symmetrical upper arms. Note the gradual decline in the ability to regenerate which occurs between 5 and 30 days of graft healing. (Redrawn from Tank & Holder, 1978.)

circumference and results in the build up of a complete new circumferential set of positional values which constitute the next most distal circular set (Bryant *et al.* 1981; French, 1981). As outgrowth continues, more and more distal sets of circumferential positional values are produced, until the most distal set is formed and the complete pattern is restored. This process of intercalation occurring in localized regions of the circumference during blastema formation emphasizes the importance of the circumference in distal outgrowth. This emphasis is made clearer still if abnormal arrangements of positional values are created at the amputation plane, and the resulting regenerated limb patterns are analysed.

Regeneration of symmetrical limbs

One simple way of altering this circular sequence is to surgically create symmetrical limbs which have two sets of positional values symmetrical about one of the transverse limb axes (Fig. 1*c–d*). Such symmetrical stumps yield remarkably different patterns of limbs depending on the level of the proximal to distal axis at which they are created, and the local conditions at the amputation site which affect cell contact between cells from each of the symmetrical regions. These conditions include the shape of the amputation plane, that is whether the amputation plane is round or oval, and the effect of the wound created by the surgery on the ability of cells from either side of the wound to contact. The amount of regeneration achieved by symmetrical limb stumps is also affected by the number of positional values present at the amputation plane in each specific construction.

These last two features, the ability of cells to contact and the number of values present, are demonstrated by examining the effect that graft healing time prior to amputation has on the degree of distal outgrowth achieved by symmetrical upper arms. A clear inverse relationship exists between the degree of distal outgrowth and the time for which grafts creating symmetrical upper arms are allowed to heal before amputation (Stocum, 1978; Tank & Holder, 1978). Figure 2 shows the amount of regeneration occurring from double anterior and double posterior upper forelimbs amputated at periods from 5 to 60 days after the operation. Several points can be gleaned from these results. Double posterior limb stumps regenerate more structure than double anterior limb stumps. If it assumed that the posterior side of the normal limb contains more than half of the circumferential set of positional values and the anterior side bears the remainder, then the creation of symmetrical limb stumps will accentuate this difference because two sets of these values are present (see Fig. 1 *c–d*). In addition, the fall off in the distal extent of regeneration of symmetrical limbs with healing time may be explained in terms of cell contacts leading to inter-calation of new positional values. We argue that in the long-healing-time groups the complete closure of the wound created by the surgery enables cells from either side of it to contact during initial blastema formation (Holder, Tank & Bryant, 1980; Bryant, Holder & Tank, 1981). In this case cells with the same or normally adjacent positional values will contact across the line of symmetry. Such cell contacts will not stimulate intercalation of new values and no cell divisions will occur. Thus the number of new cells and the number of circum-ferential positional values available at the next most distal level will be reduced. This reduction in positional values will occur at the line of symmetry (Fig. 1 *c*; Holder *et al.* 1980; Bryant *et al.* 1981). Midline loss of positional values will continue as distal outgrowth occurs until intercalation, and regeneration, ceases. In the double anterior limbs, because there are assumed to be fewer positional values to start with, midline loss of positional values will lead to the cessation of regeneration faster than in the double posterior limbs which are therefore able to achieve a greater degree of distal regeneration.

Midline loss of positional values and the premature cessation of intercalation in those limbs predicts the formation of tapering, distally incomplete symmetrical regenerates. Should the cessation of intercalation be particularly rapid, virtually no regeneration will occur. These predicted tapering spike-like limbs and regenerates which consist of only nodules of cartilage (non-regenerates) are seen in many cases (Tank & Holder, 1978; Holder *et al.* 1980) and examples are shown in Fig. 3.

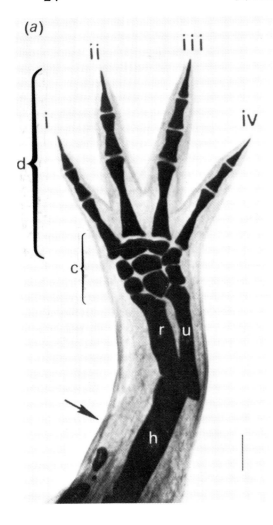

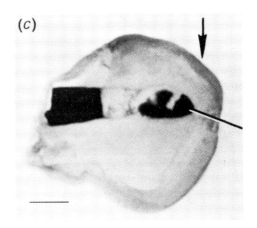

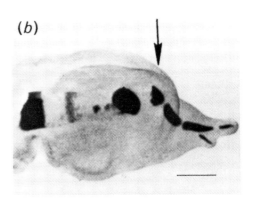

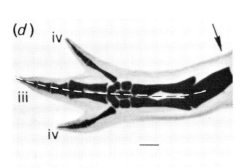

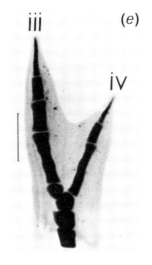

In contrast to these long-healing-time groups, short-healing-time symmetrical limbs regenerate far more structure. This is particularly evident in double posterior limbs (see Figs. 1*d*, 2); this is again presumed to reflect the unequal distribution of circumferential positional values. Immediate amputation of such limbs leads to the regeneration of double posterior symmetrical structures which vary in the extent of formation of the *a-p* axis distally and thus show a variation in their number of digits (Holder *et al.* 1980). Double anterior limbs amputated immediately predominantly regenerate tapered distally incomplete limbs. The proposed explanation for the greater regeneration potential of limbs in short-healing-time groups involves the presence of a wound existing between the symmetrical halves at the amputation plane, which results from the operation. If cells are unable to contact across this central reservation then cells with the same positional value will not contact. Therefore, these non-productive cell contacts which are assumed to be the primary cause of the early cessation of regeneration in long-healing-time groups, are prevented. The short-healing-time symmetrical limb stumps will thus be capable of more distal outgrowth (Figs. 1*d*, 3). It should be made clear, however, that local cell contacts will still cause loss of midline positional values, as is the case in the long-healing-time groups, and the outgrowths, whether double anterior or double posterior, should show a gradual midline loss of structure as distal outgrowth occurs. Analysis of the skeletal structure shows this to be the case. This midline loss will lead to more rapid cessation of regeneration in double anterior limbs than double posterior limbs but will allow greater degrees of distal outgrowth of both when compared with long healing time groups.

In addition to the various symmetrical outgrowths regenerated by these symmetrical upper arm stumps they also occasionally produce asymmetrical half limb structures. These occurred in both double anterior and double posterior constructions; the double anteriors producing half anterior limbs and the double posteriors half posterior limbs (Fig. 3) (Holder *et al.* 1980). One conclusion from the explanation for distal outgrowth of short-healing-time limbs is that each half of the surgically created stump distally transforms independently. Thus, if one side is prevented from growing, the other will regenerate regardless. In all of the cases where half limbs developed, they appeared to do so from the

Fig. 3. Victoria-blue-stained whole-mount preparations of forelimbs of the axolotl. (*a*) Normal regenerated right limb. (*b*) Spike-like outgrowth regenerated from a double posterior upper arm construction amputated 10 days after grafting. (*c*) Non-regenerate formed from a double posterior upper arm construction amputated 15 days after grafting. The leader points to the small, regenerated nodule of cartilage. (*d*) Three-digit symmetrical double posterior regenerate formed after amputation of a double posterior upper arm construction five days after grafting. The dotted line denotes the line of symmetry. Note the distally converging pattern. (*e*) A posterior half-limb formed following amputation of a double posterior upper arm construction immediately after grafting. d, digits; c, carpals; r, radius; u, ulna; h, humerus. The Roman numerals denote the digital number with i being anterior and iv posterior. The arrows denote the position of amputation. The scale bars equal 1 mm.

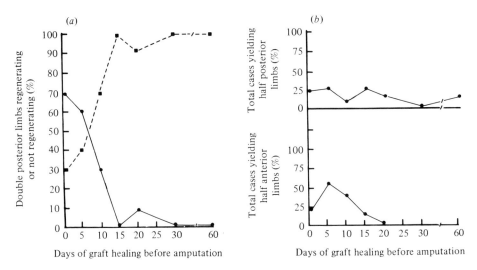

Fig. 4. Analysis of the relationship of graft healing time to degree of regeneration. (a) Regenerated limbs of either non-regenerates or spikes (dashed line), or distally complete limbs with digits (solid line) produced following amputation of double-posterior symmetrical upper arms formed following amputation at various times after grafting. Note the sudden fall in appearance of distally complete limbs between 0 and 15 days. The double anterior upper arms stumps all yielded non-regenerates or spikes and are not shown. All data taken from Tank & Holder (1978) and Holder *et al.* (1980) (see text). (b) The relationship between the appearance of half-limbs (expressed as a percentage of the total number of cases analysed in both 4 *a, b*). The top graph is half posterior limbs formed following amputation of symmetrical double posterior upper arm stumps; the bottom graph is half-anterior limbs formed following amputation of symmetrical double-anterior upper arm stumps.

host side of the symmetrical stump. It seemed therefore that the graft was the half which did not develop. This may be due to poor innervation of the grafted piece because it is well established that a local nerve trophic requirement is needed for cell division and therefore distal outgrowth to occur. Although the appearance of half limbs is consistent with the notion that each half of a symmetrical stump can regenerate independently, the polar coordinate model would require circumferential intercalation to occur on the medial edge of these half limbs during distal outgrowth. The reason why such intercalation does not occur in these cases is at present the subject of further study.

The gradual decline in regenerative ability of both double posterior and double anterior symmetrical stumps with the increase in time before amputation seems paradoxical if healing of a wound is indirectly responsible. This is because the graft and host tissues appear histologically to be completely healed by 10–15 days after the graft was performed (see histological analysis in Tank & Holder, 1978), yet the decline in structure of the regenerates occurs up to 30 days (Fig. 2). This paradox is resolved, at least in part, if the formation of half limbs is taken into account. In the light of the argument that one side of the symmetrical stump can regenerate independently if the other is unable to re-generate, it seems clear that the formation of half limbs should not bear the same clear relationship with graft healing time as regenerates forming from both sides. For this reason the regenerates produced from all graft healing time groups were re-examined in a different way (data from Tank & Holder, 1978, and Holder *et al.* 1980). Regenerates were classified into three groups: those where distal outgrowth had produced a clearly symmetrical set of digits (Fig. 3), half limbs which comprised only one forearm element and either 1 or 2 digits (Fig. 3), and distally incomplete regenerates which in fact were either spikes or non-regenerates (Fig. 3). The number of half limbs was then plotted separately in order to examine the relationship of their formation to graft healing time and the other two categories were expressed as percentages of the total of these two classes at each healing time. The results of this analysis are presented in Fig. 4, and several points emerge. The double posterior groups show a steep fall in the number of distally more complete symmetrical limbs between 0 and 15 days, which is the time during which healing between graft and host tissue is completed. This is mirrored by a steep rise in the number of spikes and non-regenerates which are formed (Fig. 4*a*). The half posterior limbs show a wider distribution which shows no distinct relationship to graft healing time (Fig. 4*b*). The double anterior groups all yield distally incomplete limbs and never produce symmetrical limbs with digits. The half anterior limbs, which do produce a complete anterior digit, are only seen during the first 20 days of healing, and do show a vague relationship with the time of graft healing. Taken together, these data strongly suggest that half limbs do not follow the same principles as symmetrical outgrowths. This supports the notion that it is the degree of inter-action between the cells derived from either side of the symmetrical stump which governs the degree of outgrowth that is achieved. Furthermore, the steep re-duction in the number of distally complete symmetrical 'double-posterior' outgrowth occurs in close association with the histological time scale of heal-ing.

The symmetrical upper arm experiments demonstrate the importance of the number and arrangement of circumferential positional values and the necessity of local cell interactions to the process of regeneration. The actual cell inter-actions that result in intercalation will be governed by the preferential healing modes of cells at or near the amputation plane. This point is best demonstrated by experiments where symmetrical limbs are amputated in the forearm region. In these cases, double posterior upper arm stumps were created and amputated immediately to ensure complete distal outgrowth in as many cases as possible (Holder *et al.* 1980). Following initial amputation through the symmetrical upper arm double posterior symmetrical regenerates which had between 3 and 6 digits were formed. Upon a second amputation through the more distal regenerated forearm region the majority of these limbs underwent a striking distal expansion of their pattern. Thus the limbs with 3 to 6 digits expanded to produce limbs with 5–7 digits (Fig. 8*a*, *b*). These remained symmetrical and in

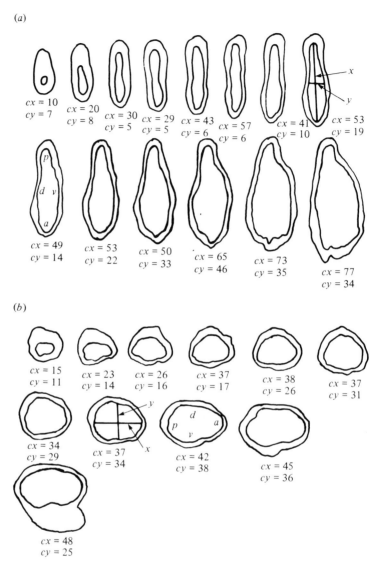

Fig. 5. Camera-lucida drawings of serial transverse sections of (*a*) a forearm (radius and ulna region) MB blastema and (*b*) an upper arm (humerus region) MB from the contralateral limb of the same animal. The sections are taken at every 100 μm and run from the blastema tip towards the stump top to bottom, left to right. The lines drawn at right angles on the blastema sections at the 800 μm level represent the measurements taken from which the ratios expressed in figure 6 were calculated. The ratio $r = (x/y)$ where x is the shorter of the two axes. *d*, dorsal; *v*, ventral; *p*, posterior and *a*, anterior. cx = the number of cells contacted by the line drawn on the x axis, cy = the number of cells contacted by the line drawn on the y axis.

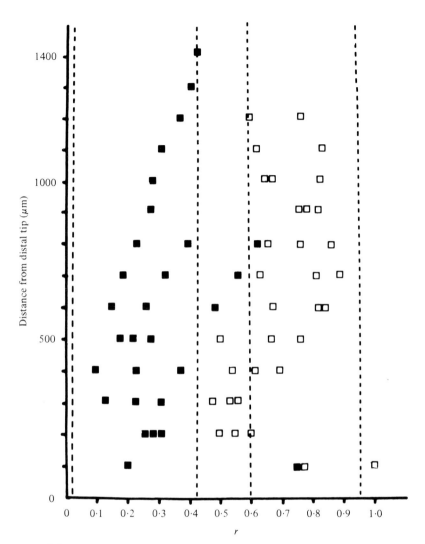

Fig. 6. The ratio of axis length represents a measure of blastemal shape (see text and Figure 5). The ordinate represents the ratio r, the abscissa the level of the section measured with 0 as the distal tip of the blastema. The open squares are individual measurements of r taken from upper arm blastemas, the solid squares are individual measurements of r taken from forearm blastemas. The dotted lines represent 95 % confidence limits (upper arm blastemas upper limit 0·95, lower limit 0·42; forearm blastemas upper limit 0·59, lower limit 0·02). The two populations show distributions that are statistically significant by the t-test at the 1 % level.

some cases where 7 digits were formed a digit 1 appeared in the centre of the digital array. Thus, in these cases the anterior positional values necessary to form a digit 1 (the most anterior digit) had been intercalated as a result of simply amputating these symmetrical limbs at a more distal level. The only position where such an intercalatory event could occur is in the midline of a double

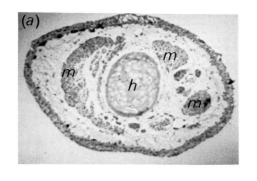

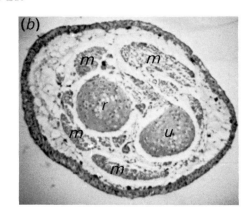

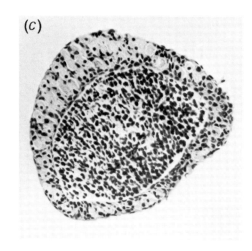

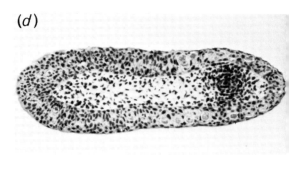

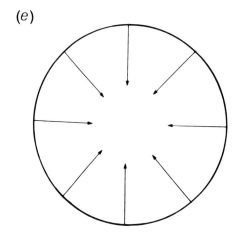

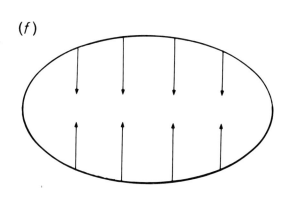

posterior symmetrical limb stump (Fig. 8). However, the previous arguments concerning normal regeneration and regeneration of symmetrical upper arms have been based on local cell contacts and intercalation, and now it seems that cells from the furthest points of the circle are interacting between the dorsal and ventral sides of the symmetrical forearm stump. The reason for this became apparent when the shape of the upper arm and forearm limb amputation planes were compared.

Blastemal shape and healing modes

The shape of the forearm and upper arm blastemas was examined in the following way. Right limbs of four larval axolotls (ranging from 68–90 mm) were amputated through the mid-upper arm (humerus) region, while the contra-lateral left limbs were amputated through the mid-forearm (radius and ulna) region. The limbs were then allowed to regenerate to the stage of Medium Bud (MB: Tank *et al.* 1976) before being removed at the shoulder, fixed in Bouin's fluid and processed for wax sectioning. Serial transverse 10 μm sections were cut at right angles to the long axis of the limb and stained with haematoxylin and eosin. Camera-lucida drawings were made of every tenth section (every 100 μm) and the axial orientation of these blastemas was determined by the muscle patterns in the stump regions of each limb (see Fig. 7*a*, *b*). The lengths of two transverse axes were measured on the drawings (dorsal to ventral and anterior to posterior) and the ratios of these distances calculated (see Fig. 6). (The ratio was calculated by dividing the length of the longest axis into that of the shortest.) Two sample sets of camera-lucida drawings are shown in Fig. 5. The ratios of axial lengths give a rough measure of the shape of the blastemal section. The ratio of a perfect circle will be 1 and this value will fall towards zero as the blastema becomes progressively more elliptical. Although these blastemas are clearly not perfect ellipses, the measurements are adequate to demonstrate quantitatively the obviously different shapes of forearm and upper arm blastemas (Figs. 5, 6). The clearcut change in shape demonstrated in Fig. 6 must occur during blastema outgrowth when dedifferentiation causes the collapse of muscle

---

Fig. 7. Blastemal shape and healing modes. (*a*) A transverse section through the mid upper arm of a normal axolotl. Note the centrally situated humerus (h) and the surrounding muscle blocks (m). The limb appears circular. Magnification: × 147. (*b*) A transverse section through the mid forearm of a normal axolotl. Note the two skeletal elements, the anterior radius (r) and the posterior ulna (u) and the surrounding muscle blocks (m). Again, the limb appears roughly circular. Magnification: × 150. (*c*) A transverse section taken 430 μm from the distal tip of a MB stage upper arm blastema. The shape is grossly circular. The ratio *r* of this section is 0·65. Magnification: × 350. (*d*) A transverse section taken 400 μm from the distal tip of MB stage forearm blastema from the contralateral limb of the same animal from which the section 7c was taken. The shape is elliptical. The ratio *r* of this section is 0·1. Magnification: × 280. (*e*) The radial healing mode proposed for short arc inter-calations in the circular upper arm blastema. (*f*) The dorsal to ventral healing mode proposed for short arc intercalation in the elliptical lower arm blastema.

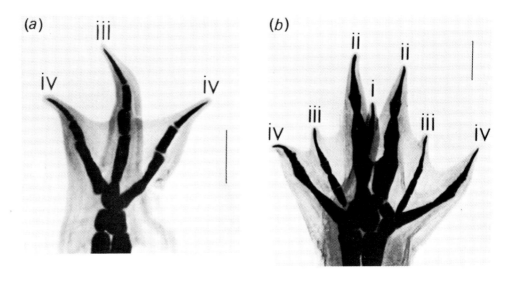

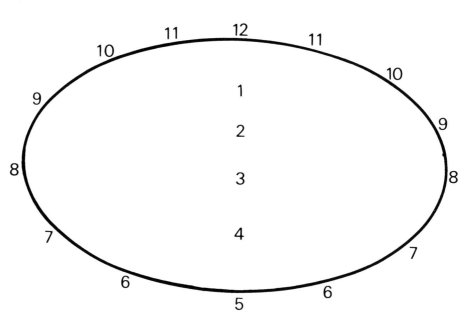

Fig. 8. (*a*) A three-digit double posterior limb formed following immediate amputation of a double posterior upper arm construction. (*b*) A seven-digit double posterior limb formed following a second amputation of the limb shown in 8*a* through the forearm region. Note the appearance of digit 1 in the midline, and the larger mass of carpal elements. Scale bar = 1 mm. (*c*) Diagrammatic representation of the proposed intercalation of anterior circumferential position values brought about by the dorsal to ventral healing mode of the elliptical double posterior forearm blastema.

blocks in the stump, leaving an array of blastemal cells surrounding the central supporting skeletal structure (see Fig. 7 a–d). Since there are-two-such skeletal structures in the forearm lying in parallel across the anterior-to-posterior axis, the shape of this axis is maintained while the dorsal to ventral sides collapse and become closer together, thus forming an ellipse. In the upper arm, by the same argument, the centrally situated humerus supports a more circular blastema.

The consequences of this clear difference in shape between the oval forearm blastema and more circular upper arm blastema are important for a model based on local interactions and intercalation. The healing modes governing which cells contact which during growth of the blastema will be grossly different in the forearm and the upper arm blastemas. Included in Fig. 5 is a series of cell counts which show the number of blastemal cells contacting a thin line drawn along the two transverse axes which were measured. These cell counts are meant only to indicate roughly the likelihood of contact between cells from the extreme ends of these axes, and have not been quantitatively analysed. For example, in Fig. 5a, the section taken 700 μm from the distal blastemal tip of this forearm blastema has 41 cells separating the anterior to posterior extremes yet only 10 lie between the dorsal and ventral sides. In the corresponding section from the contralateral upper arm blastema 37 cells separate the posterior to anterior and 31 cells the dorsal to ventral sides. It seems clear that if local cell contacts govern the establishment of new positional values then the cells from the dorsal and ventral sides of the forearm blastema are more likely to contact than cells from the anterior to posterior sides. By the same reasoning, in the upper arm, cells from different circumferential positions will have a more even chance of contacting. This kind of analysis stresses the importance of the probablistic nature of cell contacts and intercalation which is governed by healing modes.

Having established that these shape changes exist at the two proximal to distal levels examined, we can return to the regenerative ability of symmetrical limbs. Double posterior symmetrical regenerates expand their patterns when amputated in the forearm by the addition of midline structures. These extra structures will occur when, for example, 12 and 5 contact from dorsal to ventral, and produce 4, 3, 2 and 1 (Fig. 8c), the values normally lying on the anterior limb region (see Fig. 1b; see also Holder *et al.* 1980). The shape of the forearm blastema ensures that this event will occur frequently. This shape difference also makes a second prediction. If the preferential direction of healing in the forearm is dorsal to ventral, then cells from anterior and posterior sides will only contact rarely. This possibility should ensure that double anterior and double posterior limbs surgically constructed in the forearm will not show the healing time effect which is so clearly seen in the upper arm. Such symmetrical forearms should behave like short-healing-time groups from the upper arm and readily distally transform. This has indeed been shown to be the case by Krasner & Bryant (1980).

Symmetrical limb stumps produce symmetrical blastemas

Taken together, these symmetrical limb experiments demonstrate the central importance of the limb circumference to the process of distal outgrowth. This central role can be understood in terms of the number, arrangement and localized contacts that cells with different circumferential positional values make during the early stages of regeneration. Despite the considerable body of evidence which relates to the role of the circumference (see also Stocum, 1978, 1980*b, c*), it has been suggested that the healing of tissues in surgically constructed symmetrical limbs is a non-specific phenomenon (Maden, 1981). It is clear that the experiments presented here rely very much on differential directions of healing governed either by wound sites or the shape of amputation planes. The evidence which clearly shows that blastemas derived from double limb stumps bear symmetrical arrangements of circumferential position values comes from grafts where such blastemas are exchanged with comparable-staged blastemas derived from normal asymmetrical limb stumps (Holder & Tank, 1979; Stocum, 1980*b*). Double anterior and double posterior upper arms amputated after a month of healing form Medium Bud blastemas which are indistinguishable from those formed on normal amputation planes. This in itself indicates that the early phases of regeneration, that is dedifferentiation and early blastemal cell accumulation, are relatively normal in symmetrical limb stumps. Following the exchange of blastemas between symmetrical stumps and normal asymmetrical stumps, two distinct points emerge. Such grafts reliably produce supernumerary limbs when either the symmetrical blastema apposes the asymmetrical stump or when the asymmetrical blastema apposes the symmetrical stump. In addition, these extra limbs emerged at the precise circumferential positions and have the handedness that would be expected from the polar coordinate model (Holder & Tank, 1979). For example, a blastema derived from a double posterior stump produces a supernumerary limb on the anterior side of a normal stump and this supernumerary is of stump handedness. This can only occur if the grafted blastema comprises a set of posterior positional values which interact with the adjacent anterior stump values. That is, the blastema truly is a double posterior construction. The same observation applies to double anterior blastemas. The second important point to emerge from these experiments is that the symmetrical blastemas are not rescued by the asymmetrical stump upon which they are grafted. Once the stage of MB has been reached, these blastemas always cease regenerating and produce distally incomplete structures. Therefore it is clear that blastemas derived from long-healing-time symmetrical stumps are symmetrical in terms of the positional values which they bear and that this symmetrical arrangement of positional values is likely to be responsible for the early cessation of regeneration and growth.

CONCLUSIONS

This paper attempts to draw together the evidence from experiments involving symmetrical limb tissues which strongly indicate that local cell-to-cell interactions and intercalation govern the process of pattern regulation in the regenerating limb. The eventual pattern of structures regenerated from symmetrical or asymmetrical normal stumps depends on three major factors; the number of circumferential positional values present at the amputation plane, the spatial arrangement of these values around the circumference and the constraints of blastemal shape upon the directions of healing which bring cells from specific regions of the circumference together during blastemal outgrowth. The role of local cell contact is further emphasized when cells on either side of a symmetrical midline are prevented from contacting by a wound.

The symmetrical limb experiments discussed in this paper also demonstrate that the circumference of the limb plays a major role in determining the degree of distal outgrowth and the nature of the pattern of structures within the regenerate (see also Stocum, 1980*b*, *c*). The importance of local cell interactions and the intimate association between the amount of proximodistal regeneration and the circumferential properties of the amputation plane ensure that the polar coordinate model remains as a powerful theoretical tool with which to further our understanding of epimorphic regeneration and the link between growth and spatial patterning mechanisms.

It is a pleasure to thank Susan Bryant and Patrick Tank for their support in many ways and for sunny days in Irvine which they shared. I am also indebted to Susan Reynolds for all sorts of help with the manuscript and my colleagues in the Anatomy Department at King's College London for innumerable favours. I also thank Charleston Weekes for his expert technical assistance and his patience, and Philip Batten for the photography. My own work reported in this paper was supported financially by the Science Research Council.

REFERENCES

BRYANT, S. V., HOLDER, N. & TANK, P. W. (1981). Cell–cell interaction and distal outgrowth in amphibian limbs. *Am. Zool.* (in the press).

BRYANT, S. V., FRENCH, V. & BRYANT, P. J. (1981). Distal regeneration and symmetry. *Science* **212**, 993–1002.

CHALKLEY, D. T. (1954). A quantitative histological analysis of forelimb regeneration in *Triturus viridescens. J. Morph.* **94**, 21–71.

FRENCH, V. (1981). Pattern regulation and regeneration. *Phil. Trans. Roy. Soc.* (in the press).

FRENCH, V., BRYANT, P. J. & BRYANT, S. V. (1976). Pattern regulation in epimorphic fields. *Science* **193**, 969–981.

HOLDER, N. & TANK, P. W. (1979). Morphogenetic interactions occurring between blastemas and stumps after exchanging blastemas between normal and double half forelimbs in the axolotl, *Ambystoma mexicanum. Devl Biol.* **68**, 271–279.

HOLDER, N., TANK, P. W. & BRYANT, S. V. (1980). The regeneration of symmetrical fore-limbs in the axolotl, *Ambystoma mexicanum. Devl Biol.* **74**, 302–314.

ITEN, L. E. & BRYANT, S. V. (1973). Forelimb regeneration from different levels of amputation in the newt, *Notophthalamus viridescens*: length, rate and stages. *Wilhelm Roux Archiv. EntwMech. Org.* **173**, 263–282.

KRASNER, G. N. & BRYANT, S. V. (1980). Distal transformation from double half forelimbs in the axolotl, *Ambystoma mexicanum*. *Devl Biol.* **74**, 315–325.

MADEN, M. (1976). Blastemal kinetics and pattern formation during amphibian limb regeneration. *J. Embryol. exp. Morph.* **36**, 561–574.

MADEN, M. (1981). Supernumerary limbs in amphibians. *Am. Zool.* (in the press).

SMITH, A. R. & CRAWLEY, A. M. (1977). The pattern of cell division during growth of the blastema of regenerating newt forelimbs. *J. Embryol. exp. Morph.* **37**, 33–48.

STOCUM, D. L. (1978). Regeneration from symmetrical hindlimbs in larval salamanders. *Science* **200**, 790–793.

STOCUM, D. L. (1979). Stages of forelimb regeneration in *Ambystoma maculatum*. *J. exp. Zool.* **209**, 395–416.

STOCUM, D. L. (1980a). The relation of mitotic index, cell density and growth to pattern regulation in regenerating *Ambystoma maculatum* forelimbs. *J. exp. Zool.* **212**, 233–242.

STOCUM, D. L. (1980b). Autonomous development of reciprocally exchanged regeneration blastemas of normal forelimbs and symmetrical hindlimbs. *J. exp. Zool.* **212**, 361–371.

STOCUM, D. L. (1980c). Intercalary regeneration of symmetrical thighs in axolotl *Ambystoma Devl mexicanum. Biol.* **79**, 276–295.

TANK, P. W., CARLSON, B. M. & CONNELLY, T. G. (1976). A staging system for forelimb regeneration in the axolotl, *Ambystoma mexicanum. J. Morph.* **150**, 117–128.

TANK, P. W. & HOLDER, N. (1978). The effect of healing time on the proximodistal organisation of double half forelimb regenerates in the axolotl *Ambystoma mexicanum. Devl Biol.* **66**, 72–85.

J. Embryol. exp. Morph. Vol. 65 (Supplement), pp. 37–47, 1981
Printed in Great Britain © Company of Biologists Limited 1981

Sequence of regeneration in the *Drosophila* wing disc

By JANE KARLSSON[1]

From the Zoology Department, University of Edinburgh

SUMMARY

The sequence of appearance of different structures during regulation was ascertained for three wing disc fragments. It was found that structures were added sequentially from the cut edge. Regeneration in one fragment was very fast and successful, and it is suggested that this is difficult to explain in terms of wound healing being responsible for the decision to regenerate rather than duplicate. In another fragment, regeneration appeared to proceed from one part of the cut edge only, this being the region through which the anteroposterior compartment boundary runs. Compartment boundaries have been implicated in both proximodistal and circumferential regeneration, and it is suggested that regeneration is controlled at these sites. The decision to regenerate rather than duplicate may be made on the basis of which compartment boundaries are exposed at the cut edge, and their preferred direction of regeneration. The implications of this interpretation for the control of growth are discussed.

INTRODUCTION

A prerequisite for pattern formation in imaginal discs is that the correct number of cells, in the correct configuration, should be present at the time of differentiation. Since this is accomplished mainly by cell division, growth is an essential part of the pattern-forming process. Regeneration re-establishes this essential configuration, and therefore can be expected to provide information about the growth processes responsible.

If the wing disc is cut into two fragments, one regenerates during culture while the other duplicates (Bryant, 1975; Karlsson, 1981). This rule is in general very well obeyed, but there are two notable exceptions, in both of which fragments duplicate structures expected from the fate map but also regenerate others. In one, the regenerated structures are proximal (Karlsson & Smith, 1981), and in the other distal (Karlsson, 1980). The ability to regenerate distal structures while duplicating was found to be related to compartments; only fragments containing the ventral end of the anteroposterior (AP) compartment boundary were able to regenerate wing blade (Karlsson, 1980). After dissociation, fragments containing the dorsal end are also able to produce wing blade

[1] *Author's present address:* Genetics Laboratory, Department of Biochemistry, Oxford OX1 3QU, U.K.

(Wilcox & Smith, 1980). Thus it appears that the AP boundary is important in controlling proximodistal regeneration; this also seems to be the case in the leg disc (Schubiger & Schubiger, 1978), and implies that growth in this axis is also controlled at these places.

There are other observations which support this idea. Simpson (1976) has found that clones of slow-growing mutants disappear from the middle of compartments but not from the edges. This suggests that growth rules are different in the two places, a conclusion supported by the finding of Lawrence & Morata (1976) that small *engrailed* clones are capable of causing a shape distortion in the wing blade if they touch the dorsoventral compartment boundary but not otherwise. These authors suggested that this was because the compartment boundary was somehow instrumental in controlling growth, and that the reason internal clones caused no shape distortion was that they came under the control of the wild-type boundary.

Recently, further evidence for the involvement of compartment boundaries in regeneration has been reported (Karlsson, 1981). It was found that of two complementary wing disc fragments, the physically larger is not always the one which regenerates; in terms of the polar coordinate model (French, Bryant & Bryant, 1976), the smaller fragment has over half the circumferential positional values. Testing of a large number of fragment pairs led to the conclusion that the values are very unevenly spaced indeed, half of them being tightly clustered round the ends of the AP boundary. Since it is this boundary which affects proximodistal regeneration, this seemed to suggest a connexion between the ability to regenerate distally and the ability to regenerate rather than duplicate circumferentially, and that both phenomena were somehow connected to compartments.

It seemed possible that more information could be gained about this relationship by studying the sequence in which different structures were produced during regulation. For instance, it could be that the different parts of the cut edge contribute unequally to the regenerate, and it could also be that distal structures are produced in different ways from fragments containing different ends of the AP boundary. Two fragments were therefore chosen for detailed study, both having 8 of the 12 circumferential positional values and each containing a different end of the AP boundary. A third fragment was also studied; this was one which duplicated circumferentially and regenerated distally.

MATERIALS AND METHODS

Host and donor flies were of *ebony-11* genotype, and were raised at 25 °C on standard corn meal/agar/syrup medium seeded with live yeast. Wing discs were removed from late third-instar larvae in insect Ringer. Fragments were cut with tungsten needles and implanted into the body cavities of well-fed 1- to 3-day-old fertilized adult females, where they remained for varying periods.

They were then removed and re-implanted into the body cavities of late third-instar larvae, using the method of Ephrussi & Beadle (Ursprung, 1967). When the hosts emerged as adults, the metamorphosed implants were removed, mounted in Gurr's Hydramount, and scored for the structures shown in Fig. 1. Implants cultured for 0 days (Tables 1–3) were implanted directly into late third-instar larvae.

Terminology

For convenience, the four ends of the two compartment boundaries are referred to in the text as D-AP, V-AP (dorsal and ventral ends of the antero-posterior boundary) and A-DV, P-DV (anterior and posterior ends of the dorsoventral boundary).

RESULTS

Figure 1 shows a fate map of the wing disc, the fragments tested, and the day on which each structure first appeared (or was duplicated) at high frequency. Tables 1–3 show the frequencies with which the different structures were present.

The data show that structures were produced sequentially from the cut edge in a very orderly way. No structure far from the cut edge appeared until intervening structures were present. The amount of wing blade rose steadily with increasing culture periods, and the same was true for the numbers of wing margin (triple row) bristles which were counted in fragments (*a*) and (*b*). If a given structure appeared for the first time on one day, it was almost always present at maximum frequency the next. However, there were considerable differences in the behaviour of the two regenerating fragments.

Fragment (*a*)

The most noticeable difference between this fragment and fragment (*b*) was that wing blade and wing margin structures (triple row, double row and posterior row) appeared considerably sooner in fragment (*a*). Some wing blade was present in all implants injected directly into larvae (0 days of culture); that this was the result of regeneration is indicated by its complete absence from fragment (*b*) after 0 days, since the two fragments were cut at exactly the same distance from the edge of the disc.

It seems to be the case that the time of appearance of structures close to a particular part of the cut edge is indicative of whether it contributes to the regenerate (Abbott, Karpen & Schubiger, 1981). Thus the very early appearance of wing blade in fragment (*a*) indicates that the ventral half of this fragment, which lies next to the wing blade, contributes from the start of regeneration. In fact, since no wing blade can be produced without the ventral end of the AP boundary (V-AP), it is probably this point, at the extreme ventral end of the

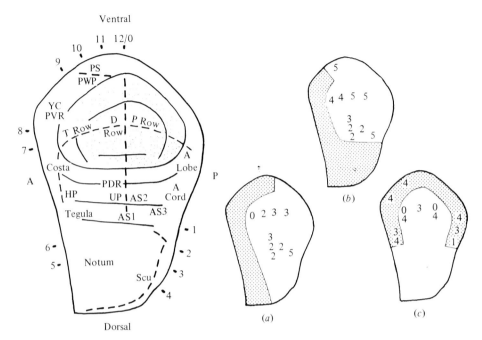

Fig. 1. Fate map of the wing disc, the fragments tested, and the sequence of regeneration.

Left: fate map of the wing disc (after Bryant, 1975). Presumptive wing blade is stippled, and compartment boundaries are shown in broken lines. The dorsoventral boundary runs along the wing margin. The position of the anteroposterior boundary was ascertained by Brower, Lawrence & Wilcox, 1981, using clones of a succinic dehydrogenase mutant. Positional values 1–12/0 from Karlsson (1981).

Right: fragments (a), (b) and (c). The numbers outside the fragments denote the first day on which a structure was present at a frequency of 30 % or more. Along the wing margin, in fragments (a) and (b), the first number refers to the day on which triple row appeared at a frequency of 30 %, and the next to the day on which the average number of triple row bristles reached 10.

The numbers inside fragment (c) denote the first day on which a structure was duplicated more often than not. Along the wing margin, the upper figure refers to the first day on which a structure was present at a frequency of 30 % and the lower figure to the first day on which it was more often duplicated than not.

Abbreviations:

Scu, scutellum; HP, humeral plate; UP, unnamed plate; AS 1, 2, 3, first, second and third axillary sclerites; T, D, Rows, triple and double rows of chaetes; P Row, posterior row of hairs; PDR, proximal dorsal radius; A Lobe, alar lobe; A Cord, axillary cord; YC, yellow club; PVR, proximal ventral radius; PS, pleural sclerite; PWP, pleural wing process.

Table 1. *Structures differentiated by implants of fragment (a)*

Implants were either injected directly into larvae (0 days of culture), or cultured in adult females for varying periods before transfer to larvae for metamorphosis. The figures refer to the percentage of implants where the marker was completely or partially present, singly or in duplicate. The amount of wing blade (WB) was estimated by eye on an arbitrary scale of 1–5, where 5 is a completely regenerated wing blade, and the numbers of bristles in the medial triple row counted; standard errors are shown. Abbreviations as in legend to Fig. 1.

Days of culture...	0	1	2	3	4	5
n...	20	16	11	15	13	23
Notum	100	100	100	80	100	96
Scu				7	15	17
Tegula	70	100	91	67	92	87
HP	45	44	73	47	92	61
UP	—	—	45	53	54	70
AS 1	—	—	45	60	38	74
AS 2	—	—	36	60	38	70
AS 3	—	—	—	7	15	57
Costa	90	94	64	67	100	78
T row	30	44	45	80	77	91
D row	—	—	—	53	46	70
P row	—	—	—	53	38	78
PDR	—	—	9	60	69	74
A lobe	—	—	—	13	8	22
A cord	—	—	—	7	—	4
YC/PVR	100	100	82	73	92	65
PS/PWP	100	94	82	73	92	52
WB	95	100	100	100	100	100
Amount of WB	0.8 ± 0.1	1.2 ± 0.1	2.8 ± 0.4	3.3 ± 0.1	3.3 ± 0.3	3.8 ± 0.1
No. of T row bristles	1.9 ± 1.0	3.9 ± 1.5	9.8 ± 3.8	15.7 ± 4.0	15.0 ± 4.6	24.7 ± 4.3

fragment, at which regeneration starts. This can only be inferred since all parts of the wing blade appear the same in implants.

Whether the dorsal half of the fragment contributes is less clear. The structures of the dorsal hinge appeared somewhat later than the wing blade and could possibly have been produced from it; however, if this were so, the first hinge structure to appear should be the proximal dorsal radius (see Fig. 1), and in fact this appeared one day later than the remaining ones. Thus it seems likely that some of the dorsal edge does contribute. The extreme dorsal end possibly does not, since scutellum, which lies close to it, never appeared at high frequency.

The above observations suggest that a wave of regeneration spreads along the cut edge from ventral to dorsal.

Regeneration was very successful in this fragment. Few implants failed to regenerate, and only six had duplicated structures, a single one in each case (two cases each of pleural sclerite, humeral plate and third axillary sclerite).

Regeneration could be completed in as little as 3 days; the only structure to continue increasing in frequency after this time was third axillary sclerite. Axillary cord, alar lobe and scutellum never appeared at high frequency; Bryant (1975) has cultured a similar fragment for 7 days, after which these structures were still present at low frequency, so it seems likely that in many cases they never appear.

Fragment (b)

Wing blade appeared in this fragment considerably later than in fragment (a). The first structures to appear, on day 2, were those of the dorsal hinge, which lie close to the edge in the dorsal half of the fragment. It seems very likely that these must be produced before wing blade can appear, as virtually all (18/22) implants having wing blade after 2 and 3 days also had one or more of these structures. Thus wing blade is probably produced at first entirely from this region. The ventral half, which lies close to triple row, appears not to contribute at least until day 4, which is the first day on which this structure appeared in any quantity. It seems quite likely that the very ends of the fragment do not con-

Table 2. *Structures differentiated by implants of fragment (b)*

Implants were either injected directly into larvae (0 days of culture), or cultured in adult females for varying periods before transfer to larvae for metamorphosis. For other details see legend to Table 1. Abbreviations as in legend to Fig. 1.

Days of culture...	0	1	2	3	4	5
n...	13	19	21	20	17	14
Notum	100	95	90	90	94	100
Scu	69	58	38	30	29	50
Tegula	77	42	90	70	65	64
HP	62	53	68	40	47	64
UP	—	11	33	35	41	36
AS 1	—	21	38	40	41	43
AS 2	—	11	33	35	41	50
AS 3	—	—	5	5	18	50
Costa	92	68	81	55	71	79
T row	15	5	19	20	41	71
D row	—	—	—	5	24	50
P row	—	—	—	—	24	71
PDR	—	—	24	30	29	71
A lobe	—	—	—	—	12	14
A cord	—	—	—	—	—	7
YC/PVR	77	63	71	55	47	50
PS/PWP	—	—	—	5	12	50
WB	8	16	57	55	65	79
Amount of WB	0	0	0.9 ± 0.2	1.0 ± 0.3	1.5 ± 1.5	3.1 ± 1.7
No. of T row bristles	0.8 ± 0.5	0.2 ± 0.3	1.9 ± 0.9	1.8 ± 1.2	11.5 ± 4.3	23.6 ± 5.0

tribute at all, since structures lying very close to the edge at these sites did not appear until day 5, and yellow club, at the extreme ventral end, was even duplicated in many cases.

A high frequency of duplication is the other major difference between this fragment and fragment (*a*); a total of 26 implants had duplicated structures compared with 6 in fragment (*a*). These were mostly yellow club (18 cases of yellow club, 7 each of humeral plate and notum, 2 each of scutellum and tegula, 1 each of costa, unnamed plate, and first and second axillary sclerites). Seven of the 26 had regenerated structures as well.

Thus it seems that the region of the dorsal hinge, through which the AP boundary runs, produces most and possibly all of the regenerate, while the rest of the cut edge is either doing nothing or duplicating. Indeed, many implants did nothing whatsoever and differentiated so poorly that they could not be scored and were not included in the data. This is always true of a few implants, but seldom to the extent seen in this fragment.

Fragment (c)

Most implants of this fragment regenerated distally and duplicated circumferentially; about 20 % regenerated completely with little or no evidence of duplication. Wing blade was produced very early as in fragment (*a*). Wing blade lies closer to the edge of the disc in the posterior than in the anterior, and approximately 1 unit of wing blade (on a scale of 1–5) was expected from the fate map (Karlsson, 1980). In fact, 2·5 units were present in implants injected directly into larvae. It appears likely that the whole of the distal-facing cut edge contributes to the production of wing blade, since the amounts of triple row and posterior row both increased on days 1 and 2, and double row, which lies at the distal wing tip, only appeared in implants having virtually complete triple and posterior rows. It was not possible to count the numbers of triple row bristles in implants of this fragment because this row was often duplicated and the bristles lay on top of one another.

Duplication appeared to spread inwards from the ends of the fragment; third axillary sclerite, at one end, was always duplicated after only 1 day of culture. Tegula, at the other end, was according to Table 3 not duplicated to any extent until day 4, but this is probably misleading since only part of the tegula was present and duplication in this part can be very difficult to detect. Humeral plate, which lies next to the tegula, was duplicated more often than not on day 3, and this was only true for those structures furthest from the ends after day 4. Thus duplication spreads from the posterior end and probably from the anterior end as well. Wing margin structures (triple and posterior rows) were only duplicated to any great extent by day 4, indicating that duplication of proximal structures is followed or accompanied by duplication of the newly regenerated wing blade.

Table 3. *Structures differentiated by implants of fragment* (*c*)

Implants were either injected directly into larvae (0 days of culture), or cultured in adult females for varying periods before transfer to larvae for metamorphosis. The figures refer to the percentage of implants where the marker was completely or partially present. The amount of wing blade (WB) was estimated by eye on an arbitrary scale of 1–5; standard errors are shown. +, structure present. D, structure present in duplicate. Abbreviations as in legend to Fig. 1.

Days of culture...	0		1		2		3		4		5	
n...	14		12		14		12		12		19	
	+	D	+	D	+	D	+	D	+	D	+	D
Tegula	29	—	8	—	43	—	25	—	17	25	42	5
HP	64	7	33	8	29	14	25	33	8	25	26	26
AS 3	57	—	—	83	—	86	—	67	8	67	16	58
Costa	100	—	83	—	71	—	92	—	33	58	63	26
T row	36	—	50	—	43	—	75	8	17	58	58	26
D row	—	—	—	—	29	—	50	—	33	—	47	—
P row	100	—	92	—	93	7	83	—	42	42	89	5
A lobe	93	—	92	8	79	7	67	17	42	58	47	47
A cord	64	—	76	—	86	—	17	33	50	42	37	26
YC/PVR	100	—	92	—	86	7	67	33	8	75	21	68
PS/PWP	100	—	100	—	71	21	67	33	8	83	16	74
WB	100	—	100	—	100	—	100	—	100	—	100	—
Amount of WB	$2 \cdot 5 \pm 0 \cdot 2$		$3 \cdot 4 \pm 0 \cdot 2$		$4 \cdot 1 \pm 0 \cdot 1$		$4 \cdot 1 \pm 0 \cdot 2$		$3 \cdot 7 \pm 0 \cdot 3$		$4 \cdot 5 \pm 0 \cdot 1$	

DISCUSSION

The results show that pattern regulation in the fragments studied proceeds sequentially and in an orderly way from the cut edge, and suggest that the regenerate can be produced from one part of the edge only. Both these results have been obtained by Abbott and co-workers (1981) in the leg disc; these authors have shown by means of clonal analysis that a part of the cut edge which appears, according to the sequence of structures produced, not to contribute, does indeed not do so.

It is not wholly expected that regeneration should proceed sequentially from the cut edge, since it is thought to be intercalary, and therefore to resemble an averaging process. Averaging should produce those structures furthest from the cut edge first, since these would have positional values intermediate between those at either end. Thus this result suggests that once the initial decision to regenerate or duplicate is made, the rest of the regenerative process proceeds autonomously until all cells have their correct nearest neighbours again.

Since therefore wound healing does not appear to be necessary in the later stages of regulation, one might ask whether it is necessary in the initial stages when the decision to regenerate or duplicate is made. If it is, the behaviour of fragment (*a*) is not easy to explain. At one end of this fragment, three positional values are crammed into a very small space; without this small piece the fragment would duplicate (Karlsson, 1981). This means that the rest of the fragment has to be aware of the presence of this small piece. If wound healing is to be responsible for the decision to regenerate rather than duplicate, it must somehow ensure that this small piece is in contact with the whole of the rest of the cut edge. Implants of this fragment inspected after 1 day of culture do appear to have curled up and therefore probably healed, but in a rather random fashion, some like the letter S, and others like P, C or O (Karlsson, unpublished observations). Nevertheless, this fragment is extremely successful at regenerating and only very rarely produces duplicated structures. One suspects that other mechanisms besides wound healing might be responsible for providing the signal which leads to the decision to regenerate or duplicate. Since no wing blade can be produced in this type of fragment without the ventral end of the AP boundary (V-AP), the initial signal must come from this region; the sequence of appearance of the different structures suggests that cell division spreads from this part of the edge. Thus the signal might in this case be quite simply an advancing front of cell division.

The spacing of positional values suggests that the AP boundary plays a crucial role in the decision to regenerate or duplicate. There also appears to be some value clustering in the region of the A-DV; fragments which contain more than a certain minimal amount of this boundary, plus either end of the AP boundary, are able to regenerate (Karlsson, 1981). From this it seems possible that the decision is made on the basis of which boundaries are contained in the fragment, certain pairs being able to co-operate in stimulating regeneration.

A possible clue to the role of the AP boundary in regeneration comes from its behaviour in isolation. While the ventral end has a strong tendency to regenerate distally (Karlsson, 1980), the dorsal end appears to have the reverse tendency. Thus many duplicating fragments which lack presumptive notum, through which the AP boundary runs, regenerate notum at high frequency. Notal structures are added in sequence from distal to proximal, and regeneration stops once most of the notum is present (Karlsson & Smith, 1981). That the boundary is somehow responsible seems very likely because fragments lacking any part of it are unable to regenerate notum, and the structures produced lie along the boundary. Thus the two cases of simultaneous regeneration and duplication seem to indicate that new proximodistal positional values can be added at compartment boundaries, and that the two ends of the AP boundary preferentially add values in opposite directions.

These observations lead to the suggestion that the involvement of compartments in regeneration means that new proximodistal values must be laid down at the boundaries before they can be added elsewhere. It might then be fruitful to consider whether the regeneration/duplication decision could be entirely the result of this process. A fragment containing two boundaries which can co-operate in regenerating distally to the extent of producing half the wing blade, say, might be able to continue regenerating until the entire disc was present. Exactly the same process would be happening in the complementary fragment, but here of course it would mean duplication.

Thus the very successful regeneration of fragment (*a*) in the present experiments might be the result of the strong tendency for distal regeneration of the V-AP, together with the ability of the A-DV to sustain this process. The much less successful regeneration of fragment (*b*) might result from the tendency of the D-AP to regenerate proximally, which must be overcome before distal regeneration can occur.

If the above interpretation of regeneration and duplication is correct, it provides support for the idea that growth is controlled at compartment boundaries. New proximodistal values may be laid down at these boundaries before they are laid down elsewhere. This could be the reason clones do not grow across them; if new growth emanates from these lines in both directions, clones will be unable to invade this region. If the boundary consists of a double row of cells, which can throw off daughters proximodistally and to one side but not the other, clones will follow the lines but never cross them.

Nothing as yet is known about the way in which mitosis might be used to create complex shapes. One extreme hypothesis would be that each cell, at each cell division, is given specific instructions about whether and how to divide. For the wing disc, this would mean about 50000 separate instructions, and one might ask whether such a large amount of information is really necessary to produce the not overly complex shape of the mature wing disc. Indeed, if all cells *were* given precise instructions then clone shape would be determinate, which of course it is generally not. The very fact that it is determinate at compartment boundaries suggests that cells have precise growth instructions here and not elsewhere.

An alternative, then, to giving all cells specific growth instructions is to give them only to cells at strategic positions, and allow other cells to follow passively. If this is what is happening, it provides an explanation for Simpson's (1976) finding that clones of slow-growing mutants disappear from the middle of compartments but not at the edges; if cells in the middle all have the same instruction, namely to divide when space is available, a slow-growing cell would never be permitted to divide because the space around it would always be taken up by its faster-growing neighbours.

If the above analysis is valid, it suggests that it might be fruitful to look for these lines in other organisms. It may be that this principle is a general one, used wherever mitosis is the primary means of creating complex shapes.

This work was supported by an M.R.C. studentship and a Beit Memorial Fellowship.

REFERENCES

ABBOTT, L. C., KARPEN, G. H. & SCHUBIGER, G. (1981). Compartmental restrictions and blastema formation during pattern regulation in *Drosophila* imaginal leg discs. *Devl Biol.* (In the Press.)

BROWER, D. L., LAWRENCE, P. A. & WILCOX, M. (1981). Clonal analysis of the undifferentiated wing disc of *Drosophila. Devl Biol.* (In the Press.)

BRYANT, P. J. (1975). Pattern formation in the imaginal wing disc of *Drosophila melanogaster*: fate map, regeneration and duplication. *J. exp. Zool.* **193**, 49–78.

FRENCH, V., BRYANT, P. J. & BRYANT, S. V. (1976). Pattern regulation in epimorphic fields. *Science* **193**, 969–981.

KARLSSON, J. (1980). Distal regeneration in proximal fragments of the wing disc of *Drosophila. J. Embryol. exp. Morph.* **59**, 315–323.

KARLSSON, J. (1981). The distribution of regenerative potential in the wing disc of *Drosophila. J. Embryol. exp. Morph.* **61**, 303–316.

KARLSSON, J. & SMITH, R. J. (1981). Regeneration from duplicating fragments of the *Drosophila* wing disc. *J. Embryol. exp. Morph.* (In the Press.)

LAWRENCE, P. A. & MORATA, G. (1976). Compartments in the wing of *Drosophila*: a study of the *engrailed* gene. *Devl Biol.* **50**, 321–337.

SCHUBIGER, G. & SCHUBIGER, M. (1978). Distal transformation in *Drosophila* leg imaginal disc fragments. *Devl Biol.* **67**, 286–295.

SIMPSON, P. (1976). Analysis of the compartments of the wing of *Drosophila melanogaster* mosaic for a temperature-sensitive mutation that reduces mitotic rate. *Devl Biol.* **54**, 100–115.

URSPRUNG, H. (1967). *In vivo* culture of *Drosophila* imaginal discs. In *Methods in Developmental Biology* (ed. F. H. Wilt & N. K. Wessels). New York: Crowell.

WILCOX, M. & SMITH, R. J. (1980). Compartments and distal outgrowth in the *Drosophila* imaginal wing disc. *Wilhelm Roux' Arch. devl Biol.* **188**, 157–161.

J. Embryol. exp. Morph. Vol. 65 *(Supplement)*, pp. 49–76, 1981
Printed in Great Britain © *Company of Biologists Limited, 1981*

Analysis of *vestigial*^{W} (vg^W): a mutation causing homoeosis of haltere to wing and posterior wing duplications in *Drosophila melanogaster*

By MARY BOWNES[1] AND SARAH ROBERTS[1]

From the Department of Molecular Biology, University of Edinburgh

SUMMARY

vg^W is a homozygous lethal mutation killing embryos prior to formation of the syncitial blastoderm. In heterozygous condition it causes duplications of the posterior wing, ranging from very small duplications of the axillary cord and alar lobe to large duplications including much of the wing blade and the posterior row of bristles. No anterior margin structures are ever observed. The thorax is sometimes slightly abnormal, but rarely shows large duplications. The size of the wing is related to the number of pattern elements deleted or duplicated.

Heterozygous vg^W flies also show homoeosis of the haltere to wing. This occurs in the capitellum, where wing blade is observed, but no wing margin structures are found. As with the *bithorax* (bx) mutation which transforms anterior haltere to anterior wing this aspect of the phenotype is repressed by the *Contrabithorax* (*Cbx*) mutation. The transformed haltere discs show more growth than wild-type haltere discs.

Flies heterozygous for vg^W also show a high frequency of pupal lethality, those forming pharate adults generally show the most extreme vg^W phenotype.

No cell death has been observed in the imaginal discs of third instar larvae, suggesting that if the wing defects result from cell death this must occur early in development. The homoeosis in the haltere discs and duplications of the wing disc are reflected by the altered morphology and growth of these discs.

There are some minor differences in the expressivity of the phenotype when flies are reared at different temperatures. Chromosome substitutions suggested that all aspects of the phenotype related to the vg^W mutation and that other mutations had not occurred in the stock. Cytological analysis indicated that vg^W is a deletion or inversion on the right arm of chromosome 2 from 47F/48A to 49C.

Complementation studies with various mutants thought to be located within the deletion, or inversion and which affect wing morphology have been undertaken.

Cbx causes transformations of wing to haltere; this occurs in the posterior compartment far more frequently than in the anterior compartment. *Cbx*; vg^W flies have wings where one of the duplicates is no longer present, presumably transformed to haltere, though this is difficult to identify. One copy of the axillary cord, alar lobe etc, the structures commonly duplicated in vg^W, are present, *but* they are the anterior duplicate rather than the original posterior copy of these structures. Thus *Cbx* acts upon genuine posterior structures but not those posterior structures in vg^W which form in anterior wing locations, suggesting that although these structures differentiate into posterior wing, to the *Cbx* gene product the cells are still 'anterior'.

[1] *Authors' address:* Department of Molecular Biology, University of Edinburgh, King's Buildings, Mayfield Road, Edinburgh EH9 3JR, U.K.

INTRODUCTION

There are several approaches to the analysis of growth and the development of pattern. In many systems the most helpful answers can be obtained by direct experimental manipulation of the tissues and organs being studied. This has been valuable in *Drosophila* especially in the determination of how growth and pattern are regulated during duplication and regeneration of the imaginal discs. It is, however, far more difficult to analyse how growth and pattern formation are regulated in normal *Drosophila* development. One way of at least finding some of the features involved is to analyse mutations which interfere with either growth, such as mutations causing cell death, or with pattern, such as the homoeotic mutants which have replacement of one body structure with another, or mutations causing morphological defects such as deletions and duplications of pattern elements in the larva or adult. It is hoped that detailed studies of these mutants will enable us to understand how the wild-type genotype regulates and controls growth and the establishment of pattern in *Drosophila*.

The *vestigial* (*vg*) mutations in *Drosophila* cause deletions of parts of the wing. The size and nature of the deletions depend upon the particular *vg* allele present in the stock (Lindsley & Grell, 1968; Waddington, 1940). The size of the mutant wing is related to the number of pattern elements which differentiate; smaller wings have fewer pattern elements. The *vg* wing phenotype seems to be the result of position-specific cell death in some of the *vg* mutants (Fristrom, 1968, 1969; Bownes & Roberts, 1981). The size of the presumptive wing blade is reduced in *vg* wing discs indicating altered growth characteristics in this region during larval development. The capitellum of the haltere is also reduced in size in the *vg* mutants.

Several overlapping deletions are known which include the *vg* locus, and many of these are homozygous lethals affecting embryonic development (Lindsley & Grell, 1968). A new dominant *vg* allele was recently isolated by Shukla, and named vestigial wingless. The preliminary report in Drosophila Information Service (Shukla, 1980) indicated that it not only caused a dominant wing reduction, but also caused homoeosis of haltere to wing in some offspring. As only one other *vg* mutant causes dominant wing reduction, vg^U and is a chromosomal inversion, and none of the other *vg* alleles caused homoeotic transformations, we have investigated this new mutant in more detail.

One of our first observations was that in fact the vg^W mutation causes duplication of posterior wing structures and a deletion of anterior wing structures. This duplication was in marked contrast to our studies on *vg* which caused no duplications of wing structures (Bownes & Roberts, 1981).

The paper describes the phenotype of vg^W wings, halteres and wing and haltere discs in detail and presents cytological evidence to show that it is a deletion or inversion on the second chromosome including the *vg* locus. Complementation studies were undertaken with other *vg* mutations. The chromosomal defect extends close to, or includes the *engrailed* (*en*) gene, which causes anterior wing duplications, so preliminary complementation studies with *engrailed* mutants are described (Garcia-Bellido, Ripoll & Morata, 1973; Lawrence & Morata, 1976; Morata & Lawrence, 1975, 1978). The genetics suggests that vg^W is an inversion.

Homoeotic mutations alter the state of determination in specific presumptive adult cells. This leads to a substitution of pattern elements in the adult and altered growth characteristics in the affected imaginal disc. The mutant *bithorax* (*bx*) causes homoeosis of anterior haltere to anterior wing and *postbithorax* (*pbx*) transform the posterior haltere to posterior wing. (Morata & Garcia-Bellido, 1976; Lewis, 1963, 1964, 1967, 1968, 1978). A further mutant from the bithorax complex *Contrabithorax* (*Cbx*) represses the homoeosis caused by *bx*, and itself causes transformations of wing to haltere (Lewis, 1963, 1964; Morata, 1975). Thus we finally studied the interactions between *Cbx* and vg^W in the hope that it would provide some insight into the mechanism of action of the vg^W mutation.

MATERIALS AND METHODS

Maintenance of flies

All flies were maintained on freshly-yeasted, sugar, cornmeal, yeast and agar medium at 25 °C, unless otherwise stated in the Results Section.

Mutations used

The mutations used for characterizing vg^W are listed in Table 1. For description of the symbols in marked stocks see Lindsley & Grell (1968). This table also supplies information on the phenotype of the mutants, who supplied the stock, and its chromosomal location, if it is a previously studied deficiency. Where more than one stock is listed for maintaining the mutation, the first was supplied to us, and the others were constructed in our lab to facilitate identification of chromosomes in genetic crosses.

Our wild-type stock was Oregon R, OrR.

Other stocks used were: $\dfrac{\text{Binscy}}{\text{Binscy}}$; $\dfrac{\text{rucuca}}{\text{TM3}}$.

Examination of adults

Adults resulting from various crosses were first examined using a dissecting microscope, then the thorax was mounted between two coverslips using Gurr's water mounting media. Detailed analysis of structures present was made using Zeiss Nomarski Interference optics.

Examination of embryos

Embryos were collected for 2–3 h, washed and dechorionated in 3 % sodium hypochlorite. They were lined up on double-sticky tape, covered in Voltalef oil and observed immediately and subsequently at frequent intervals until hatching, using Zeiss Nomarski interference optics.

Table 1. *Mutations used in genetic analysis of* vg^W

Mutant	Phenotype	genetic location	Supplied by	reference (not all refs to each mutant are listed)	Stocks used
vg^W	heterozygote; vestigial wing, homoeosis of haltere wing homozygote; lethal.	Df(2R) 47F–49C	Shukla	Shukla, 1980 and this manuscript.	$\dfrac{vg^W}{+}$; $vg^W/In(2L)t\ In(2R)Cy\ Roi\ cn^2\ bw$ or sp^2; $\dfrac{al\ dp\ b\ cn\ vg^W\ bw}{al\ dp\ b\ cn+bw}$
vg^B	homozygote; lethal. heterozygote; occasional wing nicks.	Df(2R) 49D3/4–50 A2/3	Bull	Bull 1966	$\dfrac{vg^B}{SM5}$; $vg^B/In(2L)t\ In(2R)Cy\ Roi\ cn^2\ bw$ or sp^2.
vg^C	homozygote; lethal. heterozygote; occasional wing nicks.	Df(2R) 49B2/3–49 E7/F1	*Cal. Tech. Stock Centre	Bull 1966	$\dfrac{DF(2R)vg^C\ vg^C}{In(2LR)Rev^B\ Rev^B}$; $\dfrac{Df(2R)vg^C\ vg^C}{In(2L)t\ In(2R)Cy\ Roi\ cn^2\ bw$ or sp^2}
vg^D	homozygote: lethal. heterozygote; sparse hairs and bristles on thorax.	Df(2R) 49C1/2–49 E2/6	Cal. Tech. Stock Centre	Bull 1966	$\dfrac{vg^D}{SM5}$; $vg^D/In(2L)t\ In(2R)Cy\ Roi\ cn^2\ bw$ or sp^2.
vg^S	homozygote; lethal. heterozygote; wing nicks in some flies.	Df(2R) 49B12/C1–49F1/50A1	Cal. Tech. Stock Centre	Bridges, Morgan & Schultz 1938	$\dfrac{vg^S}{Cy\ L4\ sp^2}$; $vg^S/In(2L)t\ In(2R)Cy\ Roi\ cn^2\ bw$ or sp^2.
vg^U	homozygote; lethal. heterozygote; vestigial	In(2R) 49C1/2–50C/2	Cal. Tech. Stock Centre		vg^U $\overline{In(2L)t\ In(2R)Cy\ Roi\ cn^2\ bw$ or $sp^2.}$

Allele	Description	Cytology	Source	Reference	Genotype
vg	homozygote; vestigial wings.	placed between 49C1/2–50C/2	Cal. Tech. Stock Centre	Bridges & Morgan 1919 Fristrom 1968	vg
Cbx	homozygote; heterozygote; homoeosis of wing → haltere.	89E1/2	Lewis	Lewis, 1978, Morata 1975	Cbx
bx^{34e}	homozygote; homoeosis of anterior haltere → anterior wing.	89D	Lewis	Lewis 1978 Morata & Garcia-Bellido 1976	bx^{34c}
en^{L1031}	homozygote; lethal abnormal segments in embryo.	48A	Lawrence	Nusslein-Volhard & Wieschaus 1980	$\dfrac{cn\ en^{L1034}\ bw\ sp}{CyO}$
en^1	homozygote: anterior wing duplications.	48A	Lawrence	Lawrence & Morata 1976	$\dfrac{lt\ stw\ en^1}{SM5};\ \dfrac{cn\ en^1}{SM5}$
en^A	homozygote; lethal.	Df(2R)47D3-48B4/5	Gubb	per. comm.	$\dfrac{w\ Df\ en^A}{w},\ \dfrac{mwh\ jv\ tra}{CyO},\ \dfrac{cn\ Df\ en^A}{TM3};\ \dfrac{}{CyO}$
en^B	homozygote; lethal.	Df(2R)47E3-48B2	Gubb	per. comm.	$\dfrac{pr\ pwn\ Df\ en^B}{CyO};\ \dfrac{cn\ Df\ en^B}{CyO}$
en^{30}	homozygote; lethal.	Df(2R)48A-48C6/8	Russell	per. comm.	$\dfrac{Df(2)en^{30}}{SM5},B^s y$

* Cal. Tech. = California Institute of Technology Stock Centre.

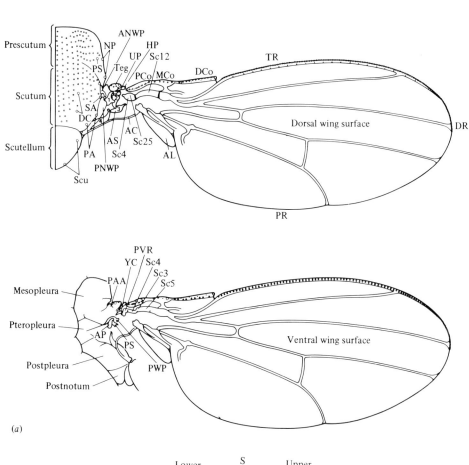

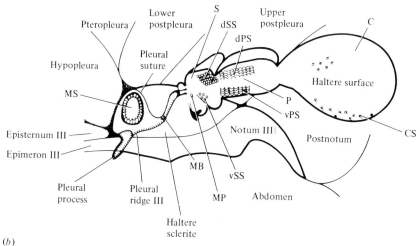

Fig. 1.(a) Drawings of the dorsal and ventral wing structures. (b) Drawing of the haltere structures. (Figure taken from Bownes & Seiler, 1977. Copyright Alan R. Liss Inc.)

Examination of imaginal discs

Imaginal discs were dissected from vg^W heterozygotes at the beginning, middle and end of the third larval instar. They were stained with a Ringer's solution (Chan & Gehring 1971) containing neutral red and trypan blue which stains dead cells blue. They were then observed using Zeiss Nomarski Interference optics.

Cytological analysis of chromosomes

Salivary glands were dissected into 3:1 ethanol:propionic acid to fix them, the nuclei were stained with aceto-orcein, and they were squashed in 45 % acetic acid to prepare the polytene chromosomes. The squash preparations were observed with Zeiss phase optics.

RESULTS AND DISCUSSION

(A) *The phenotype of* vg^W *wings and halteres*

Identifying the phenotype of abnormal wings and halteres depends upon our detailed knowledge of the structure of the wild-type wing and haltere. These are shown as line drawings, Fig. 1 (taken from Bownes & Seiler, 1977) and as photographs in Figs 2a and 3a, and it will be essential to refer to particular structures shown in these figures throughout the text.

Homozygotes

When vg^W heterozyotes are mated 27 % of the embryos die prior to migration of nuclei to form the syncitial blastoderm, and no further lethality is observed prior to larval hatching. This is consistent with vg^W homozygotes being lethal very early in development. We have not investigated whether any nuclear divisions occur in these embryos. Since a vg^W egg fertilized by a wild-type sperm

---

ABBREVIATIONS USED THROUGHOUT FIGURES

Wing abbreviations: PS, pleural sclerite; DC, dorsocentral bristles; PA, postalar bristle; Scu, scutellar bristles; SA, supraalar bristles; NP, notopleural bristles; ANWP, anterior notal wing process; Teg, tegula; AS 1–4, axillary sclerites 1–4; Sc 4, group of 4 sensilla campanoformia; AC, axillary cord; SC25, group 25 sensilla campanoformia; AL, alar lobe; PR, posterior row; DR, double row; TR, triple row; D Co., distal costa; M Co., medial costa; P Co., proximal costa; SC12, group 12 sensilla campanoformia; HP, humeral plate; UP, unnamed plate; PNWP, posterior notal wing process; PAA, prealar apophysis; YC, yellow club; PVR, proximal ventral radius; Sc4, group 4 sensilla campanoformia; Sc3, group 3 sensilla campanoformia; Sc5, group 5 sensilla campanoformia; AP, axillary pouch; PS, pleural sclerite; PWP, pleural wing process. WB, wing blade. VWS ventral wing surface; DWS dorsal wing surface.

Haltere abbreviations: MS, metathoracic spiracle; S, scabellum; d, dorsal; v, ventral; C, capitellum; CS, capitellar sensilla; P, pedicel; PS, pedicellar sensilla; SS, scabellar sensilla; MP, metathoracic papillae; MB metathoracic bristle group.

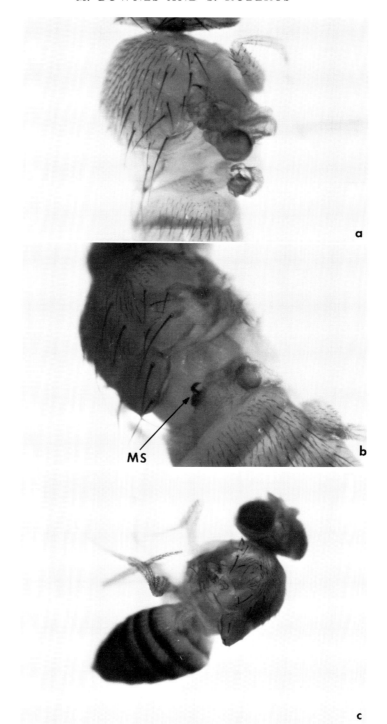

Table 2. *The phenotype of* vg^W *at 25 °C*

	Flies showing phenotype (%)
Haltere → wing homoeosis	83
Haltere absent	14
Haltere normal	1
Haltere abnormal	3
Wing-large posterior duplications (Classes 5 & 6 Fig. 5)	51
Wing-large posterior duplications and small haltere-like patches of tissue	7
Wing-small posterior duplication (Classes 1–4 Fig. 5)	46
Duplications or bristle misarrangement within thorax	10
Total number of half flies scored	140

will develop normally, yet a similar egg fertilized by a *vg^W* sperm dies early in embryogenesis, one might conclude that a gene product which is defective in *vg^W* zygotes is required very early in development for normal embryogenesis to proceed. This would be an unusual example of an early zygotic effect on development, as in general, there is no evidence for new gene activity prior to cellularization of the blastoderm (Wright, 1970; Zalokar, 1976; Anderson & Lengyel, 1979; McKnight & Miller, 1976).

Heterozygotes

Flies heterozygous for *vg^W* show a number of morphological abnormalities. They often have missing halteres or the halteres are transformed into wing. The thoraces sometimes show bristle misarrangements and small local duplications. The penetrance of these phenotypic variations is given in Table 2, and examples of the phenotype are shown in Figs. 2–4. Some flies have abnormal legs, but this aspect of the *vg^W* phenotype has not yet been analysed in detail.

Rearing the flies at 18 °C or 29 °C does not significantly alter the types of defect observed at 25 °C. At 18 °C the wings tend to be smaller than at 25 °C, which is consistent with other *vg* mutants where the wings are larger when flies are reared at higher temperatures.

It is difficult to establish which regions of the haltere are transformed to wing, since no wing margin is observed (Fig. 4b). Very occasionally *vg^W* flies show some metathorax transformation to mesothorax (Fig. 2b) suggesting that

Fig. 2. *vg^W*/+ (a) Shows a fly with a relatively small wing duplication and homoeotic transformation of haltere to wing. (b) An example where there is some mesothorax in the metathoracic segment (ms). (c) Shows a fly with abnormal bristle orientations on the mesothorax.

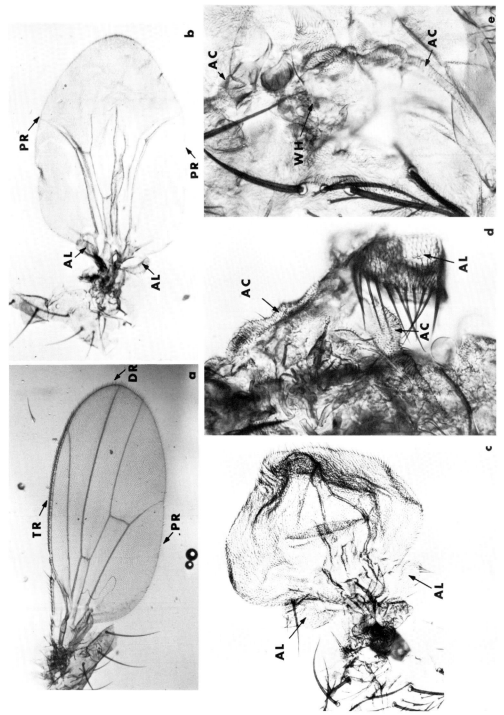

Fig. 3. Types of wing duplications found in $vg^W/+$ flies. (a) Wild-type wing. (b) Large wing duplication including posterior row and much of the wing blade. (c) Smaller wing duplication also including posterior row. (d) Small wing duplication extending to

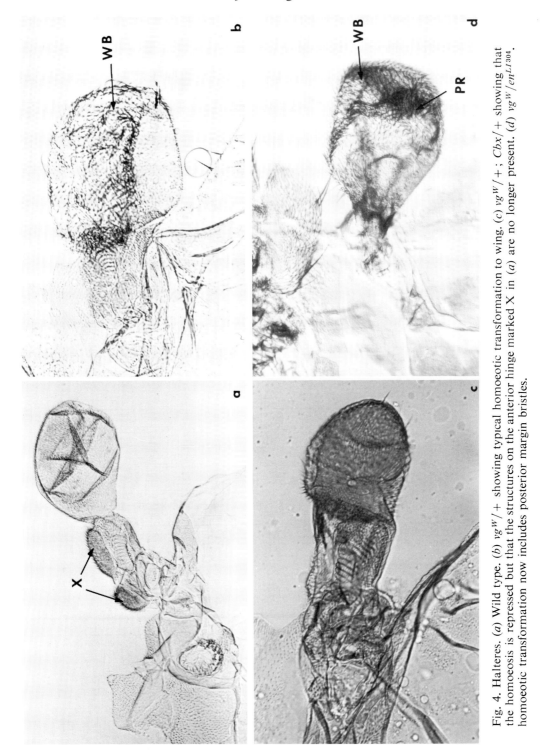

Fig. 4. Halteres. (*a*) Wild type. (*b*) $vg^W/+$ showing that the structures on the anterior hinge marked X in (*a*) are no longer present. (*c*) $vg^W/+$; *Cbx*/+ showing that the homoeosis is repressed but that the homoeotic transformation now includes posterior margin bristles. (*d*) vg^W/en^{LJ304} showing typical homoeotic transformation to wing.

anterior transformations similar to *bithorax* are occurring. The wings of vg^W flies never differentiate anterior margin structures, which may explain why they are not observed in the transformed halteres, since in other *vg* mutants both the wing and haltere are defective in similar positions. However, vg^W wings can differentiate posterior margin, yet these structures are never observed in the transformed halteres. It is likely then, that the homoeosis is genuinely similar to that caused by the *bithorax* mutation and restricted to the anterior compartment of the wing. However, because of the nature of the vg^W mutation, we cannot be sure of this, since it cannot be directly observed. The duplications observed in the posterior region of the wing were very variable in size, ranging from very large duplications, including the posterior row and most of the posterior wing blade (Fig. 3 *b–d*) to extremely small duplications, showing only duplicated axillary cord and no wing blade (Fig. 3 *e*). The structures of the ventral hinge are usually present in one copy and structures of the dorsal hinge are sometimes present, again only in one copy. The costa and anterior wing margin are never present. A classification scheme for the wing duplications was devised according to which areas of the wing are deleted, present once or duplicated. The structures actually present in 52 vg^W wings were scored. A fate map of the wild-type wing disc showing the cells which give rise to these structures is shown in Fig. 5 *a*. The fate map and nomenclature is that of Bryant (1975). We then divided the wing up into regions and these are shown on the wing disc in Fig. 5 *b*. The region marked X is a region which does not differentiate any bristles or identifiable cuticular markers and thus we cannot determine its behaviour in vg^W. The regions selected were (1) ventral hinge, (2) dorsal hinge, (3) the groups of dorsal sensilla on the proximal dorsal radius, (4) the axillary cord, (5) the alar lobe and its associated wing blade, (6) the posterior row of bristles with associated wing blade and (7) the costa, anterior margin with triple and double row of bristles with associated wing blade. Thoracic defects were not included in this classification scheme. The wings were then divided into classes according to whether these regions were deleted, partially or completely, present once, or duplicated. Class 1 has one copy of region 1 and a duplicated region 4; all the other regions are deleted (Fig. 5 *c*), Class 2 is similar to Class 1, except that region 5 is now also present and duplicated (Fig. 5 *d*). Class 3 shows again one copy of region 1 and duplicated regions 4 and 5, but regions 2 and 3 are now also present in one copy (Fig. 5 *e*). Class 4 is similar to 3 except that region 3 becomes duplicated (Fig. 5 *f*). Class 5 is similar to class 3 except that region 6 now appears in duplicated form (Fig. 5 *g*). Class 6 shows one copy of regions 1 and 2, and regions 3, 4, 5 and 6 are duplicated (Fig. 5 *h*).

The general pattern, then, is that region 7, the anterior margin and costa, is always deleted. That region 1, in the ventral hinge, is always present, but never duplicated, and that the axillary cord (region 4) is always present and always duplicated. As the duplicated wings increase in size, the alar lobe and posterior row, regions 5 and 6, appear but are always duplicated, and the dorsal hinge,

region 1, appears but is never duplicated. Only the sensillae of the dorsal radius appear sometimes in one copy and sometimes duplicated.

It is possible that such a pattern could arise by cell death followed by duplication though it is difficult to explain why the ventral hinge would always fail to duplicate. The failure of *vg^W* to regenerate the anterior margin structures of region 7 as a result of the proposed cell death, would be expected, since *vg* has been shown to have a position-specific inability to produce anterior margin structures (Bownes & Roberts, 1981).

The distribution of wings falling into the various classes is shown in Fig. 5. It should be noted that duplications do not follow the compartment boundary in any of the classes observed.

A large amount of pupal lethality was observed in the *vg^W* stock. As the homozygous embryos appeared to die very early, this lethality was presumably due to the heterozygous mutation. The fate of 190 pupae was followed and 48 % of them failed to emerge. Of these 40 % died early and the other 60 % formed pharate adults. Twenty of the pharate adults were mounted and the morphology of the 40 wings and halteres was examined in detail. Twenty-nine of the halteres showed transformations to wing and the remainder had no haltere. The transformations were similar to those observed in adults which emerged. The wing duplications fell into the same classes as those described for adults, but their distribution was different. Only 8 % fell into classes 4 and 5 with large wing duplications, whereas 50 % of the adults had been in these two classes. Thus those pupae which fail to emerge show a higher penetrance of the more extreme wing phenotypes.

The dominant effects of the *vg^W* mutation are very varied, causing pupal lethality, wing duplications and deletions and homoeosis of the haltere to wing. The mutation therefore has pleiotropic effects and disturbs many aspects of normal development.

(B) *Development of* vg^W *wing and haltere discs*

Studies of third instar *vg^W* wing discs were carried out to establish if any morphological defects could be observed and to see if cell death could be detected in the presumptive wing blade.

Haltere discs of *vg^W* were relatively normal in morphology, but were usually much larger than the wild-type haltere discs, presumably as a result of the increased cell number in the region transformed to wing.

The mature third instar *vg^W* wing discs were extremely interesting. Some were small discs with reduced presumptive wing blade areas (Fig. 6b) and presumably these give rise to the flies with very little or no wing blade. Other discs had very obvious duplications of the folds normally seen in a wild-type disc and presumably these develop into wings with large duplications. (Fig. 6c, d).

There was no indication of regions of cell death after trypan blue staining of discs although this technique will not detect individual dying cells.

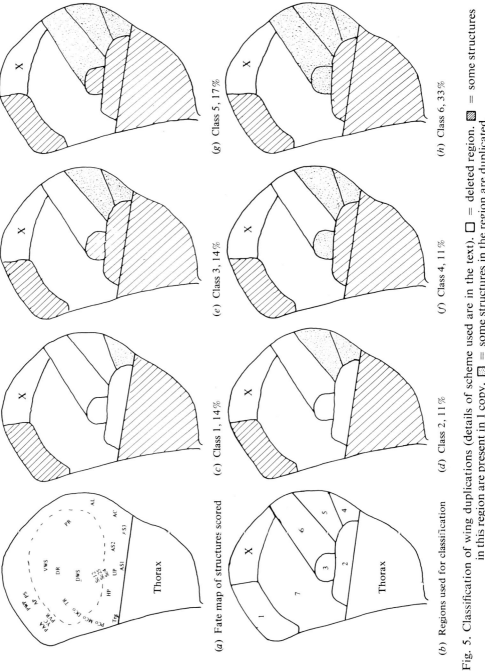

(a) Fate map of structures scored

(b) Regions used for classification

(c) Class 1, 14%

(d) Class 2, 11%

(e) Class 3, 14%

(f) Class 4, 11%

(g) Class 5, 17%

(h) Class 6, 33%

Fig. 5. Classification of wing duplications (details of scheme used are in the text). □ = deleted region. ▨ = some structures in this region are present in 1 copy. ▧ = some structures in the region are duplicated.

Mid-third- and early-third-instar larval wing discs were found showing the duplicated fold structure appearing (Fig. 6c, f) and again they showed no regions of cell death. All the discs shown in Fig. 6 had been stained for cell death.

These results are in marked contrast to findings with *vg* wing discs. These showed large areas of cell death in the presumptive wing blade region (Fristrom, 1968; Bownes & Roberts, 1981). Thus if the *vg^W* duplicated wing phenotype does result from regulation within the wing disc after cell death, this cell death must occur early in development.

(C) *Are there mutations elsewhere in the* vg^W *stock contributing to the abnormal development of the flies?*

As the *vg^W* stock showed so many defects in development and organization of the adult, it seemed possible that other mutations had been induced in the stock at the same time as the *vg* mutation, and that these were contributing to the phenotype of the stock.

We therefore systematically substituted various chromosomes from the stock, with known chromosomes and checked the phenotype.

The original stock supplied was *vg^W*/+.

We produced balanced stocks, *vg^W*/*SM1*, and *vg^W*/*In(2L)t In(2R)Cy Roi cn²* bw or sp² to do these experiments.

The 1st chromosomes were substituted with *Binscy*/*Binscy*, and the *Binscy*; *vg^W*/*SM1* stock still showed the same phenotype indicating that no mutations on the 1st chromosomes were involved.

The 3rd chromosomes were substituted with *rucuca* and *TM3*, to produce *vg^W*/+; rucuca/TM3 flies which again showed the typical *vg^W* phenotype indicating that the third chromosome was also not involved.

Finally, *vg^W* was recombined into a marked 2nd chromosome to substitute the second chromosome either side of the *vg^W* phenotype.

Thus all aspects of the *vg^W* are attributable to defective genes in the *vg* region of the second chromosome. We cannot, of course, rule out the possibility that there is more than one mutation in this region.

(D) *Cytological analysis*

As all aspects of the phenotypes were likely to be the result of one mutation, and *vg* point mutations cause only a reduction in size of the wing and haltere, or sometimes only wing nicks when heterozygous with other *vg* alleles, it seemed possible that *vg^W* may in fact be a deletion or chromosomal rearrangement.

Polytene chromosome squash preparations were made from the salivary glands of third instar *vg^W*/+ larvae. As can be seen in Fig. 7 there is a deletion or inversion from 47F/48A to 49C. Thus the affected region includes *vg*, *bic*, *scaborous* (*sca*) and some material which is also absent in some *engrailed* (*en*) deletions and thus the mutation may also include the *en* gene. Two *en* deletions from 47D3–48B4/5, Df (*en*)-A; and from 47E3–48B2, Df (*en*)–B were supplied by D. Gubb (Cytological Analysis unpublished data of D. Gubb). A further *en*

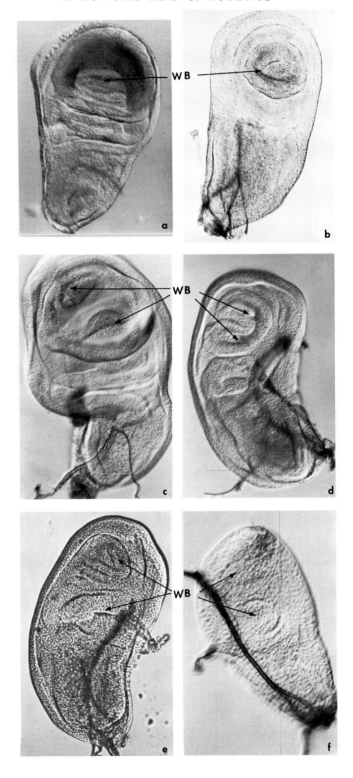

deletion from 48A–48C6/8 was supplied by M. Russell (cytological analysis unpublished data of M. Russell). Whether vg^W is a deletion or inversion will be established prior to the J.E.E.M. discussion (see page 74).

(E) *Complementation studies with other vg mutations*

The vg^W mutant was originally assigned to the vg locus because vg^W/vg heterozygotes showed no wings at all and showed an increased expressivity of the vestigial phenotype compared to either mutant alone (Shukla, 1980). There are several other deletions which include the vg gene and complementation studies with these deletions were undertaken.

The extent of these deficiencies is described in Table 1, and Fig. 8 shows which regions of the right arm of chromosome 2 are deleted by each mutation.

We found that vg^W/vg^C flies showed no wings and no halteres, but the thorax was still present. The vg^W/vg^D and vg^W/vg^B flies however, showed an even more extreme phenotype where the dorsal meso-thorax was often also deleted. These two genotypes were semi-lethal and many of the progeny died before eclosion. The vg^W/vg^S flies showed an intermediate phenotype between vg^W/vg and vg^W/vg^B. In no case was any homoeosis observed since the haltere was completely deleted and for similar reasons it was not possible to detect any of the typical vg^W wing duplications.

Only the vg^W/vg^U flies were lethal and failed to emerge as adults. This is of some interest since vg^U is the only vg mutation with a dominant vestigial wing phenotype. Analysis of $vg^U/+$ flies showed that the wings lacked margin structures and were very similar in phenotype to vg homozygotes. No duplications of any structures were seen in the vg^U wings and no homoeosis was seen in the halteres. Thus, although it has a dominant vestigial phenotype, it does not resemble vg^W in other respects. However, we did analyse $vg^U/+$ late-third-instar wing discs and discovered that they showed no cell death in the wing blade region, which is an aspect of its development which is similar to vg^W, rather than vg.

As vg^W was not lethal in combination with either vg^B, vg^C, vg^D or vg^S, yet vg^W and vg^C overlap by several bands, it seems likely that vg^W is an inversion rather than a deletion since the homozygous deletion which would be produced by vg^W/vg^C heterozygotes would be unlikely to be viable.

In combination with vg^U lethality of the vg^W/vg^U progeny was observed (vg^U is an inversion). When vg^U is combined with other vg deletions vaying results are seen. The vg^U/vg^S flies survive and are fertile, the vg^U/vg^B and vg^D/vg^U flies survive, but are sterile and the vg^U/vg^C progeny are lethal before eclosion (Seiler & Bownes, unpublished). The vg^C deletion extends further to the left

---

Fig. 6. Morphological analysis of wing discs. (*a*) Wild-type late-third-instar wing disc. (*b*)–(*d*) vg^W late-third-instar wing disc. (*e*) vg^W-mid-third-instar wing disc. (*f*) vg^W early-third-instar wing disc. All discs were stained with trypan blue to mark dead cells.

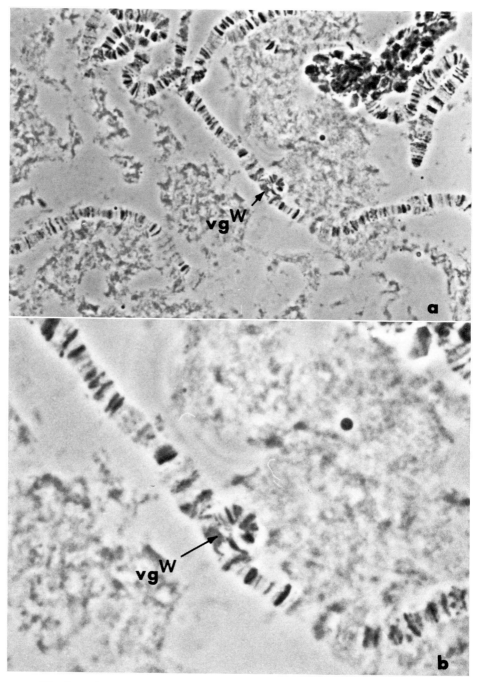

Fig. 7. Polytene chromosomes of $vg^W/+$. (*a*) Whole of second right arm.
(*b*) Detail of deleted or inverted region.

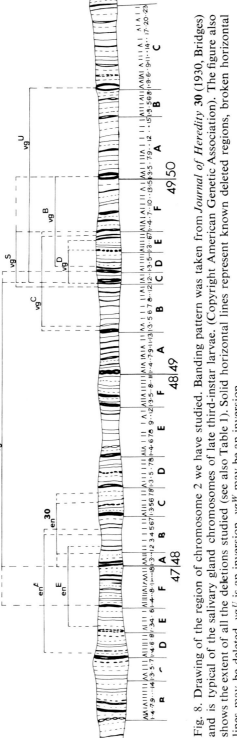

Fig. 8. Drawing of the region of chromosome 2 we have studied. Banding pattern was taken from *Journal of Heredity* **30** (1930, Bridges) and is typical of the salivary gland chromosomes of late third-instar larvae. (Copyright American Genetic Association). The figure also shows the extent of all the deletions studied (see also Table 1). Solid horizontal lines represent known deleted regions, broken horizontal lines may be deleted. vg^U is an inversion. vg^W may be an inversion.

along the chromosome than vg^B, vg^D and vg^S and thus it seems that lethality in combination with vg^U occurs when the chromosome deletions clearly include the vg^U breakpoint of the inversion.

Another interesting point from these studies is that the cytological analysis suggests that vg^W and vg^B do not in fact overlap, yet they are both vg mutants. It seems likely that careful analysis of the position of the right-hand breakpoint of vg^W and the left-hand breakpoint of vg^B could give the precise location of the vg gene itself. It is possible that the cytological analysis will show that vg^W and vg^B do just overlap, but if they do not, they may actually split the vg gene.

(F) Does vg^W delete the engrailed gene

The left-hand breakpoint of the vg^W deletion overlaps with some deletions which include the *engrailed* (*en*) gene. *Engrailed* causes anterior structures to replace posterior structures in the wing a very different phenotype to vg^W, where no anterior structures are observed and the posterior wing which is present is often duplicated. We are therefore trying to establish if vg^W also deletes the *en* gene. For this purpose we have obtained two *engrailed* deletions from Dr Gubb (unpublished, one *en* deletion, from Dr Russell (unpublished) *en¹* and a lethal *en* point mutation (*en^{LI304}*) produced by Nüsslein-Volhard & Wieschaus (1980). Unfortunately only *en^{LI304}* had the appropriate markers for the final complementation crosses to be set up immediately (see Table 1). For the other four complementation crosses the other stocks in Table 1 are being constructed. The vg^W/en^{LI304} flies survive to be adults initially suggesting that vg^W does not include *en*. However, the vg^W/en^{LI304} flies show a slightly different phenotype to $vg^W/+$. The wings, although they show duplications, are more disrupted, some forming wings with holes in, with the dorsal and ventral wing blade fused around the hole, making a doughnut shape. Wing margin was often observed on the transformed halteres (Fig. 4d). Several flies with defective legs were also observed. It seems then, that vg^W and en^{LI304} do interact. It should be noted that complementation tests with a dominant mutation never provide such clear cut results as one usually finds with recessive mutations. The best interpretation of this data would be that vg^W is an inversion with one breakpoint close to *engrailed*, but that the breakpoint does not split the *engrailed* gene.

Preliminary tests, using $Dfen^A$, $Dfen^B$ and en^1, without markers which enable us to distinguish $vg^W/balancer$ *chromosome*, from vg^W/en suggest that vg^W/en^1, $vg^W/Dfen^A$ and $vg^W/Dfen^B$ do survive. This is indicated by the number of progeny with a vg^W type phenotype compared to other genotypes of progeny expected from each cross. There is no indication that there is lethality of a class of flies carrying vg^W. In all three crosses there was a high proportion of flies with abnormal legs. There were more flies with duplications in the thorax than with $vg^W/+$, but whether these were all vg^W/en flies or due to interactions with the genetic background is not yet known. Again these results suggest that vg^W is an inversion since it is not lethal when heterozygous with quite large *en* deletions.

The complementation crosses with *en*[1] and the three *en* deletions in marked stocks should help to decide if *vg^W* does delete *en* and to localise the *en* gene by comparisons of the break points of the *en* deletions and the *vg^W* deletion. This data should be available at the J.E.E.M. meeting in three months' time where this manuscript will be presented (see page 74).

(G) *Interactions of* vg^W *with Cbx*

The mutation *Contrabithorax* (Cbx) causes dominant transformations of wing to haltere (Lewis, 1963) Fig. 9 *a*. It is 100 % effective in posterior parts of the wing, but has lower penetrance in more anterior wing regions (Morata, 1975). In combination with *bx*, the *Cbx* mutation prevents the homoeosis of haltere to wing, thus *Cbx* clearly is acting on anterior wing structures in this case too, as *bx* affects only the anterior haltere. We tested if *Cbx* would be able to inhibit the homoeosis shown by *vg^W* which would suggest that the homoeosis was occurring by similar genetic and metabolic pathways to *bx*. The interactions of *Cbx* with *vg^W* was also investigated in the wing.

Cbx completely inhibited the homoeosis in *vg^W*. As seen in Fig. 4 *c* there is no wing blade in the capitellum. It is interesting that the haltere now formed is incomplete. The rows of sensillae found in the hinge are present but the hairy regions in the anterior margin of the Scabellum and pedicel are deleted. (Compare Fig. 4 *a* and Fig. 4 *c*.) This suggests that *vg^W* does cause a failure of the anterior haltere to develop, as in *vg*, and it is this wing which prevents the appearance of anterior wing margin in the transformed halteres.

The wing phenotype of *vg^W*/ + ; *Cbx*/ + was extremely unusual and may provide insight into the mechanism of action of *Cbx* and *vg^W*. The *vg^W* phenotype usually shows a duplication of the posterior wing, the 2nd posterior being in mirror-image symmetry to the original posterior and replacing anterior structures. When *Cbx* is also present the original posterior is no longer seen. Presumably it is transformed to haltere but so little tissue is present that we cannot be sure about this. The duplicated posterior, lying in an anterior position is still present. This phenotype is shown in Figs. 9 *b–e* and was true of all *Cbx*/ + ; *vg^W*/ + wings. Thus *Cbx* is clearly acting in the original posterior wing. Since we now have none of the original posterior wing tissue to compare with the duplicate, we cannot tell if it is acting to a small extent in the duplicated posterior, but clearly it does not simply act more effectively in 'posterior'-determined wing cells. It may be that *Cbx* acts in a position-specific way, but other alternatives are also presented in the general discussion.

GENERAL DISCUSSION

The *vg^W* deletion causes several different morphological defects in the wing and haltere of *Drosophila melanogaster*. We cannot, at present, separate the various pleiotropic effects from one another. It would be interesting to know if

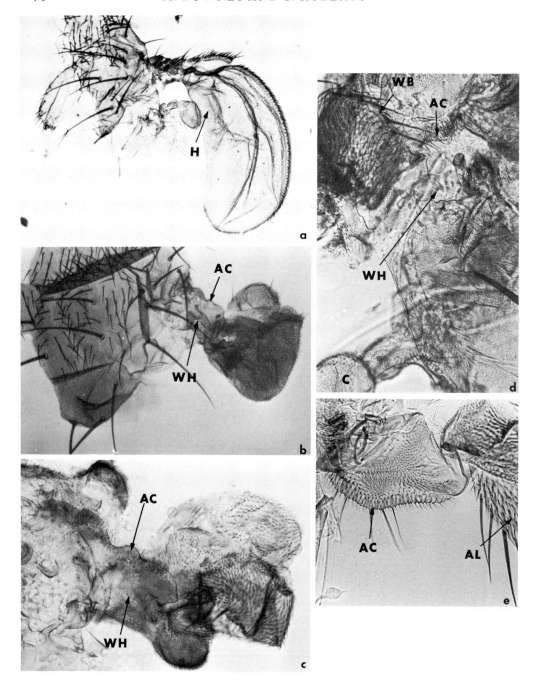

Fig. 9. Wings showing interactions with *Cbx*. (*a*) *Cbx*/+ wing and haltere. The posterior wing is transformed to haltere. (*b*)–(*d*) Are all *Cbx*/+; *vg*^{W}+. (*b*) The duplicated posterior remaining is shown in relation to the thorax for orientation. (*c*). Detail of the posterior present in reversed polarity and in an anterior location. (*d*) Detail of the remaining structures in relation to the haltere. (*e*) Normal organization and orientation of the axillary cord and alar lobe. Particularly note the location of the AC and wing hinge (WH) in these figures.

smaller deletions or point mutations in that region could be found which cause just haltere-to-wing transformations, or just the wing duplications, or if many genes must be rearranged to produce this complex pattern of developmental abnormalities.

There are three aspects of the phenotype to be discussed. (1) The wing duplications of vg^W in relation to other vg mutations. (2) The relation of *en* mutations to vg^W. (3) The interactions in the wing and haltere when both vg^W and *Cbx* are present.

(1) *Wing duplications*

It is not clear why vg^W wings show such specific patterns of duplication. The duplicate wing could arise by either incorrect specification during development of the disc by a trans-determination-like event, or by cell death and regulation phenomena. As the vg mutation shows cell death in the wing discs, the latter might be the most favoured explanation. However, third instar vg^W wing discs show no signs of cell death. It is possible that the dead cells are pushed into the lumen of the disc and therefore not detected by our staining technique though it is unlikely that we would never catch some cells before they are extruded. We shall test this by sectioning discs. The other dominant vg allele, vg^U also shows no cell death in late development of the wing discs, yet it does not duplicate.

It is possible that in both vg^U and vg^W the cell death occurs very early and in one case the nature of the wound healing is such that no duplication occurs. The phenotypes seen in vg^W wings are difficult to explain by early cell death and regeneration since ventral hinge structures never duplicate and they certainly should have time to do so. Early X-irradiation-induced cell death does lead to duplication of wing hinge (Postlethwait, 1975). Some further experiments may help to solve this problem. Regeneration studies on the vg^W disc will determine if it can duplicate hinge structure or ever regenerate anterior margin structures. Longer periods of culture of the wing discs in an adult abdomen prior to metamorphosis should enable us to determine if the regulation process is incomplete during differentiation *in vivo* and extra growth allows duplication of the wing hinge.

We shall also induce clones of marked cells early in development. Comparison of the shape and size of these clones with those found in wild-type wings should help to establish if there is cell death early in the development of the vg^W wing discs. Similar studies were undertaken with the *wingless* (*wg*) mutation (Morata & Lawrence, 1977*a*, 1978) where distal structures of the wing and haltere are replaced by a duplication of the thorax. In some ways vg^W is a slightly less extreme, but related phenotype to *wg*. In this case the clonal analysis provided no evidence for early cell death and regeneration so there is a precedent for mutations with large duplications arising by mechanisms other than cell death and regulation.

(2) *The engrailed mutation*

The *engrailed*[1](*en*[1]) mutation causes anterior wing structures to appear in the posterior compartment of the wing. (Garcia-Bellido & Santamaria, 1972; Morata & Lawrence, 1975; Lawrence & Morata, 1976). The relationship of homoeotic genes and compartments has been reviewed by Morata & Lawrence (1977*b*). It has been proposed that the insect is not just divided into segments, but is further divided into compartments or polyclones (Garcia-Bellido *et al.* 1973, 1976; Crick & Lawrence, 1975). These are goups of clonally related cells which remain segregated during differentiation. As development proceeds the polyclones are further subdivided into smaller compartments. In the wing the first clonal separation is into anterior and posterior, and subsequently there is a separation to dorsal and ventral within those compartments. Garcia-Bellido (1975) and Morata & Lawrence (1975) proposed that the process was controlled by selector genes and that when a group of cells divides into two, a given selector gene becomes activated in one polyclone and inactivated in the other. The evidence obtained with *en*[1] suggested that it was the selector gene for anterior and posterior compartments and that the defective *en*[1] gene product causes posterior cells to become partially anterior in phenotype. The 'leaky' nature of *en*[1] was therefore held responsible for the failure of complete transformation of posterior to anterior wing. Thus the *en* gene has been built into some models for segmentation in *Drosophila* (Hayes, Girton & Russell, 1979; Deak, 1980). Several other *en* point mutations have since been found and some are embryonic lethals (Nüsslein-Volhard & Wieschaus, 1980) which disturb segmentation in the embryo. Various deletions have also been isolated in order to establish if *en* is a selector gene (Gubb, pers. comm.; Russell, pers. comm.) and the vg^W mutation described in this paper perhaps also includes *en*. If *en*[1] were 'leaky', then one might expect a hemizygous *en*[1] fly to show a more extreme phenotype. However, they show a weaker phenotype (Gubb, pers. comm.; Russell, pers. comm.). Given this information and the dominant effects of vg^W, it is difficult to predict how vg^W and *en* would interact. The vg^W/en^{L1304} combination showed an interaction which led to some leg defects and slight modifications of the dominant vg^W phenotype *but* these flies did survive even though en^{L1304} is a homozygous lethal and thus it is unlikely that vg^W deletes or has a breakpoint in *en*. These facts taken together suggest to us and to D. Gubb that the *en* mutations produce novel gene products which disturb development either in the embryo or in leg and wing discs depending upon the allele present.

There are other mutations which cause wing duplications and the replacement of anterior structures by posterior structures, such as *apterous-blot* (Whittle, 1979) and *apt*[e] (Roberts & Bownes, 1981), which have phenotypic features in common with *engrailed* but do not appear to behave as selector genes (Whittle, 1979). Perhaps *engrailed* is actually similar in action to *apt*[blt]. A comparison of *apterous* wings with *engrailed*, vg^W and *wingless* wings, along with investigations of their interactions may help us to understand how the various mutations all interfere with the organisation of the wing in different, but specific ways.

The *rudimentary* (*ru*) mutants (which also have reduced wing) have been shown to be defective in pyrimidine biosynthesis (Norby, 1973; Falk, 1976). The vestigial mutants also have disturbed nucleotide synthesis (Silber, 1980). Analysis of the metabolic deficiencies of several wing mutants, including those described above, may show common mechanisms of action at the metabolic level.

(3) *Interactions between* vestigial^W *and* Contrabithorax

Cbx is one of the dominant genes of the *bithorax* series of homoeotic mutants which cause homoeotic transformations between the mesothorax, metathorax and the abdominal segments (Lewis, 1963, 1964, 1978). When heterozygous or homozygous, *Cbx* causes transformations of the wing to haltere. This happens 100 % of the time in the posterior wing and less frequently, from 11 %–91 % in various regions of the anterior wing and thorax when homozygous. It is thought that *Cbx* is a mutation which damages a regulator site next to *Ubx* (Lewis, 1978) and that as a result of this the *bx*+ and *postbithorax*+ (*pbx*) gene products are derepressed and consequently over-produced in a constitutive fashion (Morata, 1975; Garcia-Bellido, 1977). The *bx*+ gene product is required to produce anterior haltere and the *pbx*+ gene product to produce posterior haltere. Thus *bx* mutant genes cause a transformation of anterior haltere to anterior wing and *pbx* mutant genes cause a transformation of posterior haltere to posterior wing (Lewis, 1978; Hayes *et al.* 1979). *vg*^{*W*} causes homoeosis of haltere to wing and determination of whether this is *bx*-like or *pbx*-like or both is confused by the fact that the *vg*^{*W*} also prevents differentiation of margin structures, especially the anterior margin. *Cbx* is able to repress the homoeosis found in the haltere as a result of the *bx* and *pbx* mutations. Since we have found that *Cbx* can also repress the transformation of haltere to wing in *vg*^{*W*} one could propose that *vg*^{*W*}, by a related metabolic pathway to *bithorax*, produces homoeosis of haltere to wing, but *Cbx* over-produces *bx*+ and *pbx*+ in the haltere and thus the effect of *vg*^{*W*} is overcome by a competitive type of interaction. *Cbx* would also be expected to covert the *vg*^{*W*} wing to haltere. We found that this was true in the posterior region, but that the duplicated posterior in an anterior location was not affected. The fact that *Cbx* acted to transform only the original posterior, could be explained in several ways. It is possible that *Cbx* acts in a position-specific, rather than a determination-specific fashion and thus, as in flies carrying the *Cbx* mutation alone, cells in posterior locations are affected more than cells in anterior locations. This suggests that *Cbx* could be acting after any cell death and duplication had occurred, or that no regulative processes are involved in the *vg*^{*W*} wing duplications. An alternative is that the posterior region is trans-formed to haltere, but that during duplication it gives rise to wing. This kind of process has been observed in the discs of other homoeotic mutants (Adler, 1978), although since *Cbx* is expected to affect the whole disc to some extent, it would seem more likely that after cell death the *Cbx* posterior would regulate and produce cells which would differentiate into haltere. *Cbx* does sometimes act in the anterior of the wing and convert it to haltere, this may well still be happening when it is in combination with vg^{W}, but since one no longer has the original posterior for size comparisons, we cannot detect this.

Studies of interactions of vg^{W} with other homoeotics in the *bithorax* series may not only help us to understand the mechanism of action of vg^{W}, but also the mode of action of some of the *bithorax* mutations.

We are grateful to all the colleagues who supplied us with mutants, and particularly thank Drs D. Gubb and M. Russell for sharing their recent results with us, and supplying their new *en* deletions. We would like to thank Jo Rennie for excellent photographic assistance; Les Dale for reading the manuscript, useful discussions and drawing Fig. 8; and Betty McCready and Anna Nowosielska for typing the manuscript. This research was supported by the Science Research Council.

A summary of additional information on vg^W obtained between submission of the manuscript and presentation of the paper at the meeting

Cytological analysis (reference Results page 63)

We have now established that vg^{W} is an inversion with breakpoints at 47F 15·16(48A1·2 and at 49E4/49E5 (I should like to thank David Gubb, Mike Ashburner and Chris Redfern for independently confirming these breakpoints). The accurate position of the right-hand breakpoint shows that vg^{W} and vg^{B} do overlap and hence clarifies the situation discussed on page 68.

Crosses with engrailed *mutants* (reference Results page 69)

The vg^{W} flies were crossed to the *engrailed* strains, en^{1} *Dfen*[30] *Dfen*[B], *Dfen*[A], in such a way that the en/vg^{W} progeny could be distinguished from the $vg^{W}/$ Balancer chromosome progeny by virtue of their eye colour. In all cases the vg^{W}/en progeny survived. There was some variability in the phenotypes between crosses. For example, $vg^{W} \times en^{1}$ often showed extra regions of thorax close to the wing hinge and the wings were often small; $vg^{W} \times en^{30}$ often produced flies with haemolymph between the wing surfaces but with large wing blade areas; $vg^{W} \times en^{B}$ often had very small wings with duplications and bristle rearrangements within the thorax, and finally $vg^{W} \times en^{A}$ gave progeny with very abnormally shaped wings. However, in all cases these phenotypes were found amongst the $vg^{W}/$Balancer and the vg^{W}/en progeny and thus seem to result from interactions with the genetic background rather than being due to interactions of vg^{W} and *en*.

These results indicate that vg^{W} does not actually split the *en* gene at the lefthand breakpoint of the inversion. The breakpoints do suggest, however, that *en* is included in the inverted region and is thus in a new chromosomal location.

Implications of the extra information

The inversion present in *vg^{W}* is not lethal when heterozygous with large deletions covering its breakpoints, which suggests that the homozygous lethality of *vg^{W}* is not due to the genes split at the breakpoints of the inversion. Further mutations may lie just outside or within the inversion. It may be possible, however, to separate the lethality from the interesting dominant aspects of the *vg^{W}* phenotype. It is possible that some aspects of the *vg^{W}* phenotype are the result of new or untimely gene products which result from the activation of normally inactive genes or perhaps the constitutive activation of normally tissue-specific genes due to the new location of the genes in the genome. Inversions may bring genes into close proximity to different promoter sequences or place them in differently structured regions with respect to chromatin organisation, and thus alter their normal regulation.

REFERENCES

ADLER, P. (1978). Mutants of the bithorax complex and determinative states in the thorax of *Drosophila melanogaster*. *Devl Biol.* **65**, 447–461.

ANDERSON, K. V. & LENGYEL, J. A. (1979). Rate of synthesis of major classes of RNA in *Drosophila* embryos. *Devl Biol.* **70**, 217–231.

BOWNES, M. & ROBERTS, S. (1981). Regulative properties of wing discs from the *vestigial* mutant of *Drosophila melanogaster*. *Differentiation* (in press).

BOWNES, M. & SEILER, M. (1977). Developmental effects of exposing *Drosophila* embryos to ether vapour. *J. exp. Zool.* **199**, 9–23.

BRIDGES, C. B. & MORGAN, T. H. (1919). *Vestigial (vg) Carnegie Inst. Wash. Publ.* **278**, 150–154.

BRIDGES, C. B. MORGAN, T. H. & SCHULTZ, J. (1938). *Carnegie Inst. Wash. Year Book* **37**, 304–309.

BRYANT, P. J. (1975). Pattern formation in the imaginal wing disc of *Drosophila melanogaster* fate map, regeneration and duplication. *J. exp. Zool.* **193**, 49–78.

CHAN, L. N. & GEHRING, G. W. (1971). Determination of blastoderm cells in *Drosophila melanogaster*. *Proc. natn. Acad. Sci., U.S.A.* **68**, 2217–2221.

CRICK, F. H. C. & LAWRENCE, P. A. (1975). Compartments and polyclones in insect development. *Science* **189**, 340–347.

DEAK, I. I. (1980). A model linking segmentation, compartmentalisation and regeneration in *Drosophila* development. *J. theor. Biol.* **84**, 477–504.

FALK, D. R. (1976). Pyrimidine auxotrophy and the complementation map of the *rudimentary* locus of *Drosophila melanogaster*. *Mol. gen. Genet.* **148**, 1–8.

FRISTROM, D. (1968). Cellular degeneration in wing development of the mutant *vestigial* of *Drosophila melanogaster*. *J. Cell Biol.* **39**, 488–491.

FRISTROM, D. (1969). Cellular degeneration in the production of some mutant phenotypes in *Drosophila melanogaster*. *Mol. gen. Genet.* **103**, 363–379.

GARCIA-BELLIDO, A., (1975). Genetic control of wing disc development in *Drosophila*. In *Cell Patterning. Ciba Foundation* **29**, 161–182.

GARCIA-BELLIDO, A. (1977). Homoeotic and atavic mutations in insects. *Amer. Zool.* **17**, 613–627.

GARCIA-BELLIDO, A., RIPOLL, P. & MORATA, G. (1973). Developmental compartmentalisation in the wing disc of *Drosophila*. *Nature New Biology* **245**, 251–253.

GARCIA-BELLIDO, A., RIPOLL, P. & MORATA, G. (1976). Developmental compartmentalisation in the dorsal mesothoracic disc of *Drosophila*. *Devl Biol.* **48**, 132–147.

GARCIA-BELLIDO, A. & SANTAMARIA, P. (1972). Developmental analysis of the wing disc in the mutant *engrailed* of *Drosophila*. *Genetics* **72**, 87–107.

HAYES, P. H., GIRTON, J. R. & RUSSELL, M. A. (1979). Positional information and the *bithorax* complex. *J. theor. Biol.* **79**, 1–17.

LAWRENCE, P. A. & MORATA, G. (1976). Compartments in the wing of *Drosophila:* A study of the *engrailed* gene. *Devl Biol.* **50**, 321–327.

LEWIS, E. B. (1963). Genes and developmental pathways. *Amer. Zool.* **3**, 33–56.

LEWIS, E. B. (1964). Genetic control and regulation of developmental pathways. In *The Role of Chromosomes in Development* (ed. M. Locke), pp. 232–251. New York: Academic Press.

LEWIS, E. B. (1967). Genes and gene complexes. In *Heritage from Mendel* (ed. A. Brink), pp. 17–41, Wisconsin: University of Wisconsin Press, Madison.

LEWIS, E. B. (1968). Genetic control of developmental pathways in *Drosophila melanogaster*. *Proc. 12th Int. Conf. Genetics* (Abstract pp. 96–97).

LEWIS, E. B. (1978). A gene complex controlling segmentation in *Drosophila*. *Nature* **276**, 565–570.

LINDSLEY, D. L. & GRELL, F. H. (1968). Genetic variations of *Drosophila melanogaster*. *Carnegie Inst. Wash. Pub.* No. **627**.

MCKNIGHT, S. L. & MILLER, O. L. JR. (1967). Ultrastructural patterns of RNA synthesis during early embryogenesis of *Drosophila melanogaster Cell* **8**, 305–319.

MORATA, G. (1975). Analysis of gene expression during development in the homoeotic mutant *Contrabithorax* of *Drosophila*. *J. Embryol. exp. Morph.* **34**, 19–31.

MORATA, G. & GARCIA-BELLIDO, A. (1976). Developmental analysis of some mutants of the *bithorax* system of *Drosophila*. *Wilhelm Roux' Arch. devl Biol.* **179**, 125–143.

MORATA, G. & LAWRENCE, P. A. (1975). Control of compartment development by the *engrailed* gene in *Drosophila*. *Nature* **255**, 614–617.

MORATA, G. & LAWRENCE, P. A. (1977*a*). The development of *wingless*, a homoeotic mutant of *Drosophila*. *Devl Biol.* **56**, 227–240.

MORATA, G. & LAWRENCE, P. A. (1977*b*). Homoeotic genes, compartments and cell determination in *Drosophila*. *Nature* **265**, 211–216.

MORATA, G. & LAWRENCE, P. A. (1978). Cell lineage and homoeotic mutants in the development of imaginal discs of *Drosophila*. In *The Clonal Basis of Development*. New York: Academic Press.

NORBY, S. (1973). The biochemical genetics of *rudimentary* mutants of *Drosophila melanogaster*. *Hereditas* **73**, 11–16.

NÜSSLEIN-VOLHARD, C. (1977). Genetic analysis of pattern-formation in the embryo of *Drosophila melanogaster*. Characterisation of the maternal-effect mutant *bicaudal*. *Wilhelm Roux' Arch. devl. Biol.* **183**, 249–268.

POSTLETHWAIT, J. H. (1975). Pattern formation in the wing and haltere imaginal discs after irradiation of *Drosophila melanogaster* first instar larvae. *Wilhelm Roux' Arch. devl Biol.* **178**, 29–50.

ROBERTS, S. & BOWNES, M. (1981). A new apterous allele of *D. melanogaster Dros. Inform. Service* (in press).

SHUKLA, P. T. (1980). *Vestigial wingless:* A dominant allele of *vestigial* in *D. melanogaster*. *Dros. Inform. Service* **55**, 210.

SILBER, J. (1980). Metabolism of *vestigial* mutants in *Drosophila melanogaster*. 1. Resistance of *vg* flies to inhibitors of nucleotide metabolism. *Genetika* **12**, 21–29.

WADDINGTON, C. H. (1940). The genetic control of wing development in *Drosophila*. *J. Genet.* **41**, 75–151.

WHITTLE, J. R. S. (1979). Replacement of posterior by anterior structures in the *Drosophila* wing caused by the mutation *apterous-blot*. *J. Embryol. exp. Morph.* **53**, 291–303.

WRIGHT, T. R. F. (1970). The genetics of embryogenesis in *Drosophila*. *Adv. Genetics* **15**, 261–395.

ZALOKAR, M. (1976). Autoradiographic studies of protein and RNA formation during early development of *Drosophila* eggs. *Devl Biol.* **49**, 425–437.

J. Embryol. exp. Morph. Vol. 65 (Supplement), pp. 77–88, 1981
Printed in Great Britain © Company of Biologists Limited 1981

Growth and cell competition in *Drosophila*

By PAT SIMPSON[1]

From the Laboratoire de Génétique Moléculaire
des Eucaryotes du CNRS

SUMMARY

The process of cell competition, whereby slowly dividing *Minute* cells are eliminated by faster-growing *Minute*[+] cells in mosaic compartments of the imaginal wing disc, is discussed. Evidence is presented suggesting that after completion of growth of the imaginal discs, *Minute*[+] cells no longer continue overgrowing and eliminating the *Minute* cells. The process of competition thus appears to be restricted to discs that are actively growing.

No cell competition can be detected in the histoblast cells that give rise to the adult abdomen. This observation, however, has been interpreted to be the result of an extremely long perdurance effect for the *Minute*[+] product in these cells.

INTRODUCTION

We have been studying patterns of growth in *Drosophila* imaginal discs mosaic for populations of mutant cells with different growth rates. A phenomenon of cell competition has been described in these mosaic animals (Morata & Ripoll, 1975) such that the (normally non-lethal) slower-growing cells are eliminated at the expense of the faster-growing ones. Cell competition only occurs within compartments, which are the basic units of development in *Drosophila* (Garcia-Bellido, Ripoll & Morata, 1976), and not between compartments (Simpson & Morata, 1981). It has also been shown that cell competition is a result of the differences in growth rate and that it depends upon local interactions between slow- and faster-growing cells (Simpson & Morata, 1981).

Compartments that are mosaic for cells of different growth rates are always of normal size and pattern. Therefore, although the overall amount of growth is controlled, the number of mitoses undergone by any particular cell is not important. The control of growth of the imaginal discs is intrinsic to the discs themselves, that is intact discs *in situ* never grow to be larger than normal size, even when larval life is greatly prolonged and the hormonal and nutritive factors necessary for growth are still present (Simpson, Berreur & Berreur-Bonnenfant, 1980). These observations have led to the conclusion that the feedback interactions between growth and pattern size in *Drosophila* imaginal discs probably

[1] *Author's address:* Laboratoire de Génétique Moléculaire des Eucaryotes du CNRS, Unité 184 de Biologie Moléculaire et de Génie Génétique de l'Inserm, Institut de Chimie Biologique, Faculté de Médecine, Strasbourg, France.

results from the same type of local interactions that initially regulate the pattern itself. This conclusion is supported by two other lines of evidence: firstly that it is not possible to dissociate normal pattern and normal size (for discussion, see Simpson & Morata, 1980) and secondly, intercalary regeneration of *in vivo* cultured fragments of discs leads to stable pattern configurations and a cessation of growth (Bryant, 1978).

The experiments presented in this paper were designed to answer two questions. Firstly, does cell competition continue after a disc, mosaic for slow- and faster-growing cells, has completed its growth? Secondly, does cell competition occur in the histoblasts which form the adult abdomen in *Drosophila*?

MATERIALS AND METHODS

Flies were grown on corn meal–yeast–agar medium, and were allowed to lay eggs in split bottles for 2 h periods. They were X-irradiated with a dose of 1000 R (Theta X-ray machine, Machlett tube, 1 mm Al filter, 40 KVP, 20 mA) in order to induce mitotic recombination (Becker, 1957). The relevant body parts were cooked in 10 % KOH and mounted in Euparal for examination under the compound microscope. All slides were coded and scored blind.

Cross 1. Gynandromorphs were produced by means of the unstable ring-X chromosome, $R(1)2w^{vC}$ (Hinton, 1955). This chromosome is frequently lost during the early divisions in the egg, but remains stable throughout larval development. Ring-X-bearing females, also carrying *multiple wing hairs* (*mwh*), were mated to $l(1)ts\ 504\ sn^3\ l(1)ts5697;\ M(3)i^{55}/TM1$ males. $l(1)ts504$ (Simpson & Schneiderman, 1975) and $l(1)ts5697$ (Arking, 1975), hereafter abbreviated to *504* and *5697*, are sex-linked recessive lethals. These mutations have temperature-sensitive periods extending throughout the entire larval period. They both cause extensive cell death in the imaginal discs at 30 °C, the restrictive temperature. The mutations *singed* (sn^3) and *mwh* were used as cell markers; they affect bristles or trichomes respectively. sn^3 is on the *X* chromosome and served to reveal the male patches on the gynandromorphs. *mwh* is on the third chromosome and was used to mark clones induced by mitotic recombination. The use of a *Minute* mutation ($M(3)i^{55}$) leads to the production of clones with an advantaged growth rate (Morata & Ripoll, 1975). *TM1* was used to balance $M(3)i^{55}$. See Lindsley & Grell (1968) for a description of mutations used. Animals were grown at 23 °C, irradiated at 72 h after egg laying (AEL) and subjected to a heat pulse at 30 °C between 72 and 120 h AEL. The time of eclosion of the flies was recorded, once each day at the same time. Gynandromorphs bearing $M(3)i^{55}$ were selected under the dissecting microscope. The female sn^+ wings were examined for the presence of *mwh* M^+ clones. Clones were drawn on to standard wing diagrams and their size estimated, taking into account the variable cell density in different parts of the wings (Garcia-Bellido & Merriam, 1971 a). Recombination in the left arm of chromosome 3 will give rise to *mwh M*[+] clones and also *mwh M(3)i*[55] clones in those cases where the mitotic exchange takes place between *mwh* and *M(3)i*[55]. A bimodal distribution of clone sizes was observed by Morata & Ripoll (1975) after irradiation at any developmental stage. In the present experiment, clones smaller than 1000 cells were excluded; it was assumed that the remaining clones all belonged to the *mwh M*[+] class.

Crosses 2, 3 and 4 were performed as follows.

Cross 2: *wild-type* ♀ × *f*[36a]; *pwn/SM5*; *mwh* ♂
Cross 3: ♀ *FM6/T(1; 2)Bld/ +* × ♂ *f*[36a]; *pwn/SM5*; *mwh*
Cross 4: ♀ *FM6/M(1)o*[Sp] × ♂ *yw*[a]*f*[36a]

These led to the production of the following females: *f*[36a]*/ +* ; *pwn/ +* ; *mwh/ +* (genotype A), *f*[36a]*/T(1; 2)Bld/pwn*; *mwh/ +* (genotype B) and *yw*[a]*f*[36a]*/M(1)o*[Sp] (genotype C). In these flies *forked* (*f*[36a]) and *pawn* (*pwn* – see Garcia-Bellido & Dapena, 1974), were used to mark clones of bristles resulting from mitotic recombination. A description of the genotypes of these clones is given in the text where relevant. Embryos were irradiated 3 h after the middle of the egg-laying period and some at 72 h AEL. Tergites 2 to 6 inclusive were scored for the presence of clones.

RESULTS AND DISCUSSION

Cell competition after the completion of growth in the imaginal discs

In the gynandromorphs resulting from cross 1, the female discs have a wild-type phenotype, whereas the XO male discs are mutant for the marker *sn*[3] and the two temperature-sensitive cell-lethal mutations. This means that during the 48 h heat pulse, cell death occurs only in the male discs. The lesions are then repaired by the process of regeneration which causes pupariation to be delayed (Russell, 1974; Simpson & Schneiderman, 1975). Pupariation is retarded for a variable number of days depending upon the amount of male mutant tissue (Simpson *et al.* 1980). Since the female discs are unaffected by the heat pulse, they therefore undergo metamorphosis a considerable time after they have pre-sumably completed growth. In the present experiment the emergence of the *Minute* gynandromorphs varied from day 13 to day 18 AEL. Earlier observa-tions have shown that the delay in eclosion of the imagos is a consequence of a similar delay in pupariation; little or no variation in the length of the pupal period is seen (Simpson & Schneiderman, 1975).

Exclusively female wings from these flies were scored for the presence of large overgrowing *mwh M*[+] clones (see Materials and Methods). Clones from flies emerging from day 13 to day 15 (early eclosing, EE) have been pooled and also those from flies emerging from day 16 to day 18 (late eclosing, LE). Forty-one clones were obtained from the EE series and 26 from the LE series. There is no detectable difference in clone size between the two series (EE: 3800 ± 600; LE: 4100 ± 800). It will be remembered that the larvae were all irradiated at the same stage, prior to the heat pulse and subsequent cell death in the male discs.

During the growth of mosaic M/M^+ discs, the M^+ cells normally divide extensively and eliminate the M cells quite rapidly (Morata & Ripoll, 1975; Simpson, 1979). M^+ clones induced early in the larval stage often fill an entire developmental compartment (Simpson & Morata, 1981). In the present experiment, if, after the disc has reached its full size, the M^+ cells were to continue to divide and eliminate the M cells at a similar rate, then one would expect most of the clones in the LE series to fill an entire compartment. In fact only one of them did. The results suggest that after the completion of growth, the M^+ cells stop dividing and do not continue to eliminate the M cells. However, a very slow continuous overgrowth might not have been detected. This conclusion is of course based on the assumption that in the late pupariating gynandromorphs, the female discs complete their growth long before pupariation intervenes, and are not retarded due to the presence of the regenerating male discs. It is not proven that this assumption is valid, but preliminary unpublished observations (Berreur & Simpson) suggest that it is.

Analysis of Minute clones in the tergites

Description of growth

Clonal analyses performed by previous workers have established that after an initial one or two divisions during embryogenesis (Wieschaus & Gehring, 1976; Lawrence, Green & Johnston, 1978) no further cell divisions take place in the histoblasts until pupariation (Garcia-Bellido & Merriam, 1971 b; Guerra, Postlethwait & Schneiderman, 1973). This pattern of growth has been confirmed by histological observations which have also shown that during the larval period the cells increase in volume about 57-fold but do not become polytene or polyploid (Madhavan & Schneiderman, 1977). After pupariation, histoblast growth can be divided into two phases (Roseland & Schneiderman, 1979). During the first phase, from 3 to 15 h after pupariation, the histoblasts undergo about four rapid cell divisions in a manner resembling cleavage. During this time, they remain in their original area, that is in two groups of cells, continuous with the larval epithelium, situated anteriorly and posteriorly in each hemitergite. During the second phase, from 15 to 36 h after pupariation, the histoblast nests enlarge, the cells continue division and migrate over the larval polytenal cells which they replace.

Clone size and frequency

Clones were induced by irradiation at blastoderm (3 h AEL) and at 72 h AEL, in animals of genotypes A: $f^{36a}/+$; $pwn/+$; $mwh/+$, B: $f^{36a}/T1$; $2)Bld/pwn$; $mwh/+$ and C: $yw^a f^{36a}/M(1)o^{Sp}$. Clones of f^{36a} and pwn bristles of animals of genotypes A served as controls. The genetic constitution of females of genotype

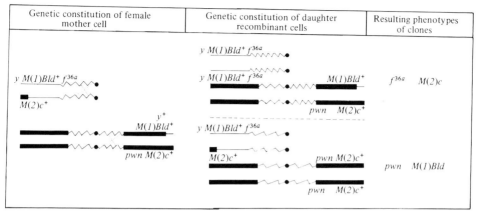

Genetic constitution of female mother cell	Genetic constitution of daughter recombinant cells	Resulting phenotypes of clones

Fig. 1. The genetic constitution of *Minute*[+] females (B) $f^{36a}/T(1;2)$ *Bld/pwn*; *mwh/ +*, and the genotype of clones produced by mitotic recombination. Only one of the two chromatids per chromosome is shown in the mother cell. The thin solid line represents the *X* chromosome and the solid bar chromosome 2. Irradiation leads to the production of f^{36a} *M(2)c* clones after recombination in the *X* chromosome and *pwn M(1)Bld* clones after recombination in chromosome 2.

B and of the clones that were induced in them is depicted in Fig. 1. These females carried two doses of *M(2)c*[+] and *M(1)Bld*[+] and therefore developed at the rate for wild-type animals. Mitotic recombination in chromosome 1 resulted in f^{36a} clones bearing only one dose of *M(2)c*[+] and therefore mutant for that *M* due to the haplo-insufficiency of the *M* loci. Similarly recombination in 2R led to *pwn M(1)Bld* clones (*mwh* clones were ignored as *mwh* does not mark bristles). The results are given only for f^{36a}*M(2)c* clones since *pwn* did not appear to be a reliable marker for abdominal bristles.* Recombination in chromosome 1 of females of genotype C led to the production of $yw^a f^{36a} M^+$ clones in a $M(1)o^{Sp}$ background (*y* clones resulting from recombination distal to f^{36a} were not scored).

The average sizes of clones obtained are presented in Table 1. In the control cross (animals of genotype A) clones induced at 3 h AEL are larger than those induced at 72 h AEL. This confirms the observations of Wiechaus & Gehring (1976) and Lawrence *et al.* (1978) that some cell division occurs in embryogenesis.

It can be seen that whereas *M(2)c* clones induced by irradiation at 72 h AEL do not differ from the controls, those induced at 3 h AEL are considerably smaller than the controls ($P < 0.001$). Therefore only clones induced at 3 h expressed the *M* phenotype of slow growth. The distribution of clone sizes is given in Fig. 2. The lower mean clone size for *M(2)c* clones induced at blastoderm is due to a decrease in the number of large clones: only 7 % of the clones

* Frequently, many *pwn* bristles were found clustered closely together and many *pwn* clones were associated with empty bristle sockets. Either *pwn* is not a reliable marker for abdominal bristles, or else the *pwn* chromosome used here had accumulated a deleterious mutation affecting bristle differentiation.

Table 1. *Average size, in number of bristles, of* f^{36a} *clones induced in flies of genotypes A*: f^{36a}/+ ; pwn/+ ; mwh/+ , *B*: f^{36a}/T(1; 2)Bld/pwn; mwh/+ *and C*: yw^{a}f^{36a}/M(1)o^{Sp}

Time at irradiation (h AEL)	Genotype					
	A		B		C	
	Control	n	Minute	n	Minute+	n
3	6·6 ± 0·5	217	2·7 ± 0·2	197	5·7 ± 0·4	257
	(12·4 ± 1·2)	87	(9·9 ± 1·4)	14	(13·9 ± 1·4)	77
72	2·8 ± 0·2	54	2·4 ± 0·2	44	2·9 ± 0·3	52
Non-irradiated	2·0 ± 0·1	146	1·4 ± 0·1	85	1·8 ± 0·1	174

n, Number of clones. Figures in parentheses represent clones composed of more than five bristles.

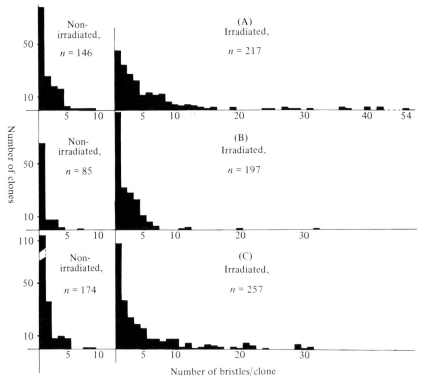

Fig. 2. Distribution of the different sizes of f^{36a} clones produced by irradiation at 3 h AEL of flies of genotype A: f^{36a}/+ ; *pwn*/+ : *mwh*/+ , B: f^{36a}/*T(1; 2)Bld/pwn*; *mwh*/+ and C: *yw^{a}f^{36a}/M(1)o^{sp}*.

Table 2. *Frequency of* f^{36a} *clones per abdomen in flies of genotypes A*: f^{36a}/ + ;
pwn/ + ; mwh/ +, *B*: f^{36a}/*T*(1; 2)Bld/pwn; mwh/ +, *and C*: yw^{a}f^{36a}/M(1)o^{Sp}

| Time at irradiation (h AEL) | Genotype | | | | | |
| | A | | B | | C | |
	Control	*n*	*Minute*	*n*	*Minute*$^+$	*n*
3	0·22	1009	0·20	985	0·31	810
72	0·54	104	0·42	105	0·76	68
Non-irradiated	0·29	494	0·18	480	0·54	330

n, Number of abdomens.

contained more than five bristles, compared with 36 % of the control clones
($P < 0.01$) and 30 % of the M^+ clones in animals of genotype C ($P < 0.01$).

The M^+ clones induced in $M(1)o^{Sp}$ flies at 3 h AEL display an average size
similar to that of the controls. Therefore, such wild-type cells undergo the same
number of divisions whether they are in an M or M^+ animal.

Spontaneous clones

The frequency of spontaneous clones has been found to vary considerably for
different stocks (Brown & Welshons, 1955; Weaver, 1960). Table 2 shows that
for the three crosses used here the frequency is quite different even for the same
marker and the same chromosome arm. The enhancing effect of the M mutations on the frequency of mitotic recombination has been well documented
(Stern, 1936; Kaplan, 1953; Ferrus, 1975). The size of such clones however
remains fairly constant (Table 1), the majority being composed of one or two
bristles as reported by earlier authors (Brown & Welshons, 1955; Weaver, 1960;
Baker, Carpenter & Ripoll, 1978; Kennison & Ripoll, personal communication). Examination of Table 2 shows that the clone frequencies in the un-
irradiated series are higher than in those irradiated at 3 h AEL. It was not
possible therefore to correct the clone sizes obtained after irradiation at 3 h
by subtraction of spontaneous clones as has been done by earlier authors.
J. Kennison and P. Ripoll (personal communication) report that, for wing discs,
a lower frequency of small clones is observed after irradiation than in non-
irradiated wings.

Perdurance in the histoblasts

Two of the observations made in the experimental series require explanation.
First, why are the M^+ clones, induced in M animals at blastoderm, no larger
than control clones, whereas M clones induced in M^+ animals at blastoderm
are smaller than control clones? Second, why are all clones induced at 72 h

AEL the same size? I interpret these results to be the consequence of a long perdurance effect, as follows.

The increase in cell volume occurring during larval development probably allows the accumulation of reserves enabling the initial, rapid, cleavage-like divisions to take place at a time when the animal stops feeding. If this is the case then it may have special consequences for clonal analysis, since it may result in an extremely long perdurance effect (Garcia-Bellido & Merriam, 1971 c) for mutations such as the *Minutes*. Perdurance is the amount of time that is necessary, after the recombination event leading to the production of an *M* cell, to reduce the amount of wild-type M^+ products synthesized by the cell. It has been concluded, from a comparison of clone sizes, that in the wing disc perdurance lasts about three cell divisions in the case of $M(1)Bld$ and $M(2)c$ (Simpson & Morata, 1981). In the case of the histoblasts, this perdurance effect may well involve many more cell divisions. If the irradiation is performed during the larval period the recombination event only takes place at the first division after pupariation. The initial cleavage divisions may not require synthesis of new products for division, and it is possible that a *further* three divisions after the four cleavage divisions may be required to effectively reduce the amount of M^+ product in the cell. As it has been calculated (Roseland & Schneiderman, 1979; Madhavan & Madhavan, 1980) that only about 8 doublings are required to make the adult abdominal epidermis, this would mean that the cells have virtually no time to express the *M* phenotype of slow growth and may explain why Morata & Ripoll (1975) observed only very small differences in clone sizes between M^+ clones in *M* flies and *M* clones in M^+ flies after larval irradiation. On the other hand, the developmental delay observed for *M* animals occurs almost entirely during the larval phase; very little delay is observed in the pupal period (Dunn & Mossige, 1937; Brehme, 1939; Steiner, 1976). Since *M* flies are similar in size to wild-type and bear the same number of bristles (Brehme, 1939) this must mean that the histoblasts in *M* flies divide at a similar rate to those of M^+ flies. This may mean that sufficient M^+ products are accumulated during larval life to tide the histoblasts over the entire pupal development. In the case of *M* larvae the extended larval period would allow the synthesis of similar amounts of M^+ products to those in wild-type larvae. If this were true, perdurance may well cover the whole of the pupal stage. It could be the reason, therefore, why all clones observed here were of a similar size, irrespective of their *M* constitution.

When animals are irradiated prior to the larval stage, for example at blastoderm, then the recombination event will occur before the larval non-mitotic period of growth. If the above reasoning is correct, and if a genetically *M* cell amasses the M^+ products necessary for division at a rate slower than that for wild-type cells, then an *M* clone induced at blastoderm in an M^+ animal would have insufficient time to synthesize the required amount of M^+ product and should therefore express a slower rate of cell division during the pupal period.

The converse, that of an M^+ clone in an M fly, would not necessarily result in extra growth of the M^+ clone, for even if the M^+ cells accumulate more reserves than is usual during the prolonged M larval stage, it is unlikely that they would then divide faster than is normal.

No cell competition can be detected in the tergites

Examination of Table 2 shows that there is no decrease in the frequency of $M(2)c$ clones induced at 3 h AEL in flies of genotype 3. Therefore, although these cells are dividing at a slow M rate, there is apparently no loss of clones. Loss of a small number of cells, however, would go undetected. The main reason for the lack of cell competition in the abdomen may be that there is no overgrowth of M^+ cells (as shown in M flies of genotype C). It has been demonstrated that the loss of M cells through cell competition is associated with extra growth of neighbouring M^+ cells (Simpson & Morata, 1981). This result is consistent with the observation that very little regulation is possible in the abdomen (Santamaria & Garcia-Bellido, 1972; Roseland & Schneiderman, 1979), since the duration of the pupal period varies little.

CONCLUSIONS

It can be concluded from the foregoing that no cell competition can be detected in the tergites. However, the pattern of growth of the histoblasts is quite different from that occurring in the imaginal discs, and this result may be due to an extremely long perdurance for the M^+ products.

Previous studies have shown that while M^+ cells divide rapidly and eliminate M cells with which they are in contact they do not cross the boundaries between compartments to invade M territories of neighbouring compartments (Garcia-Bellido *et al.* 1976). Although M^+ clones may grow to be very large their growth is in no way comparable to an invasive, cancerous type of growth; mosaic M/M^+ animals develop quite normally. Results presented here suggest also that M^+ cells stop dividing in mosaic M/M^+ wing discs that have reached full size. Therefore, whatever the nature of the signal telling the cells to stop dividing, M^+ cells that are still in contact with M cells remain sensitive to it.

It is probable that the factors leading to a cessation of growth are not external but are intrinsic to the imaginal discs themselves (Simpson *et al.* 1980; Bryant, 1978). In a fly in which pupariation is retarded because of lesions in only some of the imaginal discs, the other intact discs have an extended period of time for growth before the onset of metamorphosis. They do not, however, give organs or appendages that are any larger than the usual size (Simpson *et al.* 1980). The hormonal conditions necessary for growth are fulfilled none the less, since growth and regeneration proceed in the damaged discs. Experimental evidence for the production of giant adult insects by prolonged feeding due to juvenile-hormone treatment is not convincing. In all cases growth of the larval body decreases and eventually stops with each supernumerary larval instar. Less than 1 % of the large larvae metamorphose, and when they do the resulting adults are abnormal (Sláma, Romañuk & Sorm, 1974). The one reported instance of an additional moult in the Diptera was not followed by adult differentiation (Zdarek & Slama, 1973). The so-called giant mutant in *Drosophila* owes its slightly enlarged body size to an increased cell size and not to an additional number of cells (unpublished observations). In cases in which enlarged imaginal discs have been described, these discs either differentiate abnormal or duplicated patterns (Bryant & Schubiger, 1971; Martin, Martin & Shearn, 1977) or else prove incapable of differentiation (Gateff & Schneiderman, 1969; Stewart, Murphy & Fristrom, 1972; Garen, Kauvar & Lepesant, 1977). Fragmentation of imaginal discs usually leads to a regenerating and a duplicating fragment, and after regeneration or duplication growth stops, leaving normal-sized patterns (Bryant, 1978).

I thank Renée Dupré for her expert technical assistance, R. Arking for the mutation *I*(1)*ts5697* and Peter Lawrence for comments on the manuscript. This work was supported by the Centre National de la Recherche Scientifique.

REFERENCES

ARKING, R. (1975). Temperature sensitive cell lethal mutants of *Drosophila*: isolation and characterization. *Genetics* **80**, 519–537.

BAKER, B., CARPENTER, A. & RIPOLL, P. (1978). The utilization during mitotic cell division of loci controlling meiotic recombination and disjunction in *Drosophila melanogaster*. *Genet.* **90**, 531–578.

BECKER, H. J. (1957). Uber Röntgenmosaikflecken und Defektmutationen am Auge von *Drosophila* und die Entwicklungsphysiologie des Auges. *Z. Indukt. Abstamm. Vererbungsl.* **88**, 333–373.

BREHME, K. (1939). A study of the effect on development of *Minute* mutations in *Drosophila melanogaster*. *Genetics* **24**, 131–161.

BROWN, S. W. & WELSHONS, W. (1955). Maternal aging and crossing over of attached *X* chromosomes. *Proc. natn. Acad. Sci., U.S.A.* **41**, 209–245.

BRYANT, P. J. (1978). Cell interactions controlling pattern regulation and growth in epimorphic fields. In *Birth Defects: Original Article Series*, vol. XIV, number 2, pp. 529–545. The National Science Foundation.

BRYANT, P. J. & SCHUBIGER, G. (1971). Giant and duplicated imaginal discs in a new lethal mutant of *Drosophila melanogaster*. *Devl Biol.* **24**, 233.

DUNN, L. C. & MOSSIGE, J. C. (1937). The effects of the *Minute* mutations of *Drosophila melanogaster* on developmental rate. *Hereditas* **23**, 70–90.

FERRUS, A. (1975). Parameters of mitotic recombination in *Minute* mutants of *Drosophila melanogaster*. *Genetics* **79**, 589–599.

GARCIA-BELLIDO, A. & DAPENA, J. (1974). Induction, detection and characterisation of cell differentiation mutants in *Drosophila*. *Mol. Gen. Genet.* **128**, 117–130.

GARCIA-BELLIDO, A. & MERRIAM, J. R. (1971*a*). Parameters of the wing imaginal discs development of *Drosophila melanogaster*. *Devl Biol.* **24**, 61–87.

GARCIA-BELLIDO, A. & MERRIAM, J. R. (1971*b*). Genetic analysis of cell heredity in imaginal discs of *Drosophila melanogaster*. *Proc. natn. Acad. Sci., U.S.A.* **68**, 222–226.

GARCIA-BELLIDO, A. & MERRIAM, J. R. (1971*c*). Clonal parameters of tergite development in *Drosophila*. *Devl Biol.* **26**, 264–276.

GARCIA-BELLIDO, A., RIPOLL, P. & MORATA, G. (1976). Developmental segregations in the dorsal mesothoracic disc of *Drosophila*. *Devl Biol.* **48**, 132–147.

GAREN, A., KAUVAR, L. & LEPESANT, J. A. (1977). Roles of edcysone in *Drosophila* development. *Proc. natn. Acad. Sci., U.S.A.* **74**, 5099.

GATEFF, E. & SCHNEIDERMAN, H. A. (1969). Neoplasms in mutant and cultured wild-type tissues of *Drosophila*. *Natn. Cancer I. Monogr.* **31**, 365.

GUERRA, M., POSTLETHWAIT, J. H. & SCHNEIDERMAN, H. A. (1973). The development of the imaginal abdomen of *Drosophila melanogaster*. *Devl Biol.* **32**, 361–372.

HINTON, C. W. (1955). The behaviour of an unstable ring chromosome of *Drosophila melanogaster*. *Genet.* **40**, 951–961.

KAPLAN, W. D. (1953). The influence of *Minutes* upon somatic crossing over in *Drosophila melanogaster*. *Genet.* **38**, 630–651.

LAWRENCE, P. A., GREEN, S. & JOHNSTON, P. (1978). Compartmentalization and growth of the *Drosophila* abdomen. *J. Embryol. exp. Morph* **43**, 233–245.

LINDSLEY, D. L. & GRELL, E. H. (1968). Genetic variations of *Drosophila melanogaster*. *Carnegie Inst. Wash. Publ.* **627**.

MADHAVAN, M. M. & MADHAVAN, K. (1980). Morphogenesis of the epidermis of adult abdomen of *Drosophila*. *J. Embryol. exp. Morph* **60**, 1–31.

MADHAVAN, M. & SCHNEIDERMAN, H. A. (1977). Histological analysis of the dynamics of growth of imaginal discs and histoblast nests during the larval development of *Drosophila melanogaster*. *Wilhelm Roux's Arch. Devl Biol.* **183**, 269–305.

MARTIN, P., MARTIN, A. & SHEARN, A. (1977). Studies of *1(3)C43hsl*, a polyphasic, temperature-sensitive mutant of *Drosophila* with a variety of imaginal disc defects. *Devl Biol.* **55**, 213.

MORATA, G. & RIPOLL, P. (1975). *Minutes:* mutants of *Drosophila* autonomously affecting cell division rate. *Devl Biol.* **427**, 211–221.

ROSELAND, C. & SCHNEIDERMAN, H. A. (1979). Regulation and metamorphosis of the abdominal histoblasts of *Drosophila melanogaster*. *Wilhelm Roux's Arch. Devl Biol.* **186**, 235–265.

RUSSELL, M. (1974). Pattern formation in imaginal discs of a temperature-sensitive cell lethal mutant of *Drosophila melanogaster*. *Devl Biol.* **40**, 24–39.

SANTAMARIA, P. & GARCIA-BELLIDO, A. (1972). Localization and growth pattern of the tergite *anlage* of *Drosophila*. *J. Embryol. exp. Morph.* **28**, 397–417.

SIMPSON, P. (1976). Analysis of the compartments of the wing of *Drosophila melanogaster* mosaic for a temperature-sensitive mutation that reduces mitotic rate. *Devl Biol.* **54**, 100–115.

SIMPSON, P. (1979). Parameters of cell competition in the compartments of the wing disc of *Drosophila*. *Devl Biol.* **69**, 182–193.

SIMPSON, P., BERREUR, P. & BERREUR-BONNENFANT, J. (1980). The initiation of pupariation in *Drosophila*: dependence on growth of the imaginal discs. *J. Embryol. exp. Morph.* **57**, 155–165.

SIMPSON, P. & MORATA, G. (1980). The control of growth in the imaginal discs of *Drosophila*. In *Development and Neurobiology of Drosophila* (ed. Siddigi, Babu, Hall and Hall). New York: Plenum Press.

SIMPSON, P. & MORATA, G. (1981). Differential mitotic rates and patterns of growth in compartments in the *Drosophila* wing. *Devl Biol.* (In the Press.)

SIMPSON, P. & SCHNEIDERMAN, H. A. (1975). Isolation of temperature sensitive mutations blocking clone development in *Drosophila melanogaster*, and the effects of a temperature sensitive cell lethal mutation on pattern formation in imaginal discs. *Wilhelm Roux's Arch. Devl Biol.* **178**, 247–275.

SLÁMA, K., ROMAŇUK, S. & SORM, R. (1974). *Insect Hormones and Bio-analogues*. New York: Springer-Verlag.

STEINER, E. (1976). Establishment of compartments in the developing leg imaginal discs of *Drosophila melanogaster*. *Wilhelm Roux's Arch. Devl Biol.* **180**, 9–30.

STERN, C. (1936). Somatic crossing over and segregation in *Drosophila melanogaster*. *Genet.* **21**, 625–730.

STEWART, M., MURPHY, C. & FRISTROM, J. W. (1972). The recovery and preliminary characterization of *X* chromosome mutants affecting imaginal discs of *Drosophila melanogaster*.
Devl Biol. **27**, 71.

WEAVER, E. C. (1960). Somatic crossing over and its genetic control in *Drosophila*. *Genetics*
45, 345–357.

WIESCHAUS, E. & GEHRING, W. (1976). Clonal analysis of primordial disc cells in the early
embryo of *Drosophila melanogaster*. *Devl Biol.* **50**, 249–263.

ŽĎÁREK, J. & SLAMA, K. (1972). Supernumerary larval instars in cyclorhaphous *Diptera*.
Biol. Bull. mar. biol. Lab., Woods Hole **142**, 350.

J. Embryol. exp. Morph. Vol. 65 (Supplement), pp. 89–101, 1981
Printed in Great Britain © Company of Biologists Limited 1981

Catch-up growth

By J. P. G. WILLIAMS

*From the Department of Biological Sciences,
City of London Polytechnic*

SUMMARY

Catch-up growth, defined as growth velocity above the statistical limits of normality for age or maturity during a defined period of time, is distinguished from compensatory growth since it makes up for a potential loss rather than an actual loss and is seen in the whole body as opposed to specific organs. The cellular explanation for catch-up based on the work of Winick is described and a recent challenge to this explanation is briefly discussed. The mechanism of mismatch between actual size and 'planned' size suggested by Tanner is described and tested. In a series of experiments conducted in rats of different ages the degree of mismatch and the role of catch-up are compared for two different parameters, body weight and nose–rump length. It was found that the two parameters behaved differently and it is suggested that while the concept of mismatch is still acceptable the idea of a single central mechanism is not supported. It is suggested that the mismatch mechanism is a cellular phenomenon.

INTRODUCTION

Catch-up growth is a term introduced by Prader, Tanner & von Harnack (1963) to describe the increased growth velocity which occurs in children after a period of growth retardation when the cause of the growth retardation is removed. Catch-up growth may be defined as a growth velocity above the statistical limits of normality for age or maturity during a defined period of time (Williams, Tanner & Hughes, 1974a). The effect of catch-up growth is to take an organism towards, or in favourable circumstances right onto, its original pre-retardation growth curve. In the former case catch-up is said to be incomplete; in the latter, complete.

The phenomenon has been recognized since at least 1914 when Osborne & Mendel reported that during refeeding after undernutrition, rats increased their body weight at a rate equal to or greater than that expected for their size. Bohman (1955), describing the same phenomenon in cattle after different feeding regimes, employed the term 'compensating growth' since the extra growth compensated for the lack of earlier growth. The term 'compensatory growth' is still used by some workers describing catch-up growth. The term 'compensating growth' is however commonly used to describe the type of growth that occurs after the loss of an *actual* mass of tissue and may be viewed

---

[1] *Author's address:* Department of Biological Sciences, Sir John Cass School of Science and Technology, City of London Polytechnic, Old Castle Street, London E1 7NT, U.K.

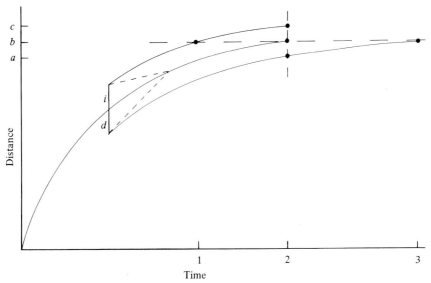

Fig. 1. Possible outcome of removing a growth decreasing, *d* or increasing, *i* stimulus. The normal growth target is *b*2. If the alteration of the growth rate is maintained then in the case of an increase the organism will reach the size target *b* at the 'wrong' time 1 or at the 'correct' time the organism will be 'too large' *c*2. If the growth rate is decreased then the organism will be 'too small' at the 'right time' *a*2 or the organism may attain the 'correct' size at a later time *b*3. The broken lines show catch-up and catch-down. In these cases the growth rate 'compensates' bringing the organism onto the original growth curve.

as being controlled by a simple feedback mechanism working on physical mass or physiological load. Catch-up growth on the other hand is rapid growth which 'compensates' for the loss of *potential* tissue and cannot be accounted for by a simple feedback mechanism. In a more abstract manner compensatory growth depends on an alteration in spatial parameters and catch-up is a response to an alteration in a temporal parameter.

In animals that attain a final adult size and cease growing before death a size target can be defined in terms of size attained at a specific time. The organism thus has a growth programme, it must reach a certain state at a certain time.

If growth is inhibited then the growth pattern or programme is altered and Fig. 1 indicates the various ways in which a time course might be changed. The target may be reached at a different time (growth delay), growth may cease at the target time thus giving rise to a small adult or the pattern may change so that the organism reaches the normal target on time, this last is catch-up. If the organism has a growth programme which responds to growth inhibition it should also respond to growth stimulation, i.e. catch-down. This is observed in the new born human where children who are born large for gestational age 'lag down' to the normal within the first year of life (Smith *et al.* 1976).

It is generally stated that the earlier the onset of inhibition the more long lasting and severe will be the consequences (Widdowson & McCance, 1960). It is also clear that the greater the deprivation the greater the period required for recovery and that the longer the growth inhibition goes on the greater will be the gap between the experimental and the control groups. Catch-up may be complete or incomplete and this seems to depend on the amount of time available for recovery and the degree of growth retardation that occurs. Failure to show complete catch-up may also be related to the mechanism for inducing the growth inhibition. Catch-up does not follow growth arrest produced by cortisone (Mosier, 1971). In catch-up after undernutrition the animals utilize food more efficiently (Miller & Wise, 1976) but after growth retardation due to cortisone rats show a 'persistent decrease in gross energy efficiency' (Mosier, 1972). Other workers however find complete catch-up in cortisone-treated rats (Mitchell, Barr & Pocock, 1978).

Following the observation of Widdowson & McCance (1963) that recovery after growth retardation depended in part at least on the age of onset of the malnutrition, Winick & Noble (1965, 1966) and Winick, Fish & Rosso (1968) undertook a series of studies to examine the cellular basis of this response.

Winick's experiments were based on the earlier study of Enesco & Leblond (1962) which examined cellular aspects of growth. These workers measured the amount of DNA present at different ages in various tissues of the rat. In addition they measured organ and tissue weights and were thus able to produce an index of cell size i.e. weight per unit DNA. With the knowledge that each cell contained 6·2 pg of DNA they were able to calculate the number of cells. Enesco & Leblond reported that during early development the number of cells ($\equiv$ amount DNA) increased rapidly and then more slowly. They also found that cell size showed nearly continuous increases in most *tissues*, such as muscle and epididymal fat. In most *organs* the increase in cell size was distinctly observed only between 17 and 48 days of age. They summarized their findings thus: 'Until about 17 days of age growth of organs and tissue is due to rapid cell proliferation with little or no change in cell size. Between about 17 and 48 days of age cell proliferation continues in all locations but at a slower rate than in the early period. Meanwhile, cell size increases in most organs and even more so in tissues. Finally after 48 days, cell proliferation slows down or even stops, while cell enlargement proceeds in most tissues but is slight or absent in organs.'

From a series of similar studies Winnick & Noble (1965) concluded that normal growth is partitioned into three periods: cell division alone; cell division with concomitant cell enlargement; and cell enlargement alone with no further increase in the number of cells. The age span of these periods varies for individual organs, but before weaning all organs grow primarily by cell division. Between weaning and about 65 days of age, beginning in the brain and lung, the pattern shifts. After 65 days growth in all organs is due primarily to cell enlargement. From these observations Winnick & Noble argued that 'the effect of any stimulus on growth inhibition may, therefore be time dependent'. They then examined the effect of undernutrition during suckling, from weaning to 42 days of age and from 65 to 86 days, with the severity of the malnutrition being kept relatively constant by maintaining the body weight without any weight gain. They reported that caloric restriction prevented the normal increase in weight, total protein and RNA in all groups of animals regardless of the time of onset of malnutrition. DNA however was only affected in the period before it had ceased to increase. There was no recovery after refeeding in those organs where cell division had been curtailed. They concluded that caloric restriction resulted in curtailment of normal growth no matter when the onset but that the effect could only be reversed if cell division had not been affected. Growth failure observed with malnutrition may thus be of two types dependent on the age of onset. Reduction in cell numbers results in permanent stunting whereas reduction in cell size results in recovery of normal size after refeeding.

The explanation for the greater severity of early undernutrition is that tissues have a finite cell number and that this is reached before the adult size. Thus growth is made up of two different phenomena, an increase in cell numbers and separately an increase in cell size. This principle has recently been challenged. Sands, Dobbing & Gratrix (1979) have reported that cell size measured as protein per unit DNA increases much earlier than had been described previously and quickly reaches a steady value. They further reported that cell multiplication continues unabated throughout tissue growth until growth itself comes to an end.

Although this challenge needs to be confirmed the consequences are of interest. The figures of Sands *et al.* (1979) only show total DNA but it seems likely that the plot of rates of accumulation or rate of synthesis would be like that shown in Fig. 2. If it is assumed that growth retardation inhibits DNA synthesis then the block is represented by the area *a, b, c, d,* in Fig. 2. To finish with the same number of cells as the control in the same period of time an increase in the rate of synthesis is needed and this would be most effective soon after growth restarts. These predictions of the rate of cell replicaton are identical to the theoretical curve of catch-up shown by Forbes (1974). Failure to catch-up completely may be due to too little time or too great an insult. Further, since cell replication and DNA synthesis can be inhibited at a number of points in the synthetic pathways failure to catch-up may be due to different sites of action of different inhibitors, (e.g. cortisone).

Malnutrition during the suckling period is not however a source of irreversible reduction in cell numbers. The adult body size of children born during the Amsterdam famine in 1945 now show a normal distribution and when rats raised in large litters until day 9 of age were suckled in small litters the number of cells returned to the normal by day 21, (Winnick, Fish & Rosso, 1968). Even more surprising perhaps is the 'catch-up' in cell numbers but not cell size

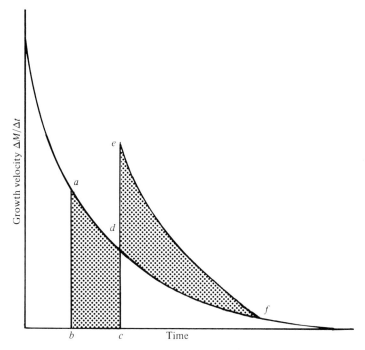

Fig. 2. Hypothetical growth velocity curve showing the effect of growth inhibition, area *abcd* and subsequent catch-up, area *def*. In the case of complete catch-up *abcd* = *def*.

observed in rat fetuses from undernourished dams. These rats are born with normal total body DNA but with 25 % reduction in body weight and total body protein (Williams & McAnulty, 1976). It may be that there is a critical time by which the total cell number must be reached and that in the examples sited here that time had not been reached.

Using the clinical cases reported by Prader *et al.* (1963), Tanner (1963) proposed a generalized mechanism to describe the events occurring during catch-up growth. Tanner's model was based on self-stabilizing systems. Essentially the model required two components, a 'flight plan' and a method for determining position. Tanner's supposition was that in the normally developing central nervous system a substance accumulated, or some cells matured, in a manner that traced out the brain's growth curve, and this was called the 'time tally'. Tanner further suggested that body growth is represented by some hypothetical substance called 'inhibitor'. The normal velocity of growth was thought to be proportional to the 'mismatch' between the two signals, in other words to the 'gap' in growth advancement between the CNS and periphery. Tanner also suggested that the effects of growth retardation were age dependent. In the case of early retardation the CNS is not mature and the 'time tally' is altered in such a way that the gap can never be fully bridged and catch-up is

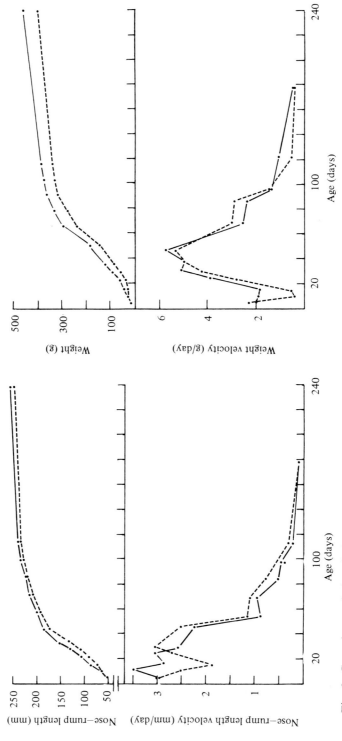

Fig. 3. Growth retardation during the suckling period followed by *ad lib* feeding from weaning at three weeks. Top panels show distance curves and the lower panels show growth velocity curves. The figures on the left show nose–rump length and the figures on the right show body weight.

incomplete. In the situation where catch-up is complete Tanner suggests that the CNS has matured and the 'time tally is fixed'. Although radiation of the brain has been reported to reduce growth (Mosier & Jansons, 1970) and it has been suggested that the appetite centre may be of importance in catch-up (Widdowson & McCance, 1975) there is no convincing evidence for the time tally in the CNS. In the period since Tanner suggested this hypothesis many growth-stimulating factors have been reported but few inhibiting agents.

A series of growth studies carried out in Tanner's laboratory have been analysed to examine parts of Tanner's hypothesis (Williams *et al.* 1974*a*, *b*; Williams & Hughes, 1975; Williams & McAnulty, 1976; Williams & Hughes, 1978).

MATERIALS AND METHODS

Three different age groups of animals were used to represent different normal growth rates and different degrees of maturation at the onset of undernutrition. The period of growth retardation was always three weeks.

The first group of animals were undernourished during suckling. This was brought about by using large (16 pup) and normal (8 pup) litters. The animals in the larger litters all survived well but were smaller.

Growth arrest was also introduced, in different groups of animals, at three weeks and at seven weeks of age. The rationale for the times was that many tissues would be producing cells predominantly during the suckling period, that during the three-week period following weaning a peak of body weight growth occurred and it was assumed that both cell number and cell size were increasing.

The group subject to zero growth after seven weeks should have made most of their cells and the subsequent catch-up would be due to cells 'filling up'. After the three weeks of reduced- or zero-growth animals in all groups were provided with food *ad lib.* Body weight, nose–rump length, and skeletal maturity were measured in the same animals throughout the experiments, i.e. they were longitudinal studies. The studies were terminated at 228 days at which time all of the animals showed 98 % bone maturity.

RESULTS AND DISCUSSION

The figures 3, 4, 5, show the effects of undernutrition at the various ages, in male rats. Table 1 shows the percentage mismatch at the end of each period of growth retardation on both male and female rats.

The percentage mismatch is always greater for body weight than for body length. Table 2 shows the time taken for male and female rats that have been undernourished at different ages to catch-up to their respective controls.

In the case of the male animals from large litters complete catch-up was not seen. Table 3 shows the rate of recovery.

Differences between the tissues and the sexes are quite clear. Body weight shows the greatest mismatch during the period 21–42 days, the period in which

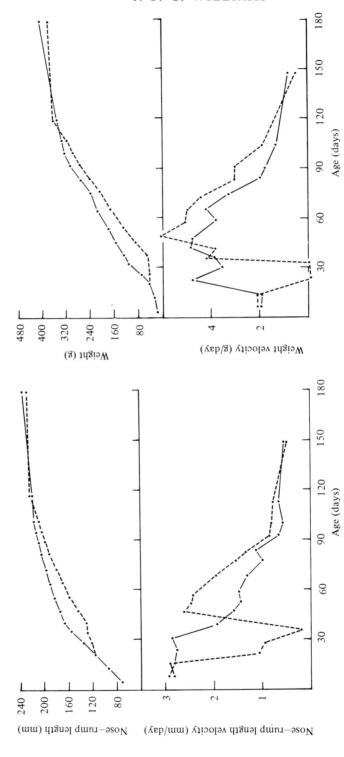

Fig. 4. Growth retardation for three weeks from three weeks of age followed by *ab lib* feeding. Top panels show distance curves and the lower panels show velocity curves. The figures on the left show nose–rump length and the figures on the right show body weight.

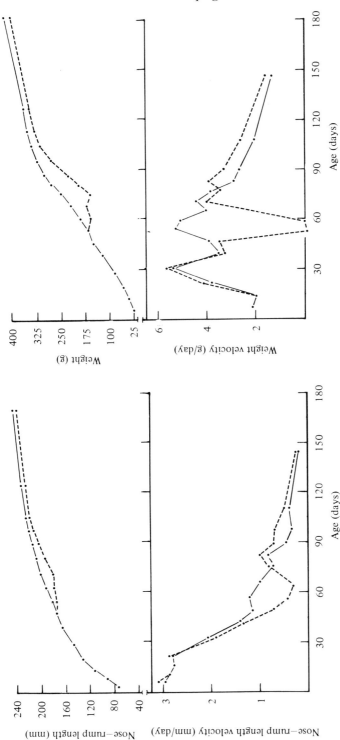

Fig. 5. Growth retardation for three weeks from seven weeks of age followed by *ab lib* feeding. Top panels show distance curves and the lower panels show velocity curves. The figures on the left show nose–rump length and the figures on the right show body weight.

Table 1. *Mismatch, percent*

Period of growth restriction days	Nose–rump length		Body weight	
	M	F	M	F
0–21	13	16	40	37
21–42	21	22	65	67
49–70	12	5	40	31

Table 2. *Time taken to obtain complete catch-up in days*

Period of growth restriction days	Nose–rump length		Body weight	
	M	F	M	F
0–21	42	63	—	56
21–42	35	56	42	28
49–70	28	28	28	21

Table 3. *Rate of recovery millimetres or grams per day*

Period of growth restriction days	Nose–rump length		Body weight	
	M	F	M	F
0–21	0·31	0·25	—	0·66
21–42	0·60	0·39	1·55	2·39
49–70	0·43	0·18	1·43	2·43

the normal body weight shows the greatest rate of increase (Figs 3–5). The sex difference in the degree of mismatch is only evident in the oldest group of animals in which the sexual difference in growth is clearly established. Sex differences in the time taken to achieve complete catch-up are most marked in the younger animals. The rate of recovery is faster in the male for nose–rump length but the female shows the fastest rate of recovery for body weight.

In Fig. 6 the velocity of catch-up is plotted against the degree of mismatch, according to the hypothesis these two measurements should be directly proportional and would therefore fall on the same line. The data for the nose–rump length appear to fall on a straight line whilst those for body weight do not. The mismatch hypothesis seems to hold for the nose–rump length but not for the body weight.

Since the two parameters, nose–rump length and body weight within the same animal behave differently it seems unlikely that a single central signal or substance or mechanism controls the phenomena of catch-up growth.

Although only body weight and nose–rump length are shown there are several cases of tissues catching up at different rates, e.g. tail and body length (Williams *et al.* 1974*a*), and different muscles (Dickerson & McAnulty, 1975). While

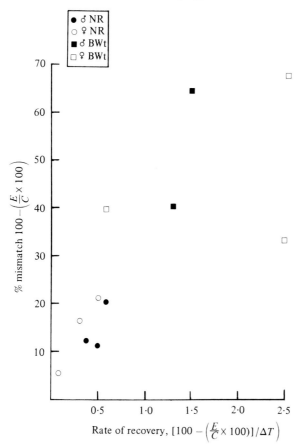

Fig. 6. Degree of mismatch plotted against the rate of recovery.

catch-up appears to be a whole body response the parts of the body seem to respond in an individual manner. Further, different tissues and organs have their normal peak growth spurts at different times. Where it has been examined in some detail (Williams & Hughes, 1975) the spurt of growth following a period of undernutrition was found to be greatly influenced by the normal pattern of growth velocity. If catch-up growth started at a time when the natural growth pattern was declining the subsequent growth rate was less than when catch-up began on an upsurge in the normal pattern of growth.

It seems then that while Tanner's idea of comparing the actual size with the programmed size may still be acceptable some other features of his thesis such as the time tally being in the CNS and the rate of catch-up being directly related to the period and intensity of the insult are now less likely.

There have been several attempts to relate the rate of normal growth and of catch-up to substances, especially hormones or growth factors (Mosier, Dearden, Jansons & Hill, 1978) in the blood but none of these correlate significantly with the rates of growth that together make up the growth of the body.

Liebel (1977) has proposed a model in which information regarding the body mass is supplied to the CNS by a 'humoral radar system' employing an insulin-like molecule as a signal. The intensity of the signal would vary as a function of the organism's total adipose cell surface area. While such a system is attractive in relation to body fat and perhaps via appetite control to total growth it does not deal with the problem posed here or by any growth study where the parts of the whole grow at different rates.

The analysis described here appears to show that there is no single mechanism regulating catch-up.

The alternative to this is a number of mechanisms. The most feasible system would be at the tissue level, with the regulation in some or all cells. This view is in accord with the increasing number of tissue-specific growth factors (Nero & Laron, 1979) and is supported by Winick's studies. The various tissues may be co-ordinated by permissive hormones such as somatotropin (growth hormone).

Completeness of catch-up depends on the normal or control. In rat experiments different workers have used different normal litter sizes and different end points. The male rats described above caught up in length but not weight. Rat experiments are also complicated by alterations in the rate of maturity (the time tally) (Williams *et al.* 1974*a*). The animal data is further complicated by age effects which must influence these often long experiments.

In the human cases complete catch-up is said to be achieved when the child reaches its predicted size or a normal size for the population. The situation is complicated by the likelihood that different growth-retarding agents behave differently e.g. cortisone. In general, in incomplete catch-up growth, cessation occurs before complete catch-up can be achieved, and this growth arrest is often associated with the onset of sexual maturity.

From the analysis of catch-up growth a generalized theory is presented. Growth regulation is a cellular phenomena in which the cell has a programme and a mechanism to recognize where it is in that programme; if it is diverted from the programme then a stabilizing mechanism will tend to return it to the right course. The cells of a given tissue type have the same programme, and are co-ordinated by tissue-specific diffusable agents. The tissues and organs are co-ordinated at the programme level and also by permissive hormones.

Experimental work reported here was carried out in Professor Tanner's Laboratories at the Institute of Child Health, London, with support from MRC. My thanks to Professor Tanner and Mr P. R. C. Hughes for their help and guidance.

REFERENCES

BOHMAN, V. R. (1955). Compensatory growth in beef cattle. The effect of hay maturity. *J. Animal Sci.* **14**, 249–255.
DICKERSON, J. W. T. & MCANULTY, P. A. (1975). The response of hind-limb muscles of the weaning rat to undernutrition and subsequent rehabilitation. *Br. J. Nutrition* **33**, 171–180.

ENESCO, M. & LEBLOND, C. O. (1962). Increase in cell number as a factor in the growth of the organs and tissues of the young male rat. *J. Embryol. exp. Morph.* **10**, 530–562.

FORBES, G. B. (1974). A note on the mathematics of 'catch-up' growth. *Pediatric Research* **8**, 929.

LEIBEL, R. L. (1977). A biological radar system for the assessment of body mass, the model of a geometry sensitive endocrine system is presented. *J. theoret. Biol.* **66**, 297–306.

MILLER, D. S. & WISE, A. (1976). The energetics of catch-up growth. *Nutritional Metabolism* **20**, 125–134.

MITCHELL, I., BARR, D. G. D. & POCOCK, S. J. (1978). Catch-up growth in cortisone-treated rats: effects of calcium and vitamin D. *Pediatric Research* **12**, 105–111.

MOSIER, H. D. JR. (1971). Failure of compensatory (catch-up) growth in the rats. *Pediatric Research* **5**, 59–63.

MOSIER, H. D. JR. (1972). Decreased energy efficiency after cortisone induced growth arrest. *Growth* **36**, 123–131.

MOSIER, H. D. JR., DEARDEN, L. C., JANSONS, R. A. & HILL, R. R. (1978). Cartilage sulfation during catch-up growth after fasting in rats. *Endocrinology* **102**, 306–392.

MOSIER, H. D., JR. & JANSON, R. A. (1970). Effect of X-irradiation of selected areas of the head of the newborn rat on growth. *Radiation Research* **43**, 92–104.

NERO, Z. & LARON, Z. (1979). Growth factors. *Amer. J. Disease Child.* **133**, 419–428.

OSBORNE, T. B. & MENDEL, L. B. (1914). The suppression of growth and the capacity to grow. *J. biol. Chem.* **18**, 95–106.

PRADER, A., TANNER, J. M. & VON HARNACK, G-A. (1963). Catch-up growth following illness or starvation. *J. Pediatrics* **62**, 646–659.

SANDS, J., DOBBING, J. & GRATRIX, C. A. (1979). Cell number and cell size: organ growth and development and control of catch-up growth in rats. *Lancet* **2**, 503–506.

SMITH, D. W., TRUGG, W., ROGERS, J. E., GREITZER, L. J., SKINNEW, A. L., McCANN, J. J. & SEDGWICK HARVEY, M. A. (1976). Shifting linear growth during infancy: illustration of genetic factors in growth from fetal life through infancy. *J. Pediatrics* **89**, 225.

TANNER, J. M. (1963). The regulation of human growth. *Child Development* **34**, 817–847.

WIDDOWSON, E. M. & McCANCE, R. A. (1960). Some effects of accelerating growth, 1 General somatic development. *Proc. R. Soc. Lond.* B **152**, 188.

WIDDOWSON, E. M. & McCANCE, R. A. (1963). The effects of finite periods of under-nutrition at different ages on the composition and subsequent development of the rat. *Proc. R. Soc. Lond.* B **152**, 188.

WIDDOWSON, E. M. & McCANCE, R. A. (1975). A review: new thoughts on growth. *Pediatric Research* **9**, 154–156.

WILLIAMS, J. P. G. & HUGHES, P. C. R. (1975). Catch-up growth in rats undernourished for different periods during the suckling period. *Growth* **39**, 179–193.

WILLIAMS, J. P. G. & HUGHES, P. C. R. (1978). Catch-up growth in the rat skull after retardation during the suckling period. *J. Embryol. exp. Morph.* **45**, 229–235.

WILLIAMS, J. P. G. & McANULTY, P. A. (1976). Foetal and placental ornithine decarboxylase activity in the rat, effect of maternal undernutrition. *J. Embryol. exp. Morph.* **35**, 545–559.

WILLIAMS, J. P. G., TANNER, J. M. & HUGHES, P. C. R. (1974a). Catch-up growth in male rats after growth retardation during the suckling period. *Pediatric Research* **8**, 149–156.

WILLIAMS, J. P. G., TANNER, J. M. & HUGHES, P. C. R. (1974b). Catch-up growth in female rats after growth retardation during the suckling period: comparison with males. *Pediatric Research* **8**, 157–162.

WINICK, M., FISH, I. & ROSSO, P. (1968). Cellular recovery in rat tissues after a brief period of neonatal malnutrition. *J. Nutrition* **95**, 623–626.

WINICK, M. & NOBLE, A. (1965). Quantitative changes in DNA, RNA and protein during prenatal and postnatal growth in the rat. *Devl Biol.* **72**, 451.

WINICK, M. & NOBLE, A. (1966). Cellular response in rats during malnutrition at various ages. *J. Nutrition* **89**, 300–306.

J. Embryol. exp. Morph. Vol. 65 (*Supplement*), pp. 103–128, 1981
Printed in Great Britain © Company of Biologists Limited 1981

The control of somitogenesis in mouse embryos

By P. P. L. TAM*

*From the MRC Mammalian Development Unit
University College London*

SUMMARY

Somitogenesis in the mouse embryo commences with the generation of presumptive somitic mesoderm at the primitive streak and in the tail-bud mesenchyme. The presumptive somitic mesoderm is then organized into somite primordia in the presomitic mesoderm. These primordia undergo morphogenesis leading to the segmentation of somites at the cranial end of the presomitic mesoderm. Somite sizes at the time of segmentation vary according to the position of the somite in the body axis: the size of lumbar and sacral somites is nearly twice that of upper trunk somites and of tail somites. The size of the presomitic mesoderm, which is governed by the balance between the addition of cells at the caudal end and the removal of somites at the cranial end, changes during embryonic development.

Somitogenesis is disturbed during the compensatory growth of mouse embryos which have suffered a drastic size reduction at the primitive-streak and early-organogenesis stages. The formation of somites is retarded and the upper trunk somites are formed at a smaller size. The embryo also follows an entirely different growth profile, but a normal body size is restored by the early foetal stage. The somite number is regulated to normal and this is brought about by an altered rate of somite formation and the adjustment of somite size in proportion to the whole body size. It is proposed that axis formation and somitogenesis are related morphogenetic processes and that embryonic growth controls the kinetics of somitogenesis, namely by regulating the number of cells allocated to each somite and the rate of somite formation.

INTRODUCTION

In the mouse embryo, mesoderm formation begins at about 6·5 days *post coitum* (*p.c.*) (Snow, 1977). At the primitive streak, epiblast cells invaginate and then spread as two lateral sheets of mesoderm in between the epiblast and the endoderm (Batten & Haar, 1979; Poelmann, 1980). During this process of gastrulation, extensive tissue relocation and marked changes in the morphology and the behaviour of cells are observed (Reinius, 1965; Revel & Solursh, 1978; Solursh & Revel, 1978; Batten & Haar, 1979; Poelmann, 1981). As the embryo develops, the mesoderm is organized into paraxial and lateral plate mesoderm along either side of the body axis. Somites, which are blocks of mesodermal cells, are representatives of the segmentation of the paraxial mesoderm and are formed in a co-ordinated cranio caudal sequence (Flint, 1977). Altogether, up to 65 pairs of somites are formed in the mouse embryo (Witschi, 1962; Rugh,

* *Present address:* Department of Anatomy, Chinese University of Hong Kong, Shatin, N.T., Hong Kong.

1968; Theiler, 1972). The somites subsequently differentiate to form the axial skeleton (vertebrae and ribs), the muscle of the trunk and the limb and the dermis of the body wall. The number of vertebrae derived from the somites is remarkably constant among individual adult mice, but some variations in number also exist in mice of different genetic backgrounds and in mutant mice showing a reduced body axis (Gruneberg, 1963; Bennett, 1975; Flint, 1977).

The capacity to regulate the normal number of somites has been demonstrated in amphibian embryos where the embryonic size or the amount of presumptive somitic material is altered. In *Xenopus* embryos, even when a large proportion of blastula cells is removed surgically, qualitatively normal neurula- and tail-bud-stage embryos with a normally proportioned body pattern are formed (Waddington & Deuchar, 1953). In such miniaturized embryos, normal numbers of somites are formed. But the somites are smaller in size and are made up of fewer cells (Cooke, 1975). A normal number of somites is also formed in haploid *Xenopus* embryos. The length of the somite is shorter than the normal diploid size, a result of the smaller size of haploid cells, although the embryo partially compensates for this small cell size by incorporating more cells into each somite (Hamilton, 1969). When the amount of somite mesoderm in the late gastrula of the newt is doubled by adding extra strips of mesodermal tissue, the resultant size of the somite is increased but apparently the total number of somites remained unchanged (Waddington & Deuchar, 1953). When the amount of somitic mesoderm is reduced by removing a longitudinal strip of tissue from the neurula-stage *Bombina* embryo, normal numbers of somites of a reduced size are formed (Cooke, 1977). Similar observations have been made in the chick embryo. The removal or addition of presumptive somitic mesoderm only alters the size but not the number of somites formed from the segmental plate (Menkes & Sandor, 1977). When the segmental plate of different developmental stages is cultured, a relatively constant number of somites is formed regardless of the initial size of the segmental plate (Packard & Jacobson, 1976). In the mutant (*amputated*) mouse embryo, despite a shorter body length, normal numbers of somites are formed by 9·5 days *p.c.* The somites in the mutant embryo are much smaller than in the normal embryo (Flint, Ede, Wilby & Proctor, 1978). All these examples show that regulation of a normal somite number is brought about by an adjustment of the somite size in accordance to the amount of precursor tissue available for somitogenesis.

In the mouse embryo, somite formation begins at a stage when most of the presumptive somitic mesoderm for the posterior somites is yet to be formed. Flint *et al.* (1978) pointed out that in such a continuously growing system, pattern formation is unlikely to be a process of proportioning of existing tissue, because it is inconceivable that the mouse embryo has a knowledge of its final body size. The control of number and size of somites is suggested to be related to the production of presomotic mesoderm by node regression, which is governed by the overall growth of the embryo. In both the mutant (*amputated*) mouse embryo and the *Xenopus* embryo formed from a small blastula, no compensatory growth was observed and therefore a regulation of whole body size did not occur (Cooke, 1975; Flint *et al.* 1978).

An upward regulation of embryonic size has been observed during the development of mouse embryos. Half-sized embryos can be produced by the destruction of one of the two blastomeres at the 2-cell stage. The embryo remains half size until about 11·5 days *p.c.* An upward regulation of size seems to have taken place between 10·5 and 11·5 days *p.c.*, presumably by an accelerated growth of the embryo (Tarkowski, 1959). Drastic reduction of embryonic size can be achieved by treating the mouse embryo at post-implantation stages with a DNA-synthesis inhibitor, Mitomycin C (MMC). Extensive cell death and cessation of cell proliferation ensue when the early-primitive-streak-stage embryo is treated with the drug transplacentally. By the late-primitive-streak stage, the treated embryo has on average only 10–15 % of the normal number of cells (Snow & Tam, 1979). The affected embryo then shows a remarkable capacity for compensatory growth so that a normal body size is restored by 13.5 days *p.c.* and the foetal weight at birth is essentially normal (Snow, Tam & McLaren, 1981). Preliminary results show that somite formation, as well as other developmental processes including the limb bud, neural tube and primordial germ cells, is disturbed during the period of compensatory growth (Snow & Tam, 1979, 1980; Snow *et al.* 1981; Tam & Snow, 1981). The MMC-treated embryo therefore provides an experimental model to study the relationship of somitogenesis to embryonic growth.

This study concerns the process of somitogenesis in mouse embryos throughout the stages of development from early-somite stage to early-foetal stage. The association between somite formation and axis formation is studied by comparing the observations made on normal embryos and on MMC-treated embryos.

MATERIALS AND METHOD

Random-bred *Q*-strain mice were used. Female mice were mated with males of proven fertility and the morning when copulation plugs were found was taken as 0·3 days *p.c.* Mating has been assumed to occur between midnight and 2 a.m.

The pregnant females were sacrificed at 8·3–13·0 days *p.c.* to obtain embryos of appropriate stages of development. These were normal embryos. Some pregnant females were each given a single intraperitoneal injection of Mitomycin C (MMC, Sigma, London; at a dose of 100–125 μg/25 g) at 6·75–7·0 days *p.c.* (Snow & Tam, 1979). Embryos were at the early- to mid-primitive-streak stage at the time of drug treatment (Snow, 1977). MMC-treated embryos were obtained from treated females sacrificed at 8·3–13·0 days *p.c.* Some females were injected with saline solution as controls. Embryos from these females

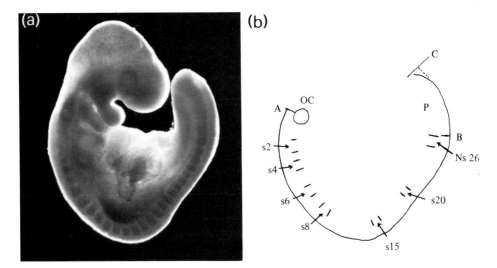

Fig. 1. (a) A 26-somite-stage mouse embryo and (b) a camera-lucida drawing of the embryo. OC, Otic capsule, S, somite, Ns, newly segmented somite, P, presomitic mesoderm.

were pooled with normal embryos since there was no discernible effect of the saline injection.

The embryos were dissected out from the decidua into phosphate-buffered saline (PB-1, Whittingham & Wales, 1969). The gross morphology and the somite number of the embryo were recorded. Camera-lucida drawings of the fresh unfixed embryos were made at $50\times$ or $100\times$ magnification using a Wild M5 dissection microscope fitted with a drawing tube. All measurements of length and size were then made on the camera-lucida drawings. Fig. 1a shows a 26-somite-stage embryo and Fig. 1b is a camera-lucida drawing of the embryo. Axial length ($A-C$) was measured from the cranial boundary of the first somite along the neural tube to the tail tip in embryos having 10 or fewer somites. In older embryos where the boundary of the first few somites was not readily discernible, axial length was measured from the centre of the optic capsule to the tail tip. This may result in an over-estimation of the axial length, but not by more than 200 μm which is equivalent to the length of about one somite. Axial length measured in this manner represents the total length of the somitic column and the presomitic mesoderm. The head was not included in such a measurement to avoid the compounding effect of head growth by brain cavity enlargement which may lead to an over-estimation of axial growth. The anterior–posterior length of the most recently segmented somite was measured midway between the dorsal and ventral side of the somite (see Flint et al. 1978). The length of the presomitic mesoderm was measured from the posterior boundary of the most recently segmented somite to the tail tip ($B-C$, Fig. 1b). Altogether, 234 normal embryos and 198 MMC-treated embryos were included.

Fifty-six normal embryos and 46 MMC-treated embryos were collected from females between 8·3 and 11·75 days *p.c.* and were fixed *in situ* within the decidua in Bouin's fixative. Serial paraffin sections of 8 or 10 μm thickness were obtained and the sections were stained with haematoxylin and eosin. At each stage, the histology of the presomitic mesoderm and the caudal end of the embryo was examined. The somite number of the embryo was determined by counting the somites in serial sections of the specimen. The cell number, cell volume and mitotic index of the newly segmented somite and those of somite 8 were estimated as follows. Camera-lucida drawings were made from all sections of the somite. Total tissue volume of the somite was then estimated from summed planimeter measurements of section areas multiplied by section thickness. For each somite, about five to six sections were selected and the number of nuclei in these sections was counted. This score was then corrected by Abercrombie's (1946) formula to give an estimate of the actual number of cells. Cell volume was then estimated by dividing the total tissue volume of the selected sections with the number of cells in that volume. Such an estimate, which does not exclude extracellular space, is close to the actual cell volume of the somitic cells which are closely packed, but may be an overestimate for the differentiating somite 8 where cells are more loosely packed. Finally, the cell number of the somite was computed from the tissue volume divided by the cell volume. Mitotic index was estimated from the total number of metaphase–anaphase figures and the cell number of the somite. For the presomitic mesoderm, mitotic index and cell volume were estimated in four randomly chosen tissue sections in different regions of the mesenchyme.

The data obtained were analysed by various statistical methods of correlation analysis, linear, non-linear and multiple linear regression, and tests for significance between groups; details of the methods are described in Meddis (1975) and Snedecor & Cochran (1967).

RESULTS

Rate of somite formation

The first pair of somites is formed at about 8·3 days *p.c.* in the *Q* strain mouse embryo. Initially, somites are formed at a fast rate of about one pair in every hour. In later stages, the rate of somite formation progressively slows down to one pair in every 2–3 h. Figure 2 shows the mean somite number for embryos at various ages. There is considerable variation in the somite number of embryos within and between litters of the same age. The somite number also overlaps among embryos belonging to litters of different age, e.g. 30-somite embryos are found in both 9·5- and 10·0-day litters. Such a variation in somite number suggests that embryos may develop at different rates even when they are in the same uterine environment. The most advanced embryo could be about 6–8 h ahead in development compared to its slowest litter-mate.

In the MMC-treated embryos, the appearance of the first pair of somites is

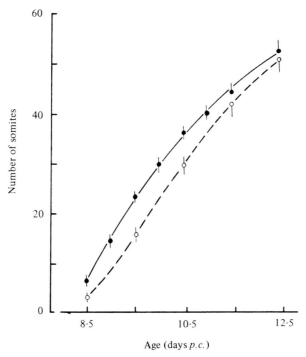

Fig. 2. The mean somite number of normal (●) and MMC-treated (○)
embryos at 8·5–12·5 days.

delayed by about 1–2 h. Subsequent somites are formed at a relatively slow
rate of one pair in every 2 h. The formation of the upper and lower trunk somites
is retarded and somites 25–35 are formed about 10–12 h later than those in normal
embryos. At 8·5–10·5 days, the MMC-treated embryos have significantly fewer
somites than the normal embryos, but by 11·5 and 12·5 days, the treated embryos
have the same somite number as the normal ones (Fig. 2).

The paraxial presomitic mesoderm

Regional differentiation of the mesoderm becomes apparent in the presomite-
stage embryo at about 8·3 days *p.c.* The mesoderm lying underneath the
cranial neural plate forms the head mesenchyme. The mesoderm in the posterior
half of the embryo forms two broad bands of paraxial and lateral plate mesoderm
on either side of the primitive streak. Figure 3 shows a longitudinal section
through the length of the paraxial presomitic mesoderm of an 8·5-day embryo
and Fig. 4 shows a similar section of a 9·5-day embryo. Somites are formed by
the segmentation of tissue at the cranial end of the presomitic mesoderm. Within
the presomitic mesoderm, and particularly near the cranial portion, mesenchymal
cells are organized into segmental units prior to overt segmentation. Figure 5
shows a magnified view of the arrangement of mesenchymal cells. Cells are
arranged radially into clusters and the boundary between clusters is marked by

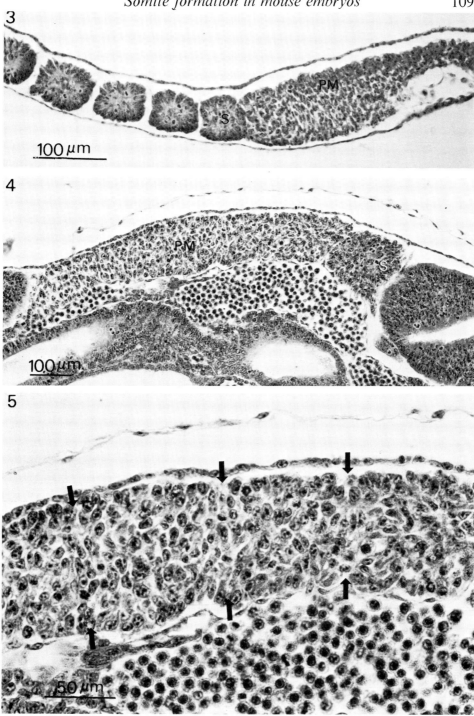

Fig. 3. Presomitic mesoderm (PM) and somitic column of a 7-somite-stage embryo. Cells in the presomitic mesoderm closest to the newly segmented somite (S) are organized into clusters.

Fig. 4. Presomitic mesoderm (PM) and newly segmented somite (S) of a 27-somite-stage embryo.

Fig. 5. The organization of cells in the presomitic mesoderm. Magnified view of Fig. 4, showing boundaries (➡) between clusters.

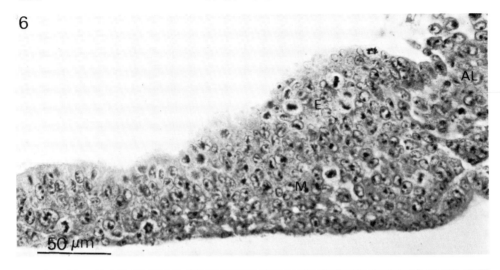

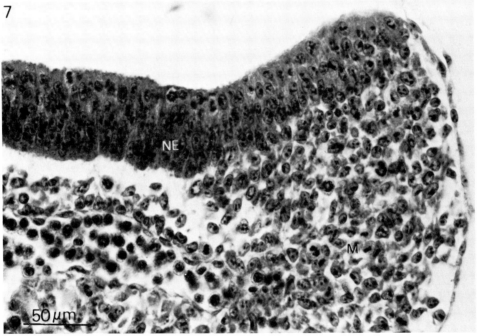

Fig. 6. Primitive streak area of a 8·5-day embryo. E, epiblast; M, mesoderm; AL, allantoic bud.

Fig. 7. Primitive streak area of a 9·5-day embryo. NE, neuroepithelium of the posterior neuropore; M, mesoderm.

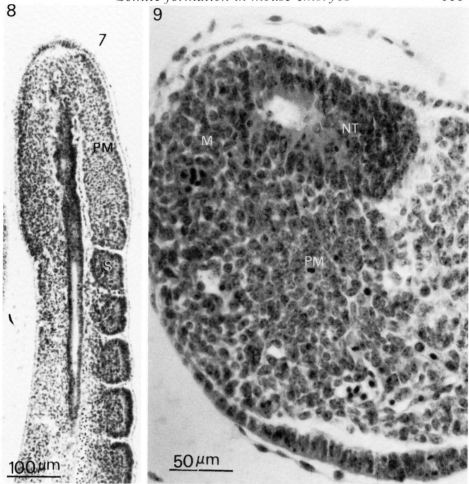

Fig. 8. A section through the tail of a 10·5-day embryo, showing presomitic mesoderm (PM) and newly segmented somite (S).

Fig. 9. The terminal zone of proliferating mesenchyme in the tail bud. NT, neural
tube; PM, presomitic mesoderm; M, mesenchyme.

the change of cellular orientation from a radial to a transverse alignment. Similar
segmental organization of cells is also found in the presomitic mesoderm closest
to the newly segmented somite in the tail (see Fig. 8).

The caudal end of the presomitic mesoderm is always associated with either
the primitive streak (Fig. 6 & 7) or the tail bud (Fig. 9). The primitive streak is first
identified in the 6·5-day early-primitive-streak-stage embryo as a localized area in
the epiblast where cell invagination occurs. The streak lengthens to 400–500 μm
in the 7·5 day late-primitive-streak-stage embryo, and it extends from the base
of the allantoic bud to the distal tip of the egg cylinder. The streak then progressively shortens during subsequent development to 200–300 μm at 8.0 days,
to 150–170 μm at 8·5 days and to 50–100 μm at 9·5 days. At 8·5 days, the

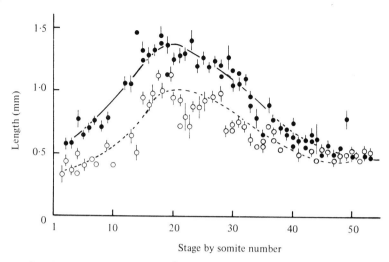

Fig. 10. Changes in the size of presomitic mesoderm during development.
Normal embryos (●) and MMC-treated embryos (○).

Table 1. *The mitotic activity and cell volume of the mesenchyme in the presomitic mesoderm of normal and MMC-treated mouse embryos*

Somite number of embryos	Mitotic index (%)		Cell volume (μm^3)	
	Normal	MMC	Normal	MMC
1–4	11·5 (1)	10·9 ± 2·2 (3)	990 (1)	1108 ± 29 (3)
5–8	10·3 ± 1·2 (9)	8·3 ± 1·1 (7)	932 ± 99 (9)	1072 ± 51 (7)
9–12	5·6 ± 0·3 (9)	7·5 ± 0·9 (7)	1012 ± 76 (9)	1134 ± 45 (7)
13–16	6·7 ± 1·1 (3)	6·5 ± 1·7 (9)	1146 ± 22 (3)	1067 ± 55 (9)
17–20	5·2 ± 0·6 (6)	7·3 (2)	1183 ± 67 (6)	1387 (2)
21–24	4·8 ± 0·3 (11)	7·4 (1)	1089 ± 30 (11)	1181 (1)
25–28	4·5 ± 0·1 (8)	8·1 ± 0·8 (7)	1096 ± 55 (8)	1134 ± 65 (7)
29–32	4·9 ± 0·5 (3)	6·6 ± 0·4 (7)	928 ± 66 (3)	1187 ± 60 (7)
33–36	4·8 (2)	6·9 ± 1·5 (3)	1106 (2)	1066 ± 94 (3)
37–41	4·6 ± 0·3 (4)	5·1 (2)	1015 ± 67 (4)	1018 (2)
Statistical test Is normal = MMC?	No, $F = 12·52$, DF. $= 1,102$ $P < 0·01$, MMC > normal		No, $F = 4·17$, D.F. $= 1,102$ $P < 0·01$, MMC > normal	

primitive streak is located immediately caudal to the neural plate of the embryo (Fig. 6) and at 9·5 days forms the posterior border of the open posterior neuropore (Fig. 7). Histological appearance of the primitive streak area suggests that invagination of cells from the overlying epiblast and the rearrangement of cells once they are in a mesodermal position are the major processes involved in the generation of paraxial presomitic mesoderm. After the closure of the posterior neuropore which takes place at about 9·6 days (26- to 29-somite stage), a terminal zone of proliferating mesenchyme is found in the tail bud (Fig. 9). No clear

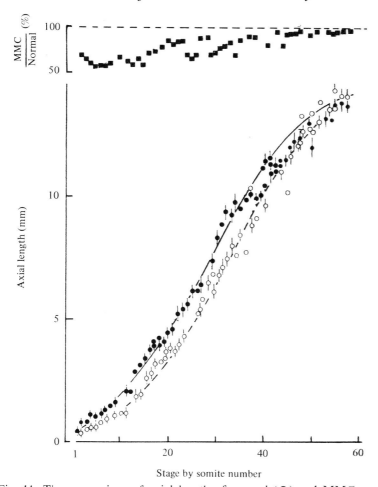

Fig. 11. The comparison of axial length of normal (●) and MMC-treated (○) embryos at different developmental stages.

distinction in morphology is found between the presumptive neural cell and the presumptive mesodermal tissue. Cranial to this terminal zone, the mesodermal cells are organized into the paraxial presomitic mesoderm on either side of the neural tube and notochord (Fig. 8).

The size of the presomitic mesoderm changes during embryonic development. It increases in length from 0·4–0·5 mm at the early-somite stage to over 1·5 mm at the 20- to 30-somite stage, then it shortens to about 0·5 mm at the 40- to 50-somite stage (Fig. 10). In the normal embryo, the mitotic activity of the mesenchyme in the presomitic mesoderm remains high until about the 20-somite stage (Table 1). This apparently coincides with the increase in the size of the presomitic mesoderm. Beyond the 20-somite stage, the mitotic activity declines. A similar change in the size of the presomitic mesoderm is also observed in MMC-treated embryos. But the size of the presomitic mesoderm is significantly smaller

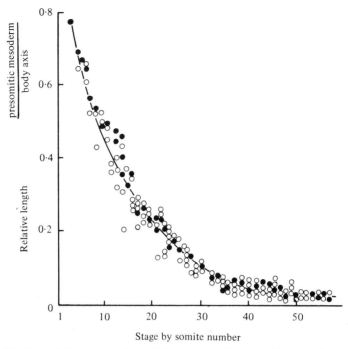

Fig. 12. The relative proportion of the length of presomitic mesoderm to whole body length. Normal embryos (●) and MMC-treated embryos (○).

than the normal size at 1- to 30-somite stage (Fig. 10). The mitotic activity of the cells is generally higher in the treated embryo (Table 1). The estimated cell volume of the mesenchymal cells in MMC-treated embryos is bigger than normal, perhaps reflecting a looser packing of cells (Table 1). This, when taken together with the reduced size of the presomitic mesoderm at 1- to 30-somite stage, suggests that many fewer cells are made available for somite formation. The enhanced mitotic activity seems also to be related to the lowering of cell density in the presomitic mesoderm.

The reduction in the size of presomitic mesoderm in MMC-treated embryos is proportional to the overall reduction in body size. Figure 11 shows the axial length measured for embryos at various somite stages. Axial length increases with the increase in somite number, but up to about the 30-somite stage the treated embryo has a significantly shorter body axis than the normal embryo of the same somite number. This information implies that the size of all the somites present in the treated embryo must be of a proportionately smaller size than those of the normal embryo. Furthermore, when the size of the presomitic mesoderm relative to the overall axis length is compared between normal and treated embryos of the same somite stage, it becomes evident that the presomitic mesoderm is always maintained at a normal proportion to the body size in the treated embryo showing a shorter body axis (Fig. 12).

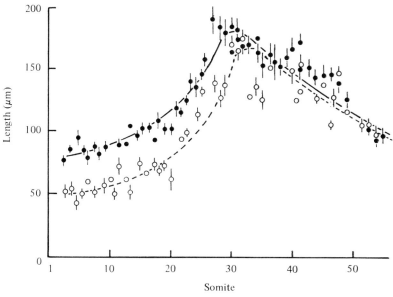

Fig. 13. The size of newly segmented somites in normal (●)
and MMC-treated (○) embryos.

Somite size at the time of segmentation

Somites are formed at different sizes according to their position in the body axis. In normal embryos, the newly segmented somites at the occipital, cervical and upper thoracic levels (somites 1–20) are about 70–110 μm in length. The size increases progressively with the more posterior somites. Lower thoracic somites (somites 21–25) are about 120–130 μm in length and upper lumbar somites (somites 26–28) are 140–150 μm and the lower lumbar somites (somites 29–31) are 150–180 μm (Fig. 13). The sacral somites (somites 32–35) are of the same size as the lower lumbar somites. The caudal somites are formed at a progressively smaller size (Fig. 13).

In the MMC-treated embryo, the occipital, cervical, thoracic and lumbar somites are all formed at a smaller size then their normal counterparts (Fig. 13). The caudal somites are formed with a normal size. The overall pattern of changes in the somite size with body level in the treated embryo is similar to that in the normal embryo.

In the normal embryo, occipital and cervical somites are formed with about 150–200 cells. More and more cells are being incorporated into thoracic, lumbar and sacral somites. About 1000–1500 cells are found in the newly segmented lumbar and sacral somites, which therefore have ten times more cells than in upper trunk somites (Table 2). In the treated embryos, somites generally have fewer cells; only about 50 % as many cells are found in the upper trunk somites, the more caudal ones have about 80–90 % of the normal number of cells (Table

Table 2. *The cell number, cell volume and mitotic activity of cells in the newly segmented somites of normal and MMC-treated mouse embryos*

Somites	Cell number		Cell volume (μm^3)		Mitotic index (%)	
	Normal	MMC	Normal	MMC	Normal	MMC
1–4	163 (1)	65 ± 8 (1)	703 (1)	978 ± 64 (3)	5·5 (1)	6·0 ± 2·0 (3)
5–8	181 ± 18 (9)	95 ± 8 (7)	859 ± 35 (9)	1021 ± 69 (7)	6·7 ± 0·5 (9)	5·0 ± 0·8 (7)
9–12	218 ± 17 (9)	189 ± 25 (7)	858 ± 42 (9)	1030 ± 61 (7)	7·3 ± 0·6 (9)	5·2 ± 0·9 (7)
13–16	435 ± 83 (3)	272 ± 45 (9)	999 ± 16 (3)	1019 ± 50 (9)	3·1 ± 0·9 (3)	4·6 ± 0·7 (9)
17–20	568 ± 39 (6)	463 (2)	1055 ± 59 (6)	1216 (2)	5·7 ± 0·6 (6)	4·4 (2)
21–24	787 ± 62 (11)	518 (1)	982 ± 42 (11)	971 (1)	4·6 ± 0·3 (11)	2·5 (1)
25–28	1051 ± 38 (8)	910 ± 144 (7)	1013 ± 66 (8)	1090 ± 53 (7)	3·8 ± 0·4 (8)	4·9 ± 1·4 (7)
29–32	1224 ± 121 (3)	1023 ± 77 (7)	969 ± 49 (3)	1068 ± 31 (7)	3·2 ± 0·4 (3)	3·1 ± 0·2 (7)
33–36	1024 (2)	903 ± 59 (3)	926 (2)	1073 ± 26 (3)	4·3 (2)	3·6 ± 0·1 (3)
37–41	1465 ± 168 (4)	824 (2)	902 ± 35 (4)	1004 (2)	3·7 ± 0·2 (4)	3·8 (2)
Statistical test	No, F = 3·9, D.F. = 1,102		No, F = 14·3, D.F. = 1,102		Yes, F = 2·3, D.F. = 1,102	
Is Normal = MMC?	$P < 0.05$, Normal > MMC		$P < 0.05$, MMC > Normal		Normal = MMC	

Table 3. *The correlation between the size of the somite and the cell number of the somite at the time of segmentation from the presomitic mesoderm*

| Somites | Normal | | MMC | |
	Cell number	Size	Cell number	Size
1–4	163	85·3	65	56·7
5–8	181	76·9	95	57·2
9–12	218	82·0	189	68·9
13–16	435	97·4	272	81·7
17–20	568	100·0	463	86·7
21–24	787	119·0	518	100·0
25–28	1051	134·0	910	129·3
29–32	1224	175·0	1023	145·5
33–36	1024	158·0	903	141·5
37–41	1465	141·0	824	137·2

Correlation analyses:
Coefficient		0·91	0·97	
Significant?		Yes, $P < 0.01$	Yes, $P < 0.01$	
Large somite has more cells?		Yes, $P < 0.01$, F = 38·1	Yes, $P < 0.01$, F = 145·8	

2). There is a good correlation between the length of the somite and the number of cells in the somite (Table 3).

The volume of somitic cells remains relatively constant among the somites in the two groups of embryos (Table 2). MMC-somitic cells have a bigger cell volume, but this is likely to be an overestimation because of the looser packing of cells in MMC somites.

In both the normal embryo and the treated embryo, the mitotic activity of the somitic cells is higher in upper trunk somites than in lower trunk somites (Table 2). There is, however, no significant difference between the same somites in the two groups of embryos. Therefore, the initial difference in somite size and the discrepancy in cell number is not modulated for some time after somite segmentation.

The growth of somite 8 is followed by counting cell numbers in the somite during the period from segmentation to sclerotome dispersion (Table 4). The somite grows in size mainly by an increase in cell number, cell growth does not seem to play a major part because cell volume does not increase during this period of growth. Compensatory growth of somites therefore does not occur prior to sclerotome dispersion.

The association between somite formation and axis growth

The axial length of the mouse embryo increases in a sigmoid relationship with the embryonic age (Fig. 14). The normal embryo accelerates growth

Table 4. *Changes in cell numbers, cell volume and mitotic activity of cells in somite 8 during its initial growth period prior to sclerotome dispersion*

Somite number of the embryo	Cell number		Cell volume (μm^3)		Mitotic index (%)	
	Normal	MMC	Normal	MMC	Normal	MMC
8	155±18 (3)	82 (1)	851±40 (3)	1233 (1)	6·6±1·1 (3)	3·7 (1)
9–10	256±53 (4)	140±38 (3)	745±75 (4)	1025±32 (3)	7·1±1·1 (4)	8·9±3·0 (3)
11–12	283±58 (5)	138±20 (4)	849±33 (5)	988±74 (4)	6·4±0·9 (5)	6·3±1·6 (4)
13–14	447±68 (5)	206±60 (4)	1166±70 (5)	949±76 (4)	6·9±1·0 (5)	6·9±0·9 (4)
15	—	302±80 (3)	—	1027±109 (3)	—	3·0±0·1 (3)

Sclerotome dispersion occurred at 14-somite stage in normal embryos, but at 15-somite stage in MMC-treated embryos. Cell number was estimated during the initial growth period form segmentation to sclerotome dispersion in the somite.

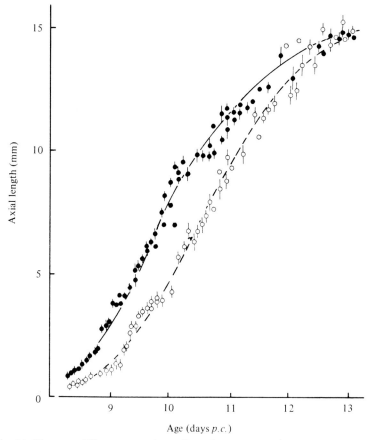

Fig. 14. The two different growth profiles of the normal (●) and the MMC-treated (○) embryo between 8·3 and 13·0 days *p.c.*

between 8·5 and 9·5 days *p.c.* and reaches a peak rate at 9·5–10.0 days (Fig. 15). The treated embryo shows an early phase of growth retardation but accelerates its growth rate after 9·5 days and it actually grows much faster than the normal embryo after 10·0 days (Fig. 15). Normal axial length is subsequently attained by 12·5 days (Fig. 14). It must be noted that the peak growth rate of the treated embryo has never exceeded that of the normal peak rate. The rate of axis growth measured in this study has taken account of both the rate of addition of tissue to the axis and the growth of the tissue already in the axis (Fig. 16). At early stages (1- to 10-somite stage), tissue addition accounts for more than 50% of the axial increase. In later stages, tissue addition consistently contributes to about 45% of the axial increase. It seems, therefore, that the rate of axial growth measured in this study is a reliable approximation of the rate of tissue incorporation at the caudal end of the embryonic axis.

Figure 17 depicts the possible relationship between somite formation and

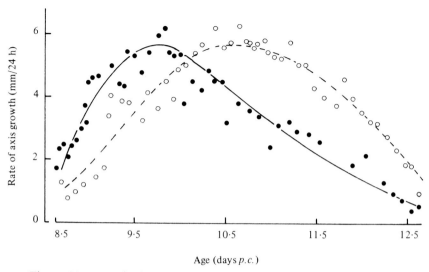

Fig. 15. The rate of axis elongation of normal (●) and MMC-treated (○) embryos at different developmental ages.

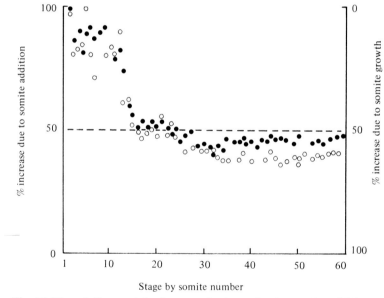

Fig. 16. The relative contribution to axis elongation by somite addition and by the growth of the tissue in the body axis at different developmental stages. Normal embryos (●) and MMC-treated embryos (○).

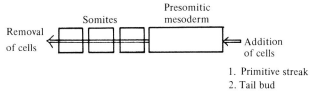

Fig. 17. The relationship between cellular addition to the caudal end of the presomitic mesoderm and the removal of cells by somite segmentation at the cranial end. Cells are added to the somitic pattern by the recruitment of epiblast cells at the primitive streak or by the proliferation of cells in the tail-bud mesenchyme. The amount of tissue removed from the presomitic mesoderm depends on somite size and the rate of somite segmentation.

axis formation during the development of mouse embryos. Presumptive somitic cells are added to the caudal end of the presomitic mesoderm. The presomitic mesoderm is a pool of cells in transit and the pool size at any instant is the balance between cellular addition at one end and the removal of somites at the other end. The amount of tissue removed depends on the size of somites and the rate of somite segmentation. The data collected on the rate of axis growth, the somite size, the rate of somite segmentation and the size of presomitic mesoderm are tested for evidence supporting a multiple linear relationship model (Table 5). The results of the regression analysis for both types of embryos suggest that the data indeed conform very well to such a model. There is also a significant correlation between the growth rate and the somite size at any stage of development (normal embryos: $r = 0.48$, D.F. $= 54$, $P < 0.01$; treated embryos: $r = 0.79$, D.F. $= 52$, $P < 0.01$). The results therefore suggest that the size of a somite is related to the amount of axis growth and the rate of somitogenesis. Despite a disturbance in the growth pattern and somitogenesis in MMC-treated embryos, a control mechanism similar to that of normal embryos is likely to be operating.

DISCUSSION

It has been reported in a study on the formation of the first 26 somites in mouse embryos that during development the size of the somite and the size of the presomitic mesoderm remain unchanged (Flint *et al.* 1978). The present study has extended the observation to a much later stage of somitogenesis and the conclusion drawn here is not in agreement with that of Flint *et al* (1978). Both the size of the somite at the time of segmentation and the size of the presomitic mesoderm change in embryos between the 1-somite and 56-somite stage. The size of the newly segmented somite increases craniocaudally and the somites are biggest at the lumbar and sacral level of the body axis. The caudal somites are progressively smaller in size. The variation in somite size is related to the number of cells being incorporated into each somite at the time of segmentation from the presomitic mesoderm. The presomitic mesoderm increases in size during the early phases of somitogenesis and gets smaller and smaller in later

Table 5. *The result of regression analyses of data obtained from normal and MMC-treated embryos, based upon a multiple linear relation*

The multiple linear relation:

$$Y = b_0 - b_1(RXS) + b_2(GR)$$

Y, size of the presomitic mesoderm;
R, rate of somite formation;
S, size of the newly segmented somite;
GR, rate of axis growth.

Regression results

	Constants			Coeff. of determination r^2	Significance of regression	Partial correlations		
	b_0	b_1	b_2			$GR:RXS$	$RXS:Y$	$GR:Y$
Normal	0·149	−0·0664	0·167	0·682 ($P < 0.01$)	$F(2,48) = 51.6$ ($P < 0.01$)	−0·46 ($P < 0.01$)	−0·49 ($P < 0.01$)	0·82 ($P < 0.1$)
MMC	0·468	0·1108	0·103	0·225 ($P < 0.01$)	$F(2,48) = 6.97$ ($P < 0.01$)	−0·80 ($P < 0.01$)	−0·29 ($P < 0.01$)	0·46 ($P < 0.01$)

phases. There is a two- to three-fold change in size of the somites and of the presomitic mesoderm during development. Since the method of measurement of sizes in this study is the same as that of Flint *et al.* the apparent conflict with those data is likely to be due to the narrow range of developmental stages and the method of data analysis used by Flint *et al.* Their study was based upon embryos at 1- to 26-somite stages whereas it is clear from the present study that the largest increase in size occurs when the embryo is at the 28- to 35-somite stage. When equivalent early somite stages are compared no significant discrepancies are found between the two studies. However, since most of the data points in Flint's study are clustered around somites 7–23 and the measured values show a large statistical variation, a linear regression of the data over a narrow range would tend to be misleading.

The variation in size of newly segmented somites in the mouse is different from that in the amphibian but similar to that in the chick. In *Xenopus* embryos, the anterior 14 somites are the biggest of all when they are formed from the paraxial mesoderm and further down the axis, the somites become smaller. This is interpreted as a manifestation of a smooth wave of cellular recruitment into the somitic pattern (Pearson & Elsdale, 1979). In the chick embryo, the newly segmented somite is always the length of two somitic cells plus a myocoel, but the actual size of the somite varies with the shape of the cells and the number of cells that are accommodated in each somite (Bellairs, 1979). The results from the measurement of newly segmented somites in chick embryos reveal that the somite size increases two-fold from 100 μm (somite 7) to 190–220 μm (for somites 20–32). The size then decreases for somites 33–50 down to about 110 μm for somite 50 (Herrmann, Schnieder, Neukom & Moore, 1951). The results in the mouse and the chick suggest that the pattern of cellular recruitment may be different from that in amphibian.

The paraxial presomitic mesoderm has been regarded as an undifferentiated tissue in which the prospective somitic cells undergo maturation and organization prior to segmentation. At the light microscopic level, no overt segmentation into presumptive somite is detected (Lipton & Jacobson, 1974). However, by using the scanning electron microscope (SEM), Bellairs (1979) has shown that tissues at the cranial-most end of the segmental plate are already structured into 'maturing somites'. Some recent findings on the organization of cells in the segmental plate of chick embryos have direct bearings on the interpretation of the result in the present study. Analysis by stereo SEM shows the paraxial mesoblast in the segmental plate to be organized into circular domains, named somitomeres, which are arranged in tandem along the cranio caudal axis (Meier, 1979). The cells in the most recently formed somitomere, which is adjacent to Hensen's node, are organized in concentric whorls, with those cells in the centre more condensed than those near the perimeter. The somitomere is divided into anterior and posterior halves along a line perpendicular to the body axis, marking the site of the future sub-division of sclerotome into the posterior and anterior components of successive vertebral bodies (Solursh, Fisher, Meier & Singley, 1979). Over a wide range of sizes of the segmental plate from embryos of different stages, a consistent number of about 10 – 11 somites is formed when the segmental plate is cultured in isolation (Packard & Jacobson, 1976; Packard, 1978). The study by stereo SEM has verified that 10 or 11 somitomeres are present in the segmental plate (Meier, 1979). Therefore, a pre-pattern of prospective somites is already specified in the segmental plate of the embryo. Extensive series of experiments have been done to show that the craniocaudal sequence of segmentation and the position of inter-somitic fissures are already determined in the segmental plate (Menkes & Sandor, 1977). The cells from the segmental plate also behave very similarly to those cells already in the somite (Bellairs, Sanders & Portsch, 1980). It has been shown in this study that in properly sectioned specimens, cells in the presomitic mesoderm of the mouse embryo are organized into segmental units. More recently, a stereo SEM study of the presomitic mesoderm of mouse embryos between 8·0 and 11·5 days *p.c.* shows that in all these stages, the presomitic mesoderm is organized into somitomeres (Tam & Meier, in preparation). Five or six somitomeres are arranged in tandem between the newly segmented somite and the primitive streak or the tail bud. The length of the presomitic mesoderm is the sum total length of all the somitomeres contained in this tissue.

Several important implications emerge from the aforementioned knowledge of the segmental plate/presomitic mesoderm. First, because a pre-pattern is already formed in the presomitic mesoderm, the specification of somite pattern must therefore have occurred at an earlier stage of development. It is also unlikely that any regulation of number or size of somite is taking place in the presomitic mesoderm as suggested by Flint *et al.* (1978). Rather, pattern formation occurs at the time of inception of presumptive somitic mesoderm as it passes through the primitive streak or leaves the zone of proliferating tail-bud mesenchyme. Secondly, since a relatively constant number of somitomeres are found in the segmental plate/presomitic mesoderm over a wide range of different sizes, it implies that the somitomere is of a different size at different phases of development. The size of the newly segmented somite may have a direct correlation with the size of its somitomere and also with the number of cells incorporated into the somitomere. However, such a correlation is complicated by the amount of growth that has taken place while the somitomere is in the presomitic mesoderm. Finally, as a result of the constancy in the number of segmental units in the segmental plate/presomitic mesoderm, the rate of segmentation of somite must follow closely the rate at which the presumptive somitic cells are organized into somitomeres at the caudal-most end of the embryo.

The regression analysis of the data obtained from the study of normal embryos and treated embryos suggest that embryonic growth controls the kinetics of somitogenesis, namely by regulating the number of cells allocated to each somite and the rate of somitogenesis. Even though the MMC-treated embryo suffers a severe reduction in size during early post-implantation period, brought about by extensive cell loss and growth retardation, it can still generate a normal number of somites by 12·5 days. This is brought about by adjusting the somite size in proportion to whole body size and by altering the rate of somitogenesis. Somite formation is mediated by the production of presumptive somitic cells from a progenitor population. This population expands by the recruitment of cells from the epiblast during gastrulation and by the proliferation of cells already residing in the population. The cells are then organized into primordia of somites (somitomeres?) which undergo morphogenesis in the presomitic mesoderm and later emerge as somites. The size of the primordium could be regulated by the quantity of precursor tissue available for somitogenesis, an idea closely in line with that of Flint *et al.* (1978). The rate of expansion of the progenitor population is probably related to the rate of axis growth. It may therefore not be coincidental that somites are largest when the embryo's in its fastest growth phase. The rate of somitogenesis, on the other hand, is a function of the rate at which cells undergo morphogenesis into primordia. It is also likely that the successful formation of a primordium depends on a minimal cell number, or tissue mass, and therefore the rate of somitogenesis is limited to some extent by the rate of expansion of the progenitor population. Extensive cell loss during the early stages of development of MMC-treated embryos may have diminished the size of the progenitor population. Subsequent expansion of the progenitor population is slowed down because of the retarded growth of the embryo. The fact that somites can still be formed but are much smaller and are made at a slower rate suggests that under normal circumstances the pool of precursor cells is considerably larger than the threshold size for somite formation. The formation of small somites could be the result of reduced growth of an initially small-sized primordium prior to segmentation but there is no evidence for a reduced growth potential of the primordium. The proliferative activity of cells in the presomitic mesoderm of treated embryos is at least the same as that of normal embryos, and may be higher at some stages. Small somites seem therefore to be the result of smaller primordia. A similar relation between the final size of the skeletal elements of the chick limb and the size of the primordia has been suggested from the study of the development of X irradiated limb bud. The successful formation of a primordium for a skeletal element also seems to require the provision of a threshold number of cells (Wolpert, Tickle & Samford, 1979).

Despite the disturbance in growth and somitogenesis, normal numbers of somites are formed in the MMC-treated embryos by 12·5 days. It is not clear how this regulation of somite number is brought about, but there are several possibilities. A regulation to the normal number can be brought about by a process whereby during early development, a fixed or quantal number of embryonic cells is set aside for somite formation regardless of the size of cell population in the whole embryo. The simplest case would be that the number of sub-groups of cells within the progenitor population of presumptive somitic mesoderm represents exactly the number of somites to be formed. By studying the heterogeneity of cell populations in the somites of mouse chimaeras, Gearhart & Mintz (1972) suggest that at least 1040 cells must be set aside in the late-primitive-streak-stage embryo in order to form all 65 pairs of somites in the mouse embryo. However, the reliability of such an estimate based on chimaeric embryos is still open to question (West, 1978). Furthermore the estimate is difficult to reconcile with the results from MMC-treated embryos where a drastic reduction of cell population occurs and normal numbers of somites are still formed. Current ideas on the regulation of somite numbers include the operation of an oscillator in the embryo, which exhibits a fixed number of oscillations during somitogenesis (Cooke, 1977; Cooke & Zeeman, 1976; Cooke & Elsdale, 1980). The same number of oscillations must be found in embryos of different sizes and with different rates of development in order to achieve a regulation of somite numbers. It further implies that only the size of somite is varied in accordance to the total amount of tissue available (Cooke, 1975, 1977). However, to account for somite number regulation in a growing system like that of the mouse embryo, Flint *et al.* (1978) suggest that regulation must be under a global control in relation to growth. If the cells designated for more posterior somites are established by increasing time or the number of rounds of cell generations in the progenitor population, then the specification of somite number could be part of the mechanism controlling embryonic axial growth. Such a mechanism for control of the number of morphological units is reminiscent of that in the chick limb bud. There appears to be a relationship between the number of divisions of mesenchymal cells in the apical growth zone of the limb bud and the number of skeletal segments laid down (Lewis, 1975). In a similar fashion, the somite number can be regulated by the number of cell generations in the progenitor population of presumptive somitic mesoderm. Even so, such a control mechanism may not be a stringent regulator of numbers because the growth profile of an embryo could be changed by intrinsic or extrinsic factors during development. Indeed, a small variation of about 3–5 % in the modal number of serially repeated units is not uncommon among individuals of the same species (Cooke, 1977; Maynard-Smith, 1960). The continuous presence of a progenitor population is crucial to the formation of somite pattern. Removal of the tail bud in amphibian embryos (Cooke, 1977) or the streak plus node area in chick embryos and in mouse embryos (Packard & Jacobson, 1976; Smith, 1964) invariably leads to a premature cessation of somitogenesis and consequently regulation of somite number becomes impossible.

I am very grateful to Dr Michael H. L. Snow for his invaluable guidance and support during the course of my work. I wish to thank Dr Antone Jacobson and Dr Stephen Meier, of the Department of Zoology, University of Texas at Austin, for reading the manuscript and for allowing me to quote the unpublished work on somitomeres. I was supported by a British Commonwealth Scholarship from the Commonwealth Scholarships Commission in the U.K.

REFERENCES

ABERCROMBIE, M. (1946). Estimation of nuclear population from microtome sections. *Anat. Rec.* **94**, 239–247.

BATTEN, B. E. & HAAR, J. L. (1979). Fine structural differentiation of germ layers in the mouse at the time of mesoderm formation. *Anat. Rec.* **194**, 125–142.

BELLAIRS, R. (1979). The mechanism of somite segmentation in the chick embryos. *J. Embryol. exp. Morph.* **51**, 227–243.

BELLAIRS, R., SANDERS, E. J. & PORTSCH, P. A. (1980). Behavioural properties of chick somitic mesoderm and lateral plate explanted *in vitro*. *J. Embryol. exp. Morph.* **56**, 41–58.

BENNETT, D. (1975). The T-locus of the mouse. *Cell.* **6**, 441–454.

COOKE, J. (1975). Control of somite number during development of a vertebrate, *Xenopus laevis*. *Nature, Lond.* **254**, 192–199.

COOKE, J. (1977). The control of somite number during amphibian development. Models and experiments. In *Vertebrate Limb and Somite Morphogenesis* (ed. D. A. Ede, J. R. Hinchliffe & M. Balls), pp. 433–448. Cambridge: Cambridge University Press.

COOKE, J. & ELSDALE, T. (1980). Somitogenesis in amphibian embryos. III. Effects of ambient temperature and of developmental stage upon pattern abnormalities that follow short temperature shocks. *J. Embryol. exp. Morph.* **58**, 107–118.

COOKE, J. & ZEEMAN, E. C. (1976). A clock and wavefront model for control of the number of repeated structures during animal morphogenesis. *J. theor. Biol.* **58**, 455–476.

FLINT, O. P. (1977). Cell interactions in the developing axial skeleton in normal and mutant mouse embryos. In *Vertebrate Limb and Somite Morphogenesis* (ed. D. A. Ede, J. R. Hinchliffe & M. Balls), pp. 463–484. Cambridge: Cambridge University Press.

FLINT, O. P., EDE, D. A., WILBY, O. K. & PROCTOR, J. (1978). Control of somite number in normal and *amputated* mouse embryos: an experimental and a theoretical analysis. *J. Embryol. exp. Morph.* **45**, 189–202.

GEARHART, J. D. & MINTZ, B. (1972). Clonal origins of somites and their muscle derivatives: evidence from allophenic mice. *Devl Biol.* **29**, 27–37.

GRUNEBERG, H. (1963). *The Pathology of Development*. Oxford: Blackwell.

HAMILTON, L. (1969). The formation of somites in *Xenopus*. *J. Embryol. exp. Morph.* **22**, 253–264.

HERRMANN, H., SCHNIEDER, M. J. B., NEUKOM, B. J. & MOORE, J. A. (1951). Quantitative data on the growth processes of the somites of the chick embryo: Linear measurements, volume, protein nitrogen, nucleic acids. *J. exp. Zool.* **118**, 243–268.

LEWIS, J. H. (1975). Fate map and the pattern of cell division: a calculation for the chick wing bud. *J. Embryol. exp. Morph.* **33**, 419–434.

LIPTON, B. H. & JACOBSON, A. G. (1974). Analysis of normal somite development. *Devl Biol.* **38**, 73–90.

MAYNARD-SMITH, J. (1960). Continuous, quantized and modal variation. *Proc. R. Soc.* B **152**, 397–409.

MEDDIS, R. (1975). *Statistical Handbook for Non-statisticians*. London: McGraw-Hill.

MEIER, S. (1979). Development of the chick mesoblast: formation of the embryonic axis and establishment of the metameric pattern. *Devl Biol.* **73**, 25–45.

MENKES, B. & SANDOR, S. (1977). Somitogenesis: regulation potencies, sequence determination and primordial interaction. In *Vertebrate Limb and Somite Morphogenesis* (ed. D. A. Ede, J. R. Hinchliffe & M. Balls), pp. 405–420. Cambridge: Cambridge University Press.

PACKARD, D. S. Jr (1978). Chick somite determination: the role of factors in young somites and the segmental plate. *J. exp. Zool.* **203**, 295–306.

5-2

PACKARD, D. S. Jr & JACOBSON, A. G. (1976). The influence of axial structures on chick somite formation. *Devl Biol.* **53**, 36–48.

PEARSON, M. & ELSDALE, T. (1979). Somitogenesis in amphibian embryo. I. Experimental evidence for an interaction between two termporal factors in the specification of somite pattern. *J. Embryol. exp. Morph.* **51**, 27–50.

POELMANN, R. E. (1980). Differential mitosis and degeneration patterns in relation to the alterations in the shape of the embryonic ectoderm of early post-implantation mouse embryos. *J. Embryol. exp. Morph.* **55**, 33–51.

POELMANN, R. E. (1981). The formation of embryonic mesoderm in the early postimplantation mouse embryo. *Anat. Embryol.* (in the Press).

REINIUS, S. (1965). Morphology of the mouse embryo from the time of implantation to mesoderm formation. *Z. Zellforsch. mikrosk. Anat.* **68**, 711–723.

REVEL, J. P. & SOLURSH, M. (1978). Ultrastructure of primary mesenchyme in chick and rat embryos. *SEM*, vol. 2, pp. 1041–1046.

RUGH, R. (1968), *The Mouse.* Minneapolis: Burgess.

SMITH, L. J. (1964). The effect of transection and extirpation on axis formation and elongation in the young mouse embryo. *J. Embryol. exp. Morph.* **12**, 787–80.

SNEDECOR, G. W. & COCHRAN, W. G. (1967). *Statistical Methods.* Iowa University Press.

SNOW, M. H. L. (1977). Gastrulation in the mouse: regionalization of the epiblast. *J. Embryol. exp. Morph.* **42**, 293–303.

SNOW, M. H. L. & TAM, P. P. L. (1979). Is compensatory growth a complicating factor in mouse teratology? *Nature, Lond.* **279**, 557–559.

SNOW, M. H. L. & TAM, P. P. L. (1980). Timing in embryological development. *Nature, Lond.* **286**, 107.

SNOW, M. H. L., TAM, P. P. L. & McLAREN, A. (1981). The control and regulation of size and morphogenesis in mammalian embryos. *Proc. Soc. Devl. Biol.* '*Levels of genetic control in development*' (in the Press).

SOLURSH, M., FISHER, M., MEIER, S. & SINGLEY, K. T. (1979). The role of extracellular matrix in the formation of the sclerotome. *J. Embryol. exp. Morph.* **54**, 75–98.

SOLURSH, M. & REVEL, J. P. (1978). A SEM study of cell shape and cell appendages in the primitive streak region of rat and chick embryos. *Differentiation* **11**, 185–190.

TAM, P. P. L. & SNOW, M. H. L. (1981). Proliferation and migration of primordial germ cells during compensatory growth in mouse embryos. *J. Embryol. exp. Morph.* (in the Press).

TARKOWSKI, A. K. (1959). Experimental studies on regulation in the development of isolated blastomeres of mouse eggs. *Acta Theriologica* **2**, 251–275.

THIELER, K. (1972). *The House Mouse.* Berlin: Springer-Verlag.

WADDINGTON, C. H. & DEUCHAR, E. M. (1953). Studies on the mechanism of meristic segmentation. I. The dimensions of somites. *J. Embryol. exp. Morph.* **1**, 349–356.

WEST, J. D. (1978). Analysis of clonal growth using chimaeras and mosaics. In *Development in Mammals*, vol. 3 (ed. M. H. Johnson), pp. 413–460. Amsterdam: North-Holland.

WHITTINGHAM, D. G. & WALES, R. G. (1969). Storage of 2-cell mouse embryos *in vitro*. *Austr. J. Biol. Sci.* **22**, 1065–1068.

WITSCHI, E. (1962). Development: Rat. I. Characteristics of stages In *Growth, including Reproduction and Morphological Development* (ed. P. L. Altman & D. S. Dittmer). *Fedn Am. Soc. exp. Biol. Washington.*

WOLPERT, L., TICKLE, C. & SAMFORD, M. (1979). The effect of cell killing by X-irradiation on pattern formation in the chick limb. *J. Embryol. exp. Morph.* **50**, 175–198.

J. Embryol. exp. Morph. Vol. 65 (Supplement), pp. 129–150, 1981
Printed in Great Britain © Company of Biologists Limited 1981

Evidence for regulation of growth, size and pattern in the developing chick limb bud

By DENNIS SUMMERBELL[1]

From The National Institute for Medical Research, London

SUMMARY

This paper examines the hypothesis that the developing chick limb bud has mechanisms for regulating the control of growth, size and pattern. The tests included: surgical removal of selected parts of the limb field, X-irradiation, temperature shock and the manipulation of known limb organizer regions (removal of the apical ectodermal ridge, or the addition of an extra zone of polarizing activity). The results strongly support the idea that there are regulatory mechanisms controlling both the pattern and the size of the limb and suggest that they involve regulation of the growth rate via control of cell division throughout the embryonic period. Possible mechanisms are discussed.

INTRODUCTION

In this communication I discuss the concepts of regulation and of regeneration, the differences between growth, size and pattern regulation, the evidence for each in the chick limb bud, and finally examine possible solutions to the problem of size regulation. The discussion is based on the concept of the limb bud as an embryonic field, explicitly as defined by Waddington (1956).

Regulative or mosaic?

Almost by convention the concept of regulation is now introduced by a reference to Driesch (1892, 1929). The very age of the reference confers a reassuring solidity. Yet the concept is not an easy one to define and it is arguable whether our understanding of the subject has advanced since the time of Driesch. He eventually succumbed to the difficulties of the problem and traced the downward path to philosophy via theoretical biology. He reasoned that the harmonious development achieved during regulation could not be explained by simple mechanistic rules and concluded that some higher spiritual component must guide the underlying life processes, this system he called *entelechy*, a return to primitive vitalism.

The debate on 'regulative or mosaic' was opened by Roux (1888) who used a hot needle to kill one of the first two blastomeres of the frog egg. The surviving

---

[1] *Author's address:* The National Institute for Medical Research, The Ridgeway, Mill Hill, London NW7 1AA, U.K.

cell produced only a half embryo during its early development. Thus it seemed that cells formed from different areas of the egg were non-uniform and therefore produced different parts of the embryo. This type of development Roux styled 'mosaic' or 'self-differentiation'.

Roux's results were soon challenged by other workers, the most compelling evidence coming from Driesch. He found that if one *separated* blastomeres of the sea urchin, each cell went on to complete embryonic development. This type of development he called regulative.

Despite Roux's experiment now being discounted, his concept has proved remarkably robust. Its meaning has widened beyond his original definition but it remains a sound empirical description of developmental phenomena. It may be simply defined as a developmental process in which part of a system can give rise only to those parts of the system that they would normally have produced had the system not been perturbed. It has subsequently been shown to be the dominant phenomenon in many systems and one could with some justice argue the generalization that all systems develop towards the mosaic state (Weiss, 1939). This generalization is seen at its most extreme in relation to cell lineage, where the division products of each cell may perform a rigidly determined programme of development (see Wilson, 1898); and the concept has been analysed recently in great depth for the nematode worm, *Caenorhabditis* (see for example Kimble, 1981).

Apart from any philosophic or mechanistic problems the concept of regulation, on the other hand, is operationally not straightforward. It is commonplace to find widely differing views on what it means. The careful author is therefore at pains to define his understanding of the concept before plunging into a discussion of ideas. A typical definition (paraphrased from Driesch, 1892) would be: a developmental process in which an abnormal complement of cells in an embryo (or part of an embryo) gives rise to a normal embryo (or part of an embryo).

The major problem with this is that while mosaic is still normally used in its absolute sense, regulative has come to describe all intermediate states between perfect mosaic and perfect regulator. The case of partial regulation may not agree with the original Driesch definition but one must suppose that they are likely to employ the same mechanisms.

There are two components that cooperate in regulative behaviour (clearly recognized by Driesch in his phrase 'harmonious equipotential system'). One concerns the intrinsic ability of the cell, the other concerns the ability of groups of cells to interact so as to modulate the intrinsic ability of the individual.

The intrinsic ability Driesch called the 'prospective potency'. In a regulative system the prospective potency of a cell is wide. Indeed, if the single cell is to produce a normal embryo, it must be all encompassing or totipotent. During development this intrinsic ability is progressively restricted so that the cell's possible mode of development is reduced. (The prospective potency is not to be confused with the 'prospective significance', today normally called the presumptive fate.) In regulative systems fate is a trivial concept. It indicates only what a cell will do during normal development. It says nothing about the cell's state or about its intrinsic abilities. A cell does not know its fate (see Summerbell, 1976 for a discussion relevant to the development of the limb bud).

The second component, the ability of cells to interact, supposes that the field recognizes the perturbation and has mechanisms that modify the behaviour of the cells so as to compensate appropriately and harmoniously. It is an article of faith with most experimental biologists that the processes active in such compensatory cell behaviour are the same as those governing normal development.

Regeneration

Regulation and regeneration have much in common. Morgan (1901) saw no reasons to distinguish between them and no compelling counter argument has been made. He suggested that there are two general ways in which regulation/regeneration may take place. Morphallaxis involves transformation of a part of the field into a different part without proliferation at a particular surface. Epimorphosis involves the appearance of a growth zone which by division gives rise to new cells that develop into the replacement part. The subject has been extensively discussed and formalized in the French Flag Model (Wolpert, 1969, 1971).

With respect to growth control there is one immediate difference between the two modes. Epimorphosis *necessarily* involves cell division, morphallaxis need not do so. Again making extreme definitions one could insist that *perfect* regulation by *truly* epimorphic means would involve restitution of the correct pattern and of the correct cell number and size. Such a system would show perfect size control with growth intimately and necessarily linked to pattern. It has been suggested that many systems approximate to this extreme (French, Bryant & Bryant, 1976; Iten & Murphy, 1980; Javois & Iten, 1981) a positional discontinuity in the field stimulates cell division locally and the new cells adopt positional values so as to regenerate those that are missing (intercalation). However, whenever the hypothesis has been tested it seems that the change in positional value takes place either without cell division (Buliere, 1972); without a localized growth zone (Cooke & Summerbell, 1980; Honig, 1981; Summerbell, 1981*a*); or without the amount of cell division correlating with the size of the positional discrepancy (Maden, 1981, this volume).

Conversely, morphallaxis involves restitution of a normal pattern without cell division; the morphology is perfect but the overall size is wrong. Examples of this may be easier to find. There is the original Driesch (1892) experiment where separated 2-cell-stage blastomeres produce normal active blastulae 'of half size'. Similarly Cooke (1981) has demonstrated that experimentally produced small tadpoles show no extra cell division and regulate their proportions to maintain size-invariant pattern. Examples such as these *may* be representative of extreme morphallaxis. However, other systems long quoted as examples of morphallaxis fail to demonstrate size invariance, e.g. *Hydra* (Bode & Bode, 1980; Wolpert, Hornbruch & Clark, 1974).

Only recently has interest been shown in intermediate cases (Maden, 1981, this volume; Summerbell, 1981 *a*) but it should be remembered that both Morgan and Wolpert were careful to state that they envisaged these two modes as extremes of a continuum.

Growth, size and pattern

When one examines a normal embryo it is clear that size is very accurately controlled. Perhaps the best example is the phenomenon of bilateral symmetry. When one compares the left and right wings of chick embryos at the end of the morphogenetic or embryonic period (day 10), then the lengths of left and right skeletal elements from the same embryo will equate very precisely (within $\pm 2\%$ in 67% of normal embryos, within $\pm 3\%$ in $95{\cdot}7\%$ of normal embryos, Summerbell & Wolpert, 1973).

This argues that during normal development either initial specification of the pattern is very accurate (Summerbell & Wolpert, 1973) and subsequent growth programmed and determinative (Lewis & Wolpert, 1976; Summerbell, 1976) or that mechanisms exist at later times to regulate or control the size of the field. The sea urchin must employ the former mechanism (at least up until blastula stage) since the cells continue to divide at the rate appropriate to the normal embryo even though it is only half size, i.e. there is regulation of pattern but not of growth or size.

I have already mentioned one extreme method of regulating size, the polar coordinate model (French *et al.* 1976), but it seems that it is not necessary to make the production of the pattern totally dependent on having a full complement of cells. It should be possible in principle to dissociate to some extent the regulation of growth, size and pattern. Such a development should not be surprising for there is ample evidence at fetal and adult stages for various mechanisms of growth control. Research in this problem has dealt almost exclusively with density-dependent control of cell division or with a search for various growth factors, humoral agents exerting either a positive or negative feedback on cell division (Summerbell & Wolpert, 1972; Stoker, 1978; Smith, 1981, this volume), while the phenomenon of catch-up growth is well documented at later stages (Williams, 1981, this volume).

However research on the embryonic period *in vivo* is scanty. Lawrence (1972) and Wolpert (1969) have both made suggestions for mechanisms controlling absolute size. More recently I have suggested a mechanism for growth control in chick limb development. Snow & Tam (1979) and Tam (1981, this volume) have direct evidence for compensatory growth control in mouse embryos.

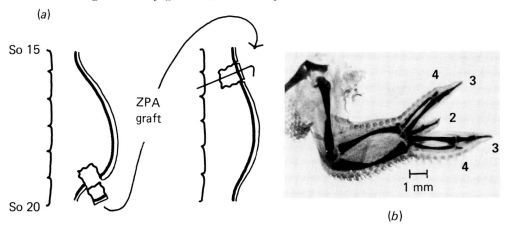

Fig. 1. The result of grafting the posterior limb organizer region (ZPA) to an anterior site on a host limb. (*a*) The operation. (*b*) Alcian-green-stained whole mount of resulting limb at day 10. The hand contains the digits **4 3 2 3 4**. In this example the 'normal' posterior digit is ~98 % the contralateral control length, and the anterior supernumerary digit is ~85 % the contralateral control length.

EVIDENCE FOR GROWTH CONTROL

This paper is based on data that has already been presented in a number of my publications, but the analysis shown in the figures is new. In some experiments I have added additional cases to those described in the original paper. Methods are not described in detail as they are already published. I include more detailed information where appropriate in the text and in the figure legends.

The zone of polarizing activity

The best direct evidence for growth control in the chick limb bud comes from experiments in which a polarizing region (ZPA) is grafted to the anterior margin of a host limb (Saunders & Gasseling, 1968). Such a graft alters the pattern so that it forms a mirror image reduplication of the host limb field causing the formation of a supernumerary limb of contralateral polarity (Fig. 1). The most immediate effect is an alteration in the growth rate, and this has now been documented at several levels.

The earliest indication (Cooke & Summerbell, 1980) is within 4–5 h after making the graft. Incubation of the embryo with [³H]thymidine for 1 h shows a significant enhancement of the labelling index in the responding tissue adjacent to the graft. This is followed somewhere between 9–17 h by a significant increase in the mitotic index (number of nuclei in mitosis per 100 nuclei). The graft appears to have shortened the cell cycle time in responding tissue so that the tissue is growing very much faster.

I have examined growth of a defined limb field under the influence of the ZPA signal by grafting two polarizing regions to a host limb, one opposite somites 16/17 and the other at a measured distance away posteriorly (Summerbell, 1981 *a*). By measuring the distance between the grafts at successive times it is possible to estimate the rate of elongation of the anteroposterior axis of the developing mirror image supernumerary pair of limbs. The width starts to increase about 6 h after operating: the intrinsic rate of growth or widening (the rate of change of length per unit length per unit time) rapidly rises to a maximum about 12–16 h after operating and thereafter declines back to a level similar to the base rate in unoperated limbs. The data on growth fits well with data on cell cycle with respect to both growth rates and timing. Smith & Wolpert (1981) using a less sensitive measure of change of width confirm these results and also show that there is no particular sensitive stage. The limb responds similarly to the graft through stages 18–21 (about 24 h).

One might question the extent to which this is an organized and controlled response to a potential change in pattern rather than a non-specific enhancement of cell division by a mitogen-like substance coincidentally present in the graft. There are three separate lines of evidence that argue in favour of specific and controlled enhancement of growth.

(1) The number of digits formed between the two ZPA grafts can be predicted very accurately by the initial distance between the grafts. Using the sample in Summerbell (1981 a), the prediction would be within ± 1 digit 95 % of the time, or $\pm \frac{1}{2}$ digit 67 % of the time. Much of this error probably lies in estimating the distances involved. The subsequent rate of change of length per unit length for the anteroposterior axis is the same for all initial distances between grafts (except for those $< 200 \mu$m) so subsequent measurements of the distance between grafts gives an equally accurate estimate of the number of digits that will develop. Initial size, growth and final pattern are closely correlated.

(2) The length of the principal axis of the proximodistal dimension remains the same as on the contralateral control limb (data from Smith & Wolpert, 1981). The change of growth rate does not cause an indiscriminate increase in the length of this dimension.

(3) The proximodistal lengths of the extra skeletal elements produced approach, but never exceed the lengths of the equivalent elements on the contra-ateral control side (Summerbell, 1974a). The earlier the stage at which the ZPA graft is made then the more closely the lengths of the reduplicated elements approach the control side (Fig. 2). This again suggests that the enhanced cell division is not uncontrolled but tends to be at a rate that will produce skeletal elements of the correct length for the host cells from which they were derived. It also suggests that the mechanism is progressive requiring time. The gradual loss of accuracy at later stages is most compatible with systems that involve negative feedback between the controlling system and the size of the field.

The apical ectodermal ridge

There is a thickened ridge of ectoderm running along the distal rim of the developing limb bud called the apical ectodermal ridge (AER). The development of the correct pattern of tissue along the proximodistal axis seems to be

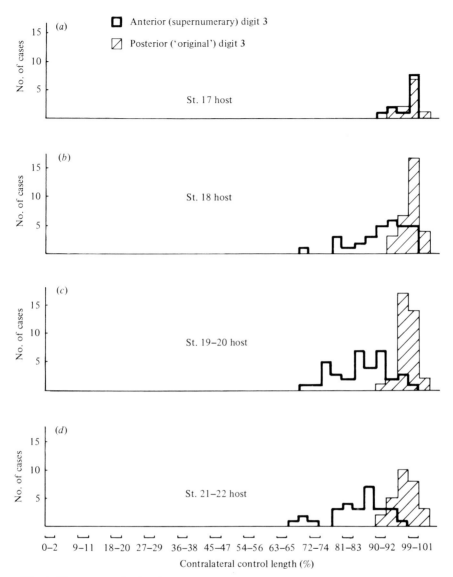

Fig. 2. Frequency distribution showing the number of cases that were measured as being the given percentage shorter than the contralateral control limb. The percentages have been grouped in blocks of 3 %. Each distribution shows both the anterior 'supernumerary' digit **3** and the posterior 'normal' digit **3**, each compared with the same contralateral control. (*a*) St. 17 host: anterior digit **3** ($\bar{x} = 99, s = 2\cdot4$); posterior digit **3** ($\bar{x} = 99, s = 2\cdot6$). $n = 11$, or which 9 were published in Summerbell (1974*a*). (*b*) St. 18 host: anterior digit **3** ($\bar{x} = 91, s = 7\cdot5$); posterior digit **3** ($\bar{x} = 99, s = 2\cdot5$). $n = 30$, of which 19 were published in Summerbell (1974*a*). (*c*) St. 19–20 host: anterior digit **3** ($\bar{x} = 86, s = 7\cdot5$); posterior digit **3** ($\bar{x} = 98, s = 2\cdot3$). $n = 36$, of which 10 were published in Summerbell (1974*a*). (*d*) St. 21–22 host: anterior digit **3** ($\bar{x} = 84, s = 8\cdot0$); posterior digit **3** ($\bar{x} = 98, s = 3\cdot3$). $n = 28$, none of which were published in Summerbell (1974*a*).

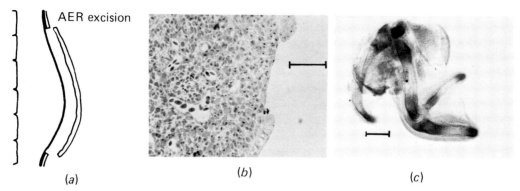

Fig. 3. Apical ectodermal ridge excision. (*a*) The operation, seen from dorsal surface. (*b*) Section of limb containing dorsoventral and anteriorposterior axes, showing missing section of ectoderm and apical ridge. This illustration is taken from Summerbell (1973). (*c*) Alcian-green-stained whole mount showing resulting limb at day 10. This illustration is taken from Summerbell (1973).

dependent on the continuing presence of the AER. If the AER is removed then the distal part of the limb will fail to develop (Saunders, 1948; Summerbell, 1974*b*; Fig. 3). The level of truncation is determined by the stage at which the operation is performed, older stages producing progressively smaller deficiencies. The cells of the AER itself do not form any durable structure and it has the general characteristics of an organizer region as defined by Spemann (1938). The only published rigorous explanation of its mode of action remains the 'progress zone model' (Summerbell, Lewis & Wolpert, 1973). Alternatively, Faber (1971) and Stocum (1975) have both suggested that it should be possible to consider the AER as the source of a morphogen, similar to that postulated for the ZPA and, by changing some of the parameters in the simulation of Summerbell (1979), it is possible to match most of the data, including the observation of truncation following AER removal. However, the loss of distal pattern does not account simply for the observed change in the rate of outgrowth of the bud. Without further assumptions neither model predicts any immediate effect on cell division or on the overall rate at which the limb grows. Furthermore, even if some adjustment is made so as to slow the overall rate of growth, both still suggest that the last programmed set of positional values will be specified over an abnormally large number of cells. Without the addition of some subsequent mechanism for growth control this would mean that the terminal element would be too big.

In practice removal of the AER causes a sharp increase in cell death and in cell cycle time (Janners & Searls, 1971; Summerbell, 1977*a*). Some cells are lost and the rate of proliferation is reduced. This appears to be a transitory effect of wounding and within 24 h the limb has resumed a programme of cell division close to normal levels, that, if extrapolated, would lead to a limb very much larger than would be appropriate for those parts of the pattern that will be present. However over the next few days there is modulation of the rate of outgrowth and the reduction is such that the total length of the operated limb is exactly equivalent to the length of the same parts on the contralateral control side (Summerbell, 1977*a*). This gradual compensatory mechanism is very similar to the growth control observed in mouse embryos after treatment with mito-mycin (Snow & Tam, 1979; Tam, 1981, this volume).

On examination, the terminal skeletal element may be found to be truncated at any point along its length. In those cases in which the element is just complete (it possesses an anatomically normal distal epiphysis but there is no indication of the next most distal element) then the length of the terminal element is within the normal limits of variation when compared to the contralateral control limb (Summerbell, 1974*b*).

This size equivalence, despite enormous perturbation and oscillations in the overall rates of growth and cell division in the operated limb can hardly be the result of pure chance. The data strongly argue the existence of mechanisms controlling at least the size and probably the growth rate.

Regulation of proximodistal deletions

The subject of regulation in the chick limb bud has been an area of high controversy for some time. Though the argument has normally been expressed in rigid rules (see for example, Wolpert, Lewis & Summerbell, 1975, with accompanying discussion by Sengel), the problems are more of detail. Embry-ology is not an exact science and survives on the generalization. A generalization, that I like, is: 'avian embryonic limb bud tissues are malleable up to a certain stage of development and can give rise to structures other than those they form in normal conditions. In other words, limb buds manifest regulative capacities' (Kieny, 1977). One can extend this generalization by saying that the younger the embryo, the more distal the position; and the smaller the amount of tissue removed, the better the regulation (Hornbruch, 1981; Summerbell, 1977*b*). When large amounts of tissue are removed from proximal levels at late stages the chick embryo limb bud behaves as a mosaic (Fig. 4*b*). There is a discrepancy in the pattern and its magnitude is linearly related to the proportion of the proximodistal axis that has been removed. Those parts of the pattern that are present are of normal size when compared to the contralateral control limb. The total length of the limb is determined by the proportion of the pattern that is present.

In contrast, when smaller amounts of tissue are removed from more distal levels or from younger embryos, then the anatomy of the skeleton at day 10 can often be normal. All of the skeletal elements are present (including the carpals) and characteristic knobs, bumps and joints are all present. There is regulation of the initial deficiency to produce a normal pattern (Fig. 4*a*). The mechanism organizing the regulation is not yet clear. The regenerated parts will include the progeny of both proximal (stump) cells and distal (tip) cells

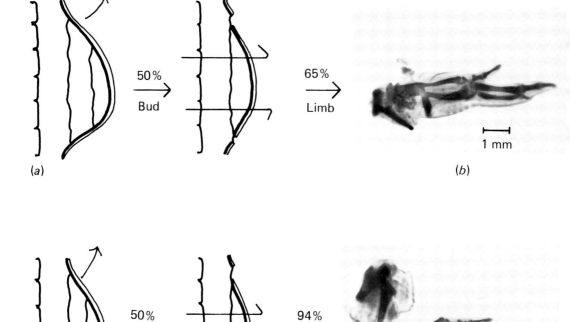

Fig. 4. Proximodistal deletion. (*a*) Removal of 50 % of the proximodistal axis of a stage-20 limb. (*b*) Alcian-green-stained whole mount of resulting limb at day 10. Result is typical of 'non-pattern regulating' cases with 65 % of the long axis of skeleton present. This limb was selected as an illustration from data first presented in Summerbell (1977*b*). (*c*) Removal of 50 % of the proximodistal axis of a stage-19 limb. (*d*) Alcian-green-stained whole mount of resulting limb at day 10. Result is typical of 'pattern regulating' cases at early stages with 94 % of the long axis of the skeleton present. This limb was selected as an illustration from data first presented in Summerbell (1977*b*).

(Kieny, 1977). Unlike the situation in amphibia or insects there is here no law of distality operating. It is possible that the regulation is purely epimorphic, but it cannot be purely morphallactic. Removal of a slice of tissue never gives a small limb with all the pattern proportionally reduced (so-called size independent). It seems most likely, extrapolating from the data on the zone of polarizing activity (ZPA, see above) that it involves something midway between the two. The regulation must involve at least some *replacement* of tissue and not only readjustment of proportions. As yet there is no direct data for changes

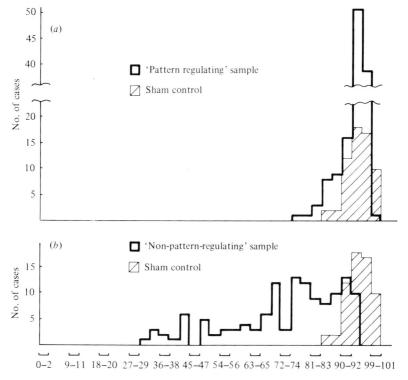

Fig. 5. Frequency distributions showing the number of cases that were measured as being given percentage shorter than the contralateral control limb. The length is the whole limb length. The percentages have been grouped in blocks of 3 %. Each distribution is compared with a set of sham control in which the tip was detached then put back without any deletion being made ($\bar{x} = 95$, $s = 3.7$). (*a*) 'Pattern regulating' sample ($\bar{x} = 93$, $s = 4.3$) $n = 119$, of which 52 were published in Summerbell (1977*b*). (*b*) 'Non-pattern' regulating sample. $n = 129$, of which 63 were published in Summerbell (1977*b*).

in the cell cycle time but it seems probable that the regulation will involve an increase in cell numbers. Again, despite the readjustment of positional values and changes in the intrinsic growth rate, no skeletal element significantly exceeded the length of the control side. Apart from restoration of the normal pattern it seems that the limb bud, despite the removal of a large proportion of the tissue present, was able to regulate the overall size of the limb fairly well. Figure 5 shows frequency distributions illustrating this regulation. The distribution shows the number of cases having a particular percentage difference from the contralateral control limb. This is compared with the results where the pattern was not regulated, and the results of a sham control.

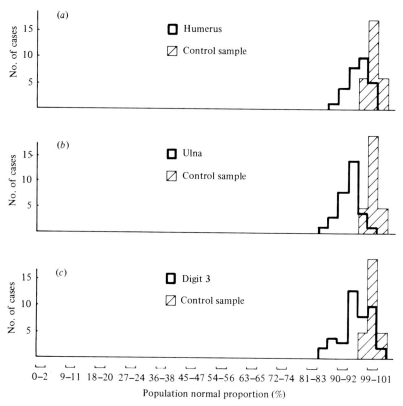

Fig. 6. Frequency distributions showing the number of cases that were measured as having a skeletal element, a given percentage shorter than the normal mean proportion of total limb length. The percentages have been grouped in blocks of 3 %. Each distribution is compared with the population distribution calculated from the mean and standard deviation of a large sample of normal embryos. The distributions contain all limbs in the teratogenic dose range that were 'pattern regulating', as defined in Summerbell (1981 b). (a) humerus ($\bar{x} = 96$, $s = 3\cdot5$) $n = 31$, all taken from Summerbell (1981 b). (b) ulna ($\bar{x} = 93$, $s = 3\cdot1$) $n = 31$, all taken from Summerbell (1981 b). (c) digit **3** ($\bar{x} = 95$, $s = 4\cdot6$). $n = 42$, all taken from Summerbell (1981 b).

X-irradiation

X-irradiation of the limb causes a complicated set of anomalies that have been explained in a number of different ways (Goff, 1962; Gumpel-Pinot, 1969; Summerbell, 1981 b; Wolff & Kieny, 1962; Wolpert, Tickle & Sampford, 1979). I will concentrate here on anomalies caused between stages 18–28 which in appearance are very similar to those caused by removing a slice of the proximodistal axis (see preceding section). Irradiation at a particular stage causes level-specific pattern defects of the skeleton. The tissue that seems to be most sensitive is the area just proximal to the distal tip, possibly the cells just emerging from the 'progress zone', Summerbell (1973, 1981 b). The problem with detailed analysis of the results of X-irradiation is that the experimental defects are normally bilateral affecting each side more or less equally. There is no contra-lateral control. Wolpert *et al.* (1979) avoided this difficulty by comparing their experimental limb population with a non-irradiated but otherwise comparably treated control population. While the method was adequate to demonstrate that the magnitude of the defect varied directly with the dose, it was too inaccurate to demonstrate the kind of small discrepancies that we begin to realize are associated with pattern regulation and growth control. One needs a very sensitive assay to detect this phenomenon, such as is provided by the analysis of propor-tions method (Summerbell, 1978). I have already shown (see above) that there is very little variation in the lengths of skeletal elements when comparing left with right wings on the same embryo. Similarly, within a wing there is accurate control of the proportions of the skeleton. In a given strain each skeletal element has a length which is within $\pm 3\%$ of that for the mean population in 67% of embryos, and within $\pm 5\%$ in 95% of embryos (Summerbell, 1978). While this is less precise than the contralateral comparison it is much more accurate than using a control sample, and certainly sufficient to demonstrate that the effects of growth control are also detectable in this experiment. When there is a visible pattern error the proportions are also severely affected (Goff, 1962; Summerbell, 1981*a*, *b*; Wolpert *et al.* 1979), but when the gross pattern appears normal there are only small, though highly significant deficiencies in size. A summary of this latter data is shown in Fig. 6. The figure is derived from examining at day 10 embryos X-irradiated at various stages. It includes data from 104 limbs, irradiated with between 10–12 Gy (1000–1200 rads.), that appeared morpho-logically normal. Each distribution contains the limbs from stages where X-irradiation selectively affected the appropriate skeletal element. The results will be presented in detail elsewhere. Meantime they demonstrate the now familiar phenomenon that disturbed limbs with normal patterns approach normal size but do not quite make it exactly.

There is one added complication when considering growth and size, and this is that the effects of the X-irradiation include killing cells and temporarily halting cell division throughout the embryo. This results in a slowing of the overall increase in cell number and hence of the growth. This is in part offset by a retardation in the rate of development, but nevertheless the whole embryo is smaller than normal when it eventually reaches stage 35. This suggests that the putative mechanism for growth control is not necessarily tied to time. It does not act to produce a normal-sized embryo by a particular time (at least not before day 10); it acts to harmoniously maintain the proportions of the embryo within the normal range of variability.

Temperature shock

A short and rather unproductive experiment consisted of cooling eggs rapidly to ~4 °C for periods of a few hours to 36 h during the period between stages 18 and 28. Development was delayed appropriately and the embryos at ten days

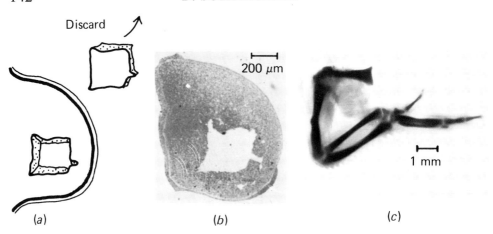

Fig. 7. Regulation of holes. A hole is cut through from dorsal to ventral surface and the tissue removed and discarded. (*a*) The operation on a stage-22 limb showing a typical position and size of hole. According to fate maps this would certainly remove the entire presumptive ulna region, and at least some of the radius, wrist and humerus. (*b*) A planar section showing an example of the profile of the hole, this illustration is taken from Summerbell (1973). (*c*) The result of a similar operation at stage 22 having normal pattern and size. This limb was selected as an illustration from data first presented in Summerbell (1973).

of incubation had limb skeletons shorter than the normal control population. However, the proportion of total wing length occupied by each skeletal element was not significantly different from the control population (*t* test on means, not significant 0·1 % level).

Regulation of holes through the dorsoventral axis

This final experiment was again designed to test the regulative abilities of the bud. In this case (Summerbell, 1973) a portion of mesenchyme was removed so as to produce a hole through the limb from dorsal to ventral surface, but without breaking the lateral or distal boundary (Fig. 7). The position of the hole was chosen so as to include predominantly areas fated to form skeleton (see fate maps of Stark & Searles, 1973; Summerbell, 1979). The size of the hole was variable, but never less than equivalent to the area of a single long bone or the whole of digit **3** as projected on to the dorsal surface as a fate map, and was often considerably larger. The total proportion of limb tissue removed averaged from approximately 30 % at stage 18 to about 10 % at stage 26. Similar experiments have been performed by Stark & Searles (1974) and by Barasa (1964) who frequently removed even larger proportions of the bud but who reported a high proportion of 'normal' wings resulting from the operation. The percentage of abnormal skeletons increased when the amount of tissue removed was increased (data from Stark & Searles), when the operations were performed on later stages (data from Stark & Searles, and Summerbell) or when it was at a more proximal level (data from Summerbell). Whenever there was a pattern abnormality then the total length of the limb and of the affected skeletal elements was shorter than on the contralateral control side. The size deficiency in these abnormal limbs could be of any magnitude. The number of operations resulting in a limb with a pattern defect increased dramatically at stage 26 (25 %) and 27 (95 %). In this consideration of size regulation I have therefore included only those cases from stages 18 to 25 (Summerbell, 1973).

In a sample of 202 limbs I found that 10 % of the cases had a pattern defect affecting one or at most two proximodistally adjacent elements. A further 70 % had a normal pattern, but one or more skeletal elements were shorter than the contralateral control (> 2 s.D.). In most of these cases the anomaly was locally restricted (one, or at most two, proximodistally *adjacent* elements affected), but in 13 cases stylopod, zeugopod and autopod were all shorter than the contralateral control. Of this last group, 2 (both from stage 18) were candidates for size independence (morphallaxis); all skeletal elements were significantly shorter than on the contralateral control side (> 2 s.D.) but each was equally reduced so that the relative proportions occupied lay within the normal limit of variation (< 1 s.D.). The remaining 20 % had normal pattern and size (within 2 s.D. of contralateral control). Frequency distributions for a given percentage deficiency are shown in Fig. 8 for humerus, ulna and digit **3**. Each distribution contained only those cases in which the original operation (as judged by the fate maps) would have affected the particular skeletal element. Skeletal elements that should *not* have been affected have been excluded but where the hole overlapped two adjacent elements both are recorded independently.

DISCUSSION

This paper discusses a number of experiments that can be divided into three groups. The removal of some of the cells from an embryonic field (proximodistal deletions, X-irradiation, holes); the slowing of development (X-irradiation, temperature shock), and the addition or removal of a field organiser (ZPA grafts, AER excision). To compare these experiments I need make the assumption that the last group involves the modification of the field so that a different pattern will be expressed. The common feature then becomes that in each case the (modified) field has an abnormal complement of cells. The first question is whether there is *any* regulation.

An embryonic field develops so as to give a pattern and size that we recognize as lying within a certain 'normal' range. If development is perturbed, then there is either a detectable variation (size and/or pattern lies outside the normal range), or the field has regulated the defect. In assessing this regulation I have used four operational criteria: anatomical pattern, relative proportion, bilateral-size symmetry and age-related normal size and proportion. It is a fair generalization that these criteria nest (Fig. 9), each lying wholly within the boundary of the previous category.

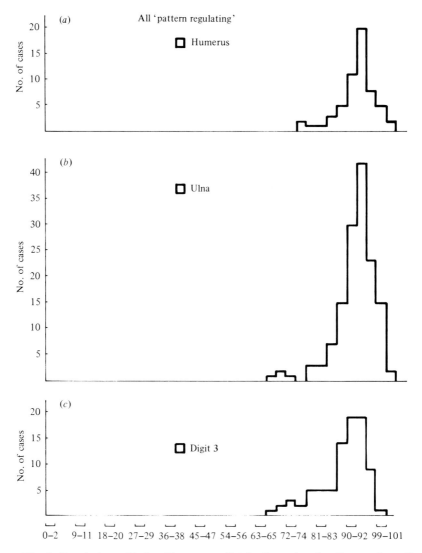

Fig. 8. Regulation of holes. Frequency distributions showing the number of cases that were measured as being given percentage shorter than the contralateral control limb. The percentages have been grouped in blocks of 3 %. Each distribution shows a particular skeletal element. The distributions contain all limbs that were pattern regulating. (a) humerus ($\bar{x} = 92\cdot6$, $s = 5\cdot7$) $n = 58$, all taken from Summerbell (1973). (b) ulna ($\bar{x} = 91\cdot8$, $s = 6\cdot5$) $n = 149$, all taken from Summerbell (1973). (c) digit 3 ($\bar{x} = 88\cdot6$, $s = 7\cdot2$) $n = 85$, all taken from Summerbell (1973).

Anatomical pattern has provided the normal criterion for regulation. It is based on a more or less scrupulous examination of the structures produced from the field. It has most often considered only the skeleton though more recently other structures are beginning to be taken more into consideration (muscle, tendons; see Shellswell, 1977; Shellswell & Wolpert, 1977; feather germs; see

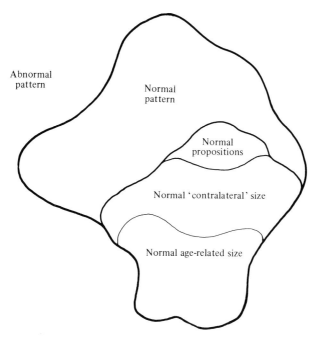

Fig. 9. Venn diagram illustrating the experiments described. The outer perimeter encloses the set of all limbs that were pattern regulating. Limbs with pattern defects lie outside this perimeter. The next lines enclose in sequence: all limbs of normal proportions, all limbs of normal bilateral size and all limbs of normal age related size. Note: (*a*) There are very few limbs with normal proportions that are not of normal bilateral size. (*b*) The sets nest, limbs of normal age related size lie inside the perimeter for normal bilateral size; limbs of normal bilateral size lie inside the perimeter for normal proportions; limbs of normal proportions lie inside the perimeter for normal pattern. Very few limbs broke this nesting rule.

McLachlan, 1980, 1981). The level of assessment is relatively arbitrary and subjective. However it seems that in the right circumstances the limb field can regulate major discrepancies, producing supernumerary limbs that involves reconstructing a large part of the skeletal pattern. This regulation is improved if occurring at early stages, or near the distal tip, or if the initial perturbation is relatively slight. If the anatomical pattern along a particular dimension is abnormal then the three size criteria will also be abnormal along the same dimension (see Fig. 9). *Pattern regulation is a prerequisite for size regulation.*

The criterion of normal relative proportions is of interest for two reasons. First, it is very useful as an adjunct to anatomical pattern markers when considering the effect of perturbations that are likely to be bilateral in their effect (X-irradiation, drugs). Second, it is the best estimate that we have for assessing the phenomenon of *size independence* (Wolpert, 1969). Size independence (the maintenance of normal pattern whatever the overall size of the field) is the main characteristic of morphallactic regeneration. If a system is regulating a defect by truly morphallactic means then the relative proportions of the systems must be reconstituted whatever the size of the field. In the experiments in which it was possible to compare the lengths of stylopod (humerus), zeugopod (ulna) and autopod (digit **3**), examples in which the proportions were regulated but in which the total length was wrong compared with the contralateral control side were extremely rare (< 1 %). The limb bud, at least along the proximodistal dimension *does not act as a morphallactic field*; the relative proportions of the pattern are not globally readjusted to compensate for a local deficiency.

The most accurate method of assaying deficiencies is to compare the perturbed limb with the unperturbed contralateral control. The rule of thumb (based on the standard deviations quote in the introduction) is that over 95 % of skeletal elements should lie with 3 % of the length of the contralateral control element (Summerbell & Wolpert, 1973). The initial length of a skeletal element at the time of differentiation is about 300 μm (Summerbell, 1976), equivalent to about 30 cell diameters. This means that whatever the details of the system, whether or not it involves size regulation, the effective length control mechanism in normal limbs is equivalent to ± 1 cell diameter (± 3 %) at the time of differentiation. In the several experiments considered here, it is obvious that most perturbed limbs are unable to modify growth or size so as to achieve *perfect* regulation. However, in each experiment the majority of limbs, despite enormous initial deficits, regulates the abnormal complement of cells so that at the end of the period of morphogenesis the limbs are within 10 % of the normal length ($\simeq 3$ cell diameter equivalents). Skeletal elements never exceeded the normal control length, even in the experiment (excision of AER) in which the field was manipulated to have an excess complement of cells for the pattern that actually appeared. *The case for some form of controlled regulation of size seems overwhelming.*

I have been unable to obtain evidence that time is an important factor in the control of growth or size. In those X-irradiated limbs that had normal proportions, the limbs were significantly smaller than the limbs of a control population that had been identically treated apart from the X-irradiation, but subjectively it seemed possible that the rate of development was retarded. This suggests that the *total size does not regulate to achieve a notional norm by a particular chronological time.* The cold shock experiment was fully compatible with this conclusion but one cannot exclude the possibility that an internal clock is stopped or slowed in step with the reduction in cell division and growth.

The essential argument of this paper is that cell division, growth and size are closely linked to pattern formation. It seems self-evident that the size of the limb and its component parts at the end of organogenesis will depend on the number of cells initially programmed to produce a specific structure and on the subsequent rate of growth. Direct studies on the cell cycle following AER excision (Summerbell, 1977a) and ZPA grafts (Cooke & Summerbell, 1980) have shown that interference with the principal organizing regions of limb field both modify the specification of pattern and concurrently the rate of cell division. The change in the cell cycle is, to a first approximation, sufficient to explain changes in the rate of growth of the field immediately afterwards (Summerbell, 1977a, 1981a). Extrapolating from these two experiments I therefore reach the tentative conclusion that *changes in the rate of growth are driven by changes in the cell cycle.*

The obvious next question is whether the link between cell cycle and pattern is intimate. Is one observing a clear case of epimorphic regeneration in which a discontinuity in the pattern stimulates local cell division and in which the new cells adopt a new programme of development that will lead to replacement of the missing part? There is not yet direct data from any of the experiments involving the removal of cells (proximodistal deletions, holes, X-irradiation), but there is abundant evidence that in the case of ZPA grafts this cannot be the mechanism. The stimulation of cell division is not restricted to the area adjacent to the graft but is widespread throughout the limb (Cooke & Summerbell, 1980; 1981 (this volume)). Nor does the supernumerary limb develop as a blastema-like outgrowth at the discontinuity between graft and host, for it clearly involves specification of pattern across $\sim 300\,\mu$m of the original field (Honig, 1981; Summerbell, 1981a; Summerbell & Honig, 1981). Thus it seems probable that *regulation in the chick wing is neither epimorphic nor morphallactic* (see also Summerbell, 1981a; Maden, 1981 this volume).

CONCLUSION

During normal development there are regulatory mechanisms that control both the pattern and size of the limb field. These do not involve an invariable and deterministic programme because experiments can still result in limbs of normal size and anatomical pattern. Nor is it an uncontrolled hypertrophy following intervention because the skeletal components of experimental animals are never too big. This harmonious interaction is not the result of morphallactic adjustment of the pattern so that it is harmoniously proportioned but smaller, nor is it an epimorphic intercalation of missing cells and pattern. It is possible that the mechanism involves an accurate programming of the correct number of cells for each part of the pattern prior to determination; but it seems more likely that there is compensatory regulation of the growth rate throughout the embryonic period.

REFERENCES

BARASA, A. (1964). On the regulative capacity of the chick embryo limb bud. *Experientia* **20**, 443–444.

BODE, P. M & BODE, H. R. (1980). Formation of pattern in regenerating tissue pieces of *Hydra attenuata*: 1. Head-body proportion regulation. *Devl Biol.* **78**, 484–496.

BULIERE, D. (1972). Etude de la régénération d'appendice chez un insecte: stades de la formation des régénérates at rapports avec le cycle de mue. *Annls Embryol. Morph.* **5**, 61–74.

COOKE, J. (1981). Scale of the body pattern adjusts to available cell number in amphibian embryos. *Nature* **290**, 775–778.

COOKE, J. & SUMMERBELL, D. (1980). Growth control early in embryonic development; the cell cycle during experimental pattern duplication in the chick wing. *Nature* **287**, 697–701.

COOKE, J. C. & SUMMERBELL, D. (1981). Control of growth related to pattern specification in the chick wingbud mesenchyme. *J. Embryol. exp. Morph.* **65** (*Suppl.*), 169–185.

DRIESCH, H. (1929). *The Science and Philosophy of the Organism.* London: Black.

DRIESCH, H. (1892). The potency of the first two cleavage cells in echinoderm development Experimental production of partial and double formation. In *Foundations of Experimental Embryology* (ed. **B.** H. Willier & J. M. Oppenheimer), pp. 38–50. New Jersey (1964): Prentice Hall.

FABER, J. (1971). Vertebrate limb ontogeny and limb regeneration: morphogenetic parallels. *Advances in Morphogenesis* **9**, 127–147.

FRENCH, V., BRYANT, P. J. & BRYANT, S. V. (1976). Pattern regulation in epimorphic fields. *Science* **193**, 969–981.

GOFF, R. A. (1962). The relation of developmental status of limb formation to X-irradiation sensitivity in chick embryos. I. Gross study. *J. exp. Zool.* **151**, 177–200.

GUMPEL-PINOT, M. (1969). Développment de L'ébauche dés membres après traitement a l'yperite azotée, irradiation aux rayons-X et culture *in vitro*. Etude comparative chez l'embryon de poulet. *Annls Embryol. Morph.* **3**, 215–234.

HONIG, L. S. (1981). Range of a positional signal in the developing chick limb. *Nature* **291**, 72–73.

HORNBRUCH, A. (1981). Abnormalities along the proximodistal axis of the chick wing bud: the effect of surgical intervention. In *Teratology of the Limbs* (ed. H.-J. Merker, H. Nau & D. Neubert), pp. 191–197. Berlin: De Gruyter.

ITEN, L. E. & MURPHY, D. J. (1980). Pattern regulation in the embryonic chick limb: supernumerary limb formation with anterior (non-ZPA) limb bud tissue. *Devl Biol.* **75**, 373–385.

JANNERS, M. Y. & SEARLS, R. L. (1971). Effect of removal of the apical ectodermal ridge on the rate of cell division in the sub-ridge mesenchyme of the embryonic chick wing. *Devl Biol.* **24**, 465–476.

JAVOIS, L. C. & ITEN, L. E. (1981). Position of origin of donor posterior chick wingbud tissue transplanted to an anterior host site determines the extra structures formed. *Devl Biol.* **82**, 329–342.

KIENY, M. (1977). Proximo-distal pattern formation in avian limb development. In *Vertebrate Limb and Somite Morphogenesis* (ed. D. A. Ede, J. R. Hinchliffe & M. Balls), pp. 87–104. Cambridge University Press.

KIMBLE, J. (1981). The role of cell lineages and cell fate in *Caenorhabditis elegans.* In *Theories of Biological Pattern Formation, Phil. Trans. R. Soc.* B. (In the press.)

LAWRENCE, P. A. (1972). The development of spatial patterns in the integument of insects. In *Development Systems – Insects* (ed. S. J. Counce & C. H. Waddington). New York: Academic Press.

LEWIS, J. H. & WOLPERT, L. (1976). The concept of non-equivalence in development. *J. theoret. Biol.* **62**, 479–490.

MADEN, M. (1981). Morphallaxis in an epimorphic system: size, growth control and pattern formation during amphibian limb regeneration. *J. Embryol. exp. Morph.* **65** (*Suppl.*), 151–167.

McLACHLAN, J. C. (1980). The effect of 6-aminonicotinamide on limb development. *J. Embryol. exp. Morph.* **55**, 307–318.

McLACHLAN, J. C. (1981). Mechanisms of limb development revealed by the teratogenic activity of nicotinamide analogues. In *Teratology of the Limbs* (ed. H.-J. Merker, H. Nau & D. Neubert), pp. 363–371. Berlin: De Gruter.

MORGAN, T. M. (1901). *Regeneration*, pp. 18–25. New York: MacMillan.

ROUX, W. (1888). Contributions to the developmental mechanics of the embryo. On the artificial production of half-embryos by the destruction of one of the first two blastomeres, and the later development (postgeneration) the missing half of the body. In *Foundations of Experimental Biology* (ed. B. H. Willier & J. M. Oppenheimer), pp. 2–37. New Jersey (1964): Prentice Hall.

SAUNDERS, J. W. (1948). The proximo-distal sequence of origin of the parts of the chick wing and the role of ectoderm. *J. exp. Zool.* **108**, 363–403.

SAUNDERS, J. W. JR. & GASSELING, M. P. (1968). Ectodermal-mesenchymal interactions in the origin of limb symmetry. In *Epithelia-mesenchymal Interactions* (ed. R. Fleischmajer), pp. 78–97. Baltimore, Maryland: Williams and Wilkins.

SHELLSWELL, G. B. (1977). The formation of discrete muscles from the chick wing dorsal and ventral muscles in the absence of nerves. *J. Embryol. exp. Morph.* **41**, 269–277.

SHELLSWELL, G. B. & WOLPERT, L. (1977). The pattern of muscle and tendon developments in the chick wing. In *Vertebrate Limb and Somite Morphogenesis* (ed. D. A. Ede, J. R. Hinchliffe & M. Balls), pp. 71–86. Cambridge University Press.

SMITH, J. C. (1981). Growth factors and pattern formation. *J. Embryol. exp. Morph.* **65** (*Suppl.*), 187–207.

SMITH, J. C. & WOLPERT, L. (1981). Pattern formation along the antero-posterior axis of the chick wing: the increase in width following a polarizing region graft and the effect of X-irradiation. *J. Embryol. exp. Morph.* **63** (in the press).

SNOW, M. H. L. & TAM, P. P. L. (1979). Is compensatory growth a complicating factor in mouse teratology? *Nature* **279**, 555–557.

SPEMANN, H. (1938). *Embryonic Development and Induction*. New Haven: Yale University Press, Reprinted New York: Hafner.

STARK, R. J. & SEARLS, R. L. (1973). A description of chick wing bud development and a model of limb morphogenesis. *Devl Biol.* **33**, 138–153.

STARK, R. J. & SEARLS, R. L. (1974). The establishment of the cartilage pattern in the embryonic chick wing and evidence for a role of the dorsal and ventral ectoderm in normal wing development. *Devl Biol.* **38**, 51–63.

STOCUM, D. L. (1975). Outgrowth and pattern formation during limb ontogeny and regeneration. *Differentiation* **3**, 167–182.

STOKER, M. G. P. (1978). The multiplication of cells. In *Paediatrics and Growth* (ed. D. Barltrop), pp. 5–14. London: The Fellowship of Postgraduate Medicine.

SUMMERBELL, D. (1973). *Growth and regulation in the development of the chick limb*. Ph.D. Thesis, University of London.

SUMMERBELL, D. (1974a). Interaction between the proximo-distal and antero-posterior co-ordinates of positional value during the specification of positional information in the early development of the chick limb bud. *J. Embryol. exp. Morph.* **32**, 227–237.

SUMMERBELL, D. (1974b). A quantitative analysis of the effect of excision of the AER from the chick limb-bud. *J. Embryol. exp. Morph.* **32**, 651–660.

SUMMERBELL, D. (1976). A descriptive study of the rate of elongation and differentiation of the skeleton of the developing chick wing. *J. Embryol. exp. Morph.* **35**, 241–260.

SUMMERBELL, D. (1977a). Reduction of the rate of outgrowth, cell density and cell division following removal of the apical ectodermal ridge of the chick limb bud. *J. Embryol. exp. Morph.* **40**, 1–21.

SUMMERBELL, D. (1977b). Regulation of deficiencies along the proximal distal axis of the chick wing bud: a quantitative analysis. *J. Embryol. exp. Morph.* **41**, 137–159.

SUMMERBELL, D. (1978). Normal and experimental variations in proportions of skeleton of chick embryo wing. *Nature* **274**, 472–473.

SUMMERBELL, D. (1979). The zone of polarizing activity: evidence for a role in normal chick limb morphogenesis. *J. Embryol. exp. Morph.* **50**, 217–233.

SUMMERBELL, D. (1981a). The control of growth and the development of pattern across the antero-posterior axis of the chick limb bud. *J. Embryol. exp. Morph* **63**, 161–180.

SUMMERBELL, D. (1981b). The effects of X-irradiation on limb development. In *Teratology of the Limbs* (ed. H.-J. Merker, H. Nau & D. Neubert), pp. 200–205. Berlin: De Gruyter.

SUMMERBELL, D. & HONIG, L. S. (1981). The control of pattern across the antero-posterior axis of the chick limb bud by a unique signalling region. *Symp. Amer. Soc. Zool., Seattle 1980, Amer. Zool.*

SUMMERBELL, D., LEWIS, J. H. & WOLPERT, L. (1973). Positional information in chick limb morphogenesis. *Nature* **244**, 492–495.

SUMMERBELL, D. & WOLPERT, L. (1972). Cell density and cell division in the early morphogenesis of the chick limb. *Nature New Biology* **238**, 24–25.

SUMMERBELL, D. & WOLPERT, L. (1973). Precision of development in chick limb morphogenesis. *Nature* **244**, 228–229.

TAM, P. P. L. (1981). The control of somitogenesis in mouse embryos. *J. Embryol. exp. Morph.*
65 (*Suppl.*), 103–128.

WADDINGTON, C. H. (1956). *Principles of Embryology*, pp. 23–28. London: George Allen
and Unwin.

WEISS, P. (1939). In *Principles of Development: a Text in Experimental Embryology*. (Facsimile
of the 1939 edition). pp. 321–341. New York (1969): Hafner.

WILLIAMS, J. P. G. (1981). Catch-up growth. *J. Embryol. exp. Morph.* **65** (*Suppl.*), 89–101.

WILSON, E. B. (1898). Cell-lineage and ancestral reminiscence. In *Foundations of Experimental
Embryology* (ed. B. H. Willier & J. M. Oppenheimer), pp. 52–72. New Jersey (1964):
Prentice Hall.

WOLFF, E. & KIENY, M. (1962). Mise en évidence par l'irradiation aux rayons-X d'un phéno-
mène de compétition entre les ébauches du tibia et du péroné chez l'embryon de Poulet.
Devl Biol. **4**, 197–213.

WOLPERT, L. (1969). Positional information and the spatial pattern of cellular differentiation.
J. theoret. Biol. **25**, 1–47.

WOLPERT, L. (1971). Positional information and pattern information. *Curr. Top. Devl Biol.*
6, 183–224.

WOLPERT, L., HORNBRUCH, A. & CLARKE, M. R. B. (1974). Positional information and
positional signalling in *Hydra*. *Amer. Zool.* **14**, 647–663.

WOLPERT, L., LEWIS, J. & SUMMERBELL, D. (1975). Morphogenesis of the vertebrate limb.
In *Cell Patterning*, Ciba Foundation Symposium **29**.

WOLPERT, L., TICKLE, D. & SAMPFORD, M. (1979). The effect of cell killing by X-irradiation
on pattern formation in the chick limb. *J. Embryol. exp. Morph.* **50**, 175–198.

J. Embryol. exp. Morph. Vol. 65 (Supplement), pp. 151–167, 1981
Printed in Great Britain © Company of Biologists Limited, 1981

Morphallaxis in an epimorphic system: size, growth control and pattern formation during amphibian limb regeneration

By M. MADEN[1]

From the National Institute for Medical Research, Mill Hill, London

SUMMARY

An essential component of most theories of pattern formation in epimorphic systems is that growth and pattern formation are strictly linked. Furthermore it has been assumed that epimorphic systems display size dependence, i.e. the pattern elements are always of a fixed size. These assumptions are challenged in the work described here on amphibian limb regeneration. The sizes of elements regenerating from small and large limbs and from normal and partially denervated limbs have been measured along with a detailed study of their subsequent growth stages. It is shown that the size of elements within one group of animals is remarkably constant even though their final, adult size is very different. But between groups of animals (large versus small or normal versus partially denervated) their sizes vary considerably. Therefore this classical epimorphic system is not size dependent, calling for a revision of current theoretical concepts. The similarities between this type of behaviour and that in morphallactic systems is discussed as well as similarities with growth control in limb development.

INTRODUCTION

During the many years that the science of experimental embryology has been in existence, the concepts derived from this line of enquiry have been categorized and defined in many different ways by different authors. None more so than the phenomenon of *regeneration* which was variously described by Roux, Barfuth, Driesch and Hertwig. The definition which seems to have been generally accepted, however, is that of Morgan (1901) who used the term regeneration to mean 'not only the replacement of a lost part but also the development of a new, whole organism or even part of an organism, from a piece of an adult, or of an embryo, or of an egg'. He further divided the phenomenon of regeneration into two modes – *epimorphosis*, 'in which proliferation of material preceeds the development of the new part' and *morphallaxis*, 'in which a part is transformed directly into a new organism without proliferation at the cut surfaces'. Morgan

[1] *Author's address:* Developmental Biology Division, National Institute for Medical Research, The Ridgeway, Mill Hill, London NW7 1AA, U.K.

also added that 'these two processes are not sharply separated and may even appear combined in the same form'.

In the years since Morgan published these definitions, but particularly since the advent of the concept of positional information (Wolpert, 1969) and the subsequent proliferation of theories of pattern formation, these terms have assumed additional meanings. Morphallactic systems of regeneration such as the early embryos of many vertebrate and invertebrate species (e.g. Cooke, 1979), *Hydra* (Wolpert, Hornbruch & Clarke, 1974) and *Dictyostelium* (Stenhouse & Williams, 1977) are now characterized as systems which show *size independence* of pattern. This means that the amount of tissue within which the pattern can be formed is variable. Since there is no cell proliferation during regeneration after removal of tissue one can therefore produce embryos, *Hydras* or *Dictyostelium* slugs of varying sizes depending on the amount of tissue remaining.

Conversely, epimorphic systems of regeneration such as insect and amphibian limbs, are now characterized as systems which show *size dependence* of pattern. This means that irrespective of how much tissue is removed, the regenerated elements will always be of the same size. This situation has arisen because most theories of pattern formation in epimorphic systems (e.g. Summerbell, Lewis & Wolpert, 1973; French, Bryant & Bryant, 1976) have assumed that the same cell interactions control both the growth and spatial pattern of differentiation within the tissue (i.e. cell division is stimulated by removal of tissue and ceases when that pattern is restored. When a greater amount of tissue is removed a correspondingly greater amount of cell division is stimulated).

In amphibian limb regeneration, the system on which the work described below was performed, just such a *size dependent* theory was proposed by Faber (1971, 1976). He assumed that within the limb field, positional information is in the form of a linear gradient of constant slope, a common concept in epimorphic systems which has its origin in Wolpert's (1969) notion of the relation between positional information and growth control. Thus regeneration blastemas from distal levels are smaller than those from proximal levels because the latter contains more pattern elements. Furthermore, this theory holds that by making distal blastemas bigger, they would produce more pattern elements than normal and, conversely, by making proximal blastemas smaller, they would produce less elements than normal.

No critical test of these specific assumptions or even of the general assumption, that epimorphic systems are *size dependent* has ever been performed. I report here two such tests on the regenerating amphibian limb. The first concerns blastemas on small and large animals to ask the question – do smaller animals produce smaller blastemas? The second involves reducing the size of blastemas growing on limbs by partial denervation to ask the question – do blastemas experimentally made smaller produce fewer elements than normal? The results reveal that in each case blastemas produce the same number of pattern elements, but the size of the elements vary according to the size of the blastema. This is typical *size independent* behaviour, occurring in an epimorphic system and it thus calls for a revision of our current theories of pattern formation.

MATERIALS AND METHODS

The first series of experiments were performed on *Pleurodeles waltl* and the second on *Ambystoma mexicanum*.

Blastemas on small and large animals: For this series a total of 56 young *Pleurodeles* were used with an average snout to tail length of 36 mm. This group of animals was selected on the basis of their similarity in size which varied from 32 to 40 mm. After anaesthetizing in 0·1 % MS222 their forelimbs were amputated through the mid-humerus and the amputated portion of the limbs fixed in neutral formalin. When the regenerates reached the following stages the limbs were reamputated and fixed in neutral formalin – late cone (on day 13), palette (days 13–15), notch (days 15–16), 3-digit (days 16–17) and 4-digit (day 19).

Both the regenerates and amputees were then stained in Alcian green and cleared in methyl salicylate, to reveal cartilage elements. Alcian green was chosen for staining (as used routinely by those working on developing chick limbs, for method see Summerbell, 1981) because it stains cartilage at an earlier stage of differentiation than Victoria blue or methylene blue (as used routinely by those working on regenerating amphibian limbs). Once stained, the lengths of the cartilage elements were determined at each stage by drawing them with a camera lucida, measuring with a ruler and converting to microns.

This same experiment was repeated when the larvae had doubled in size – average snout to tail length of 79 mm (84–72 mm). Regeneration was slower with these bigger animals, taking about 4 weeks to reach the stages of late cone to notch, rather than 2 weeks. The amputees and regenerates were fixed, stained and cartilage elements measured as described above.

Size reduction by partial denervation

Two experiments were performed in this series. In the first, 30 axolotls, 70–100 mm long had their left forelimbs partially denervated by severing spinal nerve 4 at the brachial plexus. Both forelimbs were then amputated through the mid-humerus, the right limbs serving as controls. Experimental left limbs were continually redenervated every 4 days to prevent regrowth of nerve 4 into the limb. Camera-lucida drawings of the external form of experimental and control limbs were obtained every two days to monitor blastemal growth. Regenerates were harvested at the stages of late cone, palette, notch, 3 digit and 4 digit. They were fixed and serially sectioned at 10 μm. The blastemal length, volume and stage reached were determined for each blastema and graphs drawn.

In the second experiment the above regime was repeated exactly on 35

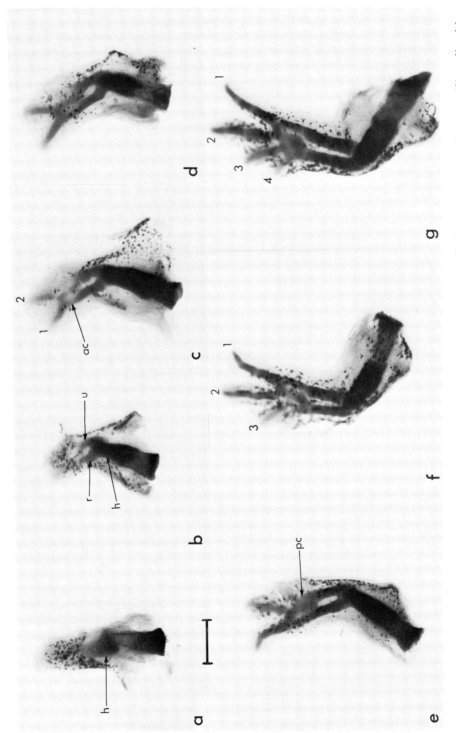

Fig. 1. (a) Stage 1 – the humerus (h) has just redifferentiated in a late cone blastema. (b) Stage 2 – now the humerus (h), radius (r) and ulna (u) are present in a palette stage blastema. The radius is larger and redifferentiates before the ulna. (c) Stage 3 – next the anterior carpals (ac) and digits 1 and 2 develop. (d) Stage 3 – slightly later than (c) in which the middle row of carpals has now redifferentiated and digits 1 and 2 are more distinct. (e) Stage 4 – the 3rd digit is just beginning to appear along with the most posterior carpals (pc). (f) Stage 4 – the 3rd digit is now much more distinct and digits 1 and 2 have begun to fragment into the metacarpals and

axolotls, 120–140 mm long. After fixing blastemas at the stages mentioned above, they were stained whole in Alcian green and the sizes of cartilage elements measured as in the first series on *Pleurodeles*.

Stages in the growth of the blastema

Growth stages from the point of view of the sequential laying down of new elements do not seem to have been studied in detail in the past. Most effort was concentrated on measuring cell densities, mitotic indices etc. on sectioned material or staging by external appearance and it has always been assumed that re-differentiation occurs in a straightforward proximodistal sequence. But this is not the case, as the following observations, which apply to both *Ambystoma* and *Pleurodeles*, reveal.

Stage 1 – humerus. The first stainable cartilage element to appear after amputation through the mid-humerus is the distal part of the humerus (Fig. 1 *a*). Externally, blastemas are at the late bud stage (Tank, Carlson & Connelly, 1976; Stocum, 1979).

Stage 2 *a* – humerus, radius. The next element(s) to redifferentiate is not as one might expect, radius and ulna together, but *only* the radius.

Stage 2 *b* – humerus, radius and ulna (Fig. 1 *b*). Then the ulna redifferentiates and remains smaller and stains less intensely than the radius throughout its early growth. Externally, blastemas have reached the palette or early redifferentiation stage (Stocum, 1979).

Stage 3 – humerus, radius and ulna, anterior wrist, digits 1 and 2 (Fig. 1 *c*, *d*). One might expect the wrist to appear next, but this could never be detected without digits 1 and 2 present. In fact the digits always stained more intensely than the wrist suggesting that the digits had redifferentiated *before* the wrist. This phenomenon can also be seen in fig. 19 of Stocum (1979). Similarly, digit 1 stains more intensely than digit 2 and thus probably differentiated before it although a 1-digit stage was never detected. Externally, the regenerates are now at the notch stage.

Stage 4 – humerus, radius and ulna, wrist, digits 1, 2 and 3. Next the posterior part of the wrist appears along with digit 3 (Fig. 1 *e*, *f*). By comparing Fig. 1 *c*, *d*, *e*, the anterior to posterior spread of redifferentiation of the wrist can be seen.

Stage 5 – humerus, radius and ulna, wrist, digits 1, 2, 3 and 4. Finally, the fourth digit appears (Fig. 1 *g*). By now the more proximal and anterior elements are well differentiated with individual anterior carpals appearing and digits 1 and 2 beginning to subdivide into metacarpals and phalanges. As reported by Smith (1978) it seems that metacarpals and phalanges appear by subdivision of a long rod of cartilage rather than by sequential addition at the distal end.

In summary, it is clear that there is an anterior to posterior spread of

Table 1. *Sizes of elements in small and large regenerates and amputated portions of the limbs*

The sizes of regenerated elements were measured at the stage when they first redifferentiated. The total length of the limb was taken to be the sum of the lengths of the humerus, radius and ulna, wrist and digit 1.

	Humerus	Radius and Ulna	Wrist	Digit 1	Digit 2	Digit 3
Small regenerate	420 μm$\pm$60	300 μm$\pm$50	330 μm$\pm$30	320 μm$\pm$40	330 μm$\pm$40	320 μm$\pm$50
% total length	31	22	24	23		
Small amputee	1110 μm$\pm$30	990 μm$\pm$40	480 μm$\pm$10	1030 μm$\pm$40	1330 μm$\pm$60	1280 μm$\pm$60
% total length	31	27	13	29		
Large regenerate	730 μm	570 μm$\pm$90	520 μm$\pm$80	480 μm$\pm$110	460 μm$\pm$100	440 μm$\pm$90
% total length	32	25	23	21		
Large amputee	2590 μm$\pm$80	2730 μm$\pm$40	1470 μm$\pm$30	1610 μm$\pm$60	2340 μm$\pm$70	2270 μm$\pm$80
% total length	31	33	18	19		

redifferentiation in each segment of the regenerating limb and a proximal to distal spread but the latter seems to 'skip a beat' at the wrist.

Size of regenerated elements from small and large animals

At each of the stages described above about ten blastemas were sampled from small and large *Pleurodeles*. The length of each element was measured from the time of its first appearance and throughout the growth to the 4-digit stage. In Table 1 the size of the elements at first appearance is recorded as well as the size of the elements amputated. In Fig. 2 the growth curves for each element are drawn.

Considering first small blastemas, what is immediately apparent is that, with the exception of the humerus, the size of the elements at first appearance is remarkably similar being 300–330 μm (Table 1). The humerus was amputated mid-way down its length so we are not considering the size of the whole element. The remaining part of the humerus in the stump could have had an effect to cause this discrepancy, a hypothesis which is currently being tested. The growth curves (Fig. 2 – dashed lines) reveal that the humerus and radius and ulna have identical growth rates, the wrist has a slower rate and the digits a faster rate. Thus although the size of the elements was originally identical, due to their own intrinsic growth rates they do not end up being of equal length in the fully grown form (see size of amputees, line 3, Table 1).

In large blastemas, again with the exception of the humerus, the size of the elements at first appearance is similar, being 440–570 μm (Table 1). Fig. 2 – solid lines reveal differences in the growth rates of individual elements. The humerus and radius and ulna have identical rates, being the fastest, the wrist is slightly slower and the digits slower still. Thus as in small blastemas, each segment has its own intrinsic growth rate and the segments are not the same length in the final form (see line 7, Table 1).

A comparison of these two sets of data immediately reveals that the size of elements at first appearance is smaller in small blastemas than in larger ones. This size difference is maintained throughout subsequent growth of the humerus, radius and ulna and wrist (Fig. 2). In the digits, however, the growth rates for small blastemas is faster than larger ones and this is reflected in the final form since the digits in small animals occupy a larger fraction of the total length of the arm than in large animals (see lines 4, 8, Table 1).

Despite the *absolute* differences in size of elements between small and large blastemas, the *proportion* of the total length that each element occupies is virtually identical (lines 2, 6, Table 1). This demonstrates that the mechanism of pattern formation which operates in regeneration seems to divide up the available tissue, whether it be a small amount or a larger amount, into the same proportions. This is typical *size-independent* behaviour, as described in the Introduction.

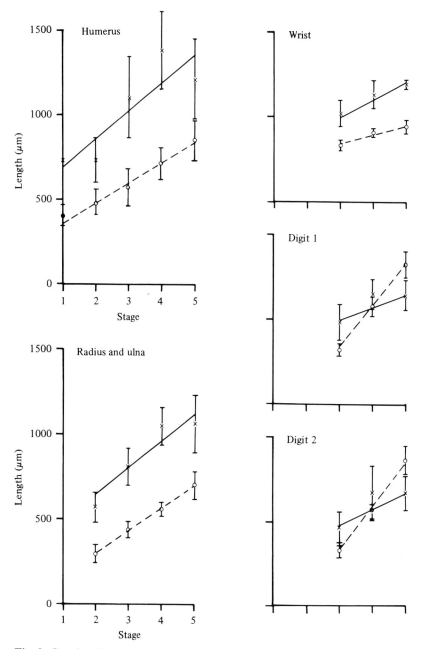

Fig. 2. Graphs of the lengths of elements (in microns) at stages 1–5 (see Fig. 1 and text for stages). *x* and *y* axes of each graph are drawn to the same scale. The element to which each graph applies is marked at the top of the graph. Solid lines are the calculated regression lines for elements from large blastemas and dashed lines those from small blastemas.

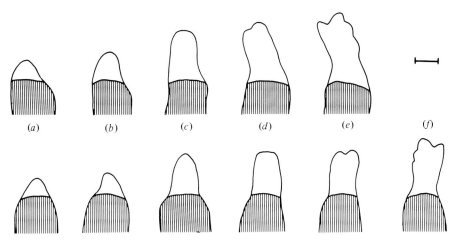

Fig. 3. Camera-lucida drawings of blastemas from control (upper row) and partially denervated (lower row) limbs. Cross hatching marks the stump, clear areas are the blastemas. (*a*) Day 12; (*b*) day 14; (*c*) day 16; (*d*) day 19; (*e*) day 21; (*f*) day 24 post amputation. Clearly partially denervated blastemas are delayed in their development and they are also smaller at equivalent stages – compare control *e* with partially denervated *f*. Bar = 1 mm.

Blastemal size in normal and partially denervated limbs

Blastemas on limbs which have had nerve 4 removed grow more slowly than normal controls (Fig. 3). Previous experiments had established that, for the purposes of this experiment, removal of nerve 4 provided the most useful test. Nerve 5 contributes the least number of nerve fibres to the limb, about 10–20 % (Singer, 1946; Karczmar, 1946) and removal of this nerve alone did not have a significant effect on blastemal growth. Nerve 3 provides 30 % of the limb inner-vation and removal of this had an effect, but it was small in magnitude. Nerve 4 provides the greatest contribution (50–60 %) and has the greatest and most consistent effect on blastemal growth.

Clearly, from external observations, growth is retarded in such partially denervated blastemas (Fig. 3), but when these blastemas get to the same stages have they caught up in size? In the first experiment normal and partially denervated blastemas were sampled and serially sectioned. For each sample, the blastemal length, volume and stage reached was recorded. Figure 4 reveals that partially denervated blastemas are both shorter in length (Fig. 4*a*) and smaller in volume (Fig. 4*b*) than controls at equivalent stages. The difference in slope between the normal and partially denervated plots is greater for blastemal volume than for blastemal length. This confirms the observation of Singer & Craven (1948) and Singer & Egloff (1949) that denervation affects blastemal volume to a greater degree than blastemal length.

The second experiment was a repeat of the first except that blastemas were stained whole for cartilage elements rather than serially sectioned. The size of elements at first appearance is recorded in Table 2. The controls revealed that just as in the previous series on *Pleurodeles*, the humerus is much larger than the other elements, which varied in size between 480 μm and 590 μm. These sizes are very similar to the data for large *Pleurodeles* (Table 1), although the axolotls used here were nearly twice as large. This suggests that there is a maximum size of element which can be produced in a developing blastema, the size does not linearly increase with size of animal. Similarly the percentage of the total length of the blastema that each element occupies (line 2, Table 2) is virtually identical to the data on *Pleurodeles*, both large and small (lines 2, 6, Table 1), attesting to the common mode of pattern formation in the two species.

Although the total length of partially denervated blastemas was smaller than normal (Fig. 4*a*) the length of individual elements as they redifferentiated was not uniformly smaller, but more variable. The humerus was 28 % smaller, the radius and ulna 19 % smaller, the wrist 14 % smaller, but the digits were the same size (Table 2, line 4). Reasons for this differential effect are considered in the Discussion. Growth curves revealed that these initial differences in size were maintained throughout subsequent growth, resulting in significant differences in size between normal and partially denervated limbs at the 4-digit stage (Fig. 5).

DISCUSSION

Normal growth of the blastema

It has, in the past, always been tacitly assumed that redifferentiation of cartilage elements within the blastema proceeds in a proximal to distal direction (e.g. Thornton, 1968; Faber, 1971). But there are two complicating factors in this oversimplified supposition.

The first is that not all the segments (humerus, radius and ulna, wrist, digits) form in a strict sequence. Certainly the humerus redifferentiates first followed by the radius and ulna, but then the digits seem to develop *before* the wrist. The phenomenon has also recently been noted during regeneration of adult *Notophthalmus* limbs (Neufeld & Settles, 1981; Settles & Neufeld, 1981) and thus seems to be common to regenerating systems. A similar anomaly, but at a different level is observed in chick limb development where the ulna and humerus appear simultaneously (Summerbell, 1976).

The second complicating factor is that imposed upon this sequence there is an anterior to posterior spread of redifferentiation. The radius develops before the ulna, the anterior wrist before the posterior wrist and digits 1 and 2 before 3 and 4. Stocum (1979) also reported this in *Ambystoma maculatum* regenerates and, again, it seems to be common to regenerating systems. We can also extrapolate this principle to Urodele limb development since the anterior digits develop before the posterior ones in both axolotls (Hinchliffe & Johnson, 1980; Maden, unpublished) and Pleurodeles (Lauthier, 1971).

Interestingly, in Anuran (Taylor & Kollros, 1946 – *Rana pipiens*; Tarin &

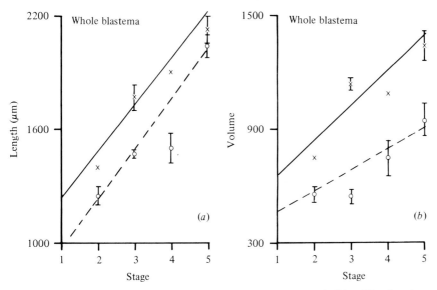

Fig. 4. Lengths (*a*) and volumes (*b*) of blastemas at stages 1–5 (see Fig. 1 and text for details of stages). Solid lines are the regression lines calculated for normal, control blastemas (crosses) and dashed lines those calculated for partially denervated blastemas (circles).

Sturdee, 1971 – *Xenopus laevis*) and chick (Summerbell, 1976) limb development, the transverse redifferentiation sequence is in the opposite direction – posterior develops before anterior. This correlation between the higher growth rate on the posterior side of the chick limb bud (Cooke & Summerbell, 1980) and the existence of the zone of polarizing activity (ZPA) in the same position provides support for the idea that the ZPA controls both growth and pattern in the developing limb. But since Urodele limb buds have a posterior organizer which controls the pattern (Slack, 1976, 1977), yet the redifferentiation sequence and thus presumably the growth rate progresses from anterior to posterior, the same principles cannot apply here. This adds weight to the proposition discussed below that growth and pattern formation are independently controlled and that the characteristics of the growth of a limb bud are not strictly related to the pattern which will appear within it.

Size regulation in normal blastemas

By examining blastemas from small and large *Pleurodeles*, it was clear that the size of elements when they first appeared was variable. Larger animals (twice the size of smaller ones) regenerated larger elements (1·3–2 × the size), but in the second series of experiments larger animals still (twice the size again) did *not* produce correspondingly larger elements. Consequently there seems to be a maximum size of element and perhaps a minimum too. However, within those constraints, element size can vary while maintaining constant propor-

Table 2. *Sizes of elements at redifferentiation in normal and partially denervated blastemas*

The total length of the limb was taken to be the sum of the lengths of the humerus, radius and ulna, wrist and digit 1.

	Humerus	Radius and Ulna	Wrist	Digit 1	Digit 2	Digit 3
Control % total length	680 μm ± 30 30	530 μm ± 20 23	510 μm ± 20 22	550 μm ± 20 24	590 μm ± 30	480 μm ± 20
Partially denervated % total length	490 μm ± 40 26	430 μm ± 20 23	440 μm ± 20 23	530 μm ± 40 28	590 μm ± 30	480 μm ± 30
% size difference between control and p.d.	28	19	14	4	0	0

tionality. Therefore it is not true that larger blastemas form more elements than smaller ones and vice versa (see Introduction), instead, there is a range of sizes. Pattern formation within the blastema does not thus depend on absolute size relationships, but on an intrinsic level-specific property of the cells which de-differentiate from the stump, combined with a proportion regulating mechanism.

Within each group of blastemas (either large or small) the size of different elements at first appearance was remarkably constant. In the smaller ones the radius and ulna, wrist and digits were each between 300 and 330 μm in length. Although their initial sizes were identical, further development of each element then continued at their own intrinsic rate to produce a fully grown form in which the elements are grossly *unequal* in size with respect to each other. Nevertheless the size of any one element in the fully grown population (measured from the amputees) was accurate to $\pm 5 \%$.

This property of constancy of size proportions in regenerating limbs is identical to that found in developing chick limb buds. In the latter the size of elements at first appearance are all the same, they then progress at their own intrinsically different growth rates (Summerbell, 1976) to produce a limb whose proportions are accurate to $\pm 5 \%$ (Summerbell & Wolpert, 1973). This co-incidence adds weight to the proposition that development and regeneration are guided by the same mechanisms. These observations are currently being extended to developing amphibian limbs to attempt further generalizations.

Size regulation in partially denervated blastemas

It has been known for many years that partially denervated limbs regenerate slower than normal (Schotté, 1923), the speed of regeneration being dependent on the number of nerve fibres present, and that the final size of the regenerate is smaller than normal (Singer & Egloff, 1949). Similarly, if the limb is totally denervated at later stages of blastemal growth, in the nerve-independent phase, the final product is smaller than normal (Schotté & Butler, 1944; Singer & Craven, 1948). But it has never been questioned whether the individual elements in such denervated regenerates are actually smaller at comparable stages or they simply grow slower to reach the same initial length. The data presented here shows that the former is indeed the case – partial denervation causes a decrease in the size of the elements when they first redifferentiate.

It is thought that the role of the nervous system in limb regeneration is to provide a neurotrophic factor which permits the cells of the blastema to divide (Singer, 1974). That is, the nerves simply provide a mitogen rather than having any instructive role in pattern formation. Clearly, this has been confirmed here – decreasing the growth rate (presumably by elongating the cell cycle time or permitting fewer cells to divide) and thereby generating a blastema of smaller size does not decrease the number of pattern elements formed. Instead each element is reduced in size.

It is interesting that the effect on element size is not uniform, but is greatest

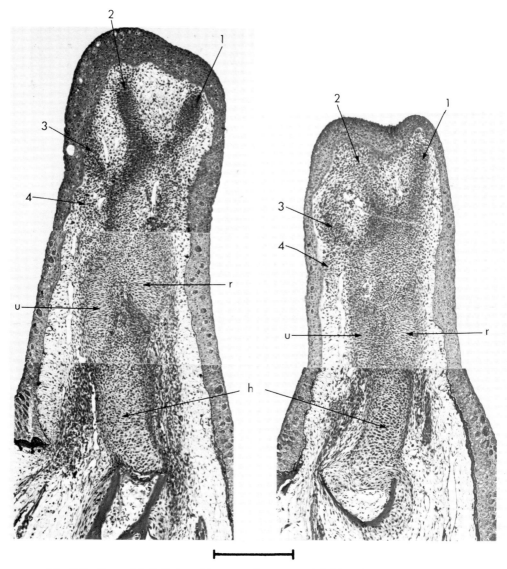

Fig. 5. Sections of 4-digit-stage blastemas from control (left) and partially denervated (right) limbs. Each is a composite photograph since the plane of section did not pass through all the elements at once. However, both sections are at the same magnification (bar = 500 µm) and it is clear that the partially denervated blastema is considerably smaller both in length and width at the same stage. h = humerus, r = radius, u = ulna, 1, 2, 3, 4 = digits.

in the humerus, decreases in the radius and ulna and wrist, and is negligeable in the digits. The reason for this can only be surmised, but seems most likely due to some aspect of neurotrophic growth control rather than an alteration of pattern formation mechanisms. For example, the regenerates could have been gradually reinnervated and thereby recovered from the decrease in growth rate.

Direct reinnervation was foreseen as a possibility and the limbs were redenervated every 4 days to prevent such an occurrence. An indirect reinnervation by collateral sprouting of the remaining spinal nerves 3 and 5 could have taken place over the 3–4 weeks duration of the experiment. Alternatively, the regenerates could have recovered from the denervation and loss of trophic factor either by starting to synthesize their own trophic factor or by gradually becoming independent of trophic factor. The latter seems to be the most likely case since total denervation at progressively later stages of blastemal development reveals just this phenomenon – a gradual appearance of resistance to the debilitating effects of denervation (Schotté & Butler, 1944). Indeed, total denervation after day 13 of regeneration in adult newts prevents any further growth in volume, but growth in length continues (Singer & Craven, 1948).

Relevance to theories of pattern formation

It is clear from the above results that growth and pattern formation in the regenerating limb are not strictly linked since elements of varying sizes can be produced by disturbing the growth rate. If the two were totally dependent upon each other then such a disturbance should not affect element size, instead it would simply take longer (or shorter) to reach the prescribed size. Thus, *size independence* has been demonstrated in an epimorphic system, a previously unrecognized behaviour which calls for a re-evaluation of contemporary theories of pattern formation (see Introduction).

What then controls the size of elements in the regeneration blastema? Although distal blastemas are smaller than proximal blastemas (Maden, 1976), the above experiments show it is not tissue mass which controls the size. One idea which would predict variable size elements is if redifferentiation occurred at a fixed absolute time after amputation. Slower growing blastemas would then have less tissue after this fixed time. However, this can be immediately dismissed since slower growing blastemas (after partial denervation or those on older animals) redifferentiate much later than normal (see Results).

Neither can size control be exerted by the ticks of a cell-cycle clock as proposed in the progress zone theory (Summerbell *et al.* 1973) as this again would predict size-dependent behaviour. Nevertheless, there are many similarities between chick limb-bud growth and that described here. For instance, the size of elements when they first redifferentiate is the same even though the size of elements in the final form is grossly different due to variable intrinsic growth rates. In both systems these growth rates are tightly controlled since the variation in the size of the final form is extremely small ($\pm 5\%$). These coincidences between a regulative system (regeneration) and a mosaic system (chick limb bud) imply a similarity of growth control and pattern formation mechanisms despite their apparent differences in response to experimental manipulation (Maden, 1981).

Not only are the similarities between development and regeneration now apparent but also those between epimorphosis and morphallaxis since both show size-independent behaviour. It seems probable, therefore, that during the evolutionary change from morphallaxis to epimorphosis, the same basic mechanisms of pattern formation were preserved, and growth was just added on top as a complicating factor. Growth did not therefore *replace* already existing mechanisms of pattern formation. Similarly, the progression during embryonic development from the morphallactic behaviour of the primary field (Cooke, 1979) to the epimorphic behaviour of secondary fields such as the limb is unlikely to involve a sudden and complete change of pattern-forming mechanisms. Rather it is more likely to produce the pattern by the same process, but now in a system which is growing. This is precisely the mode of operation of one theory already proposed for the regenerating limb (Maden, 1977) in which an early phase of morphallactic change of positional value occurs before growth begins. Hopefully this approach can be extended in the future to produce theories of pattern formation which highlight the similarities between and are relevant to development, regeneration, morphallaxis and epimorphosis.

I should like to thank Katriye Mustafa for excellent technical assistance in all aspects of this work.

REFERENCES

COOKE, J. (1979). Cell number in relation to primary pattern formation in the embryo of *Xenopus laevis. J. Embryol. exp. Morph.* **51**, 165–182.

COOKE, J. & SUMMERBELL, D. (1980). Cell cycle and experimental pattern duplication in the chick wing during embryonic development. *Nature* **287**, 697–701.

FABER, J. (1971). Vertebrate limb ontogeny and limb regeneration: morphogenetic parallels. *Advances in Morphogenesis* **9**, 127–147.

FABER, J. (1976). Positional information in the amphibian limb. *Acta biotheor.* **25**, 44–65.

FRENCH, V., BRYANT, P. J. & BRYANT, S. V. (1976). Pattern regulation in epimorphic fields. *Science* **193**, 969–981.

HINCHLIFFE, J. R. & JOHNSON, D. R. (1980). *The Development of the Vertebrate Limb.* Oxford Univ. Press.

KARCZMAR, A. G. (1946). The role of amputation and nerve resection in the regressing limbs of Urodele larvae. *J. exp. Zool.* **103**, 401–427.

LAUTHIER, M. (1971). Étude descriptive d'anomalies spontanées des membres postérieurs chez *Pleurodeles waltlii* Michah. *Ann. d'Embryol. Morphog.* **4**, 65–78.

MADEN, M. (1976). Blastemal kinetics and pattern formation during amphibian limb regeneration. *J. Embryol. exp. Morph.* **36**, 561–574.

MADEN, M. (1977). The regeneration of positional information in the amphibian limb. *J. theor. Biol.* **69**, 735–753.

MADEN, M. (1981). Experiments on Anuran limb buds and their significance for principles of vertebrate limb development. *J. Embryol. exp. Morph.* **63**, 243–265.

MORGAN, T. H. (1901). *Regeneration.* Macmillan.

NEUFELD, D. A. & SETTLES, H. E. (1981). The pattern of mesenchymal condensations prior to chondrogenesis in regenerating forelimbs of the adult newt, *Notophthalmus viridescens. Am. Zool.* (abstract) (in press).

SCHOTTÉ, O. E. (1923). La suppression partielle de l'innervation et la régénération des pattes chez les Tritons. *C. R. Seanc. Soc. Phys. Hist. Nat. Genève* **40**, 160–164.

SCHOTTÉ, O. E. & BUTLER, E. G. (1944). Phases in regeneration of the Urodele limb and their dependence upon the nervous system. *J. exp. Zool.* **97**, 95–121.

SETTLES, H. E. & NEUFELD, D. A. (1981). Morphogenesis of cartilaginous skeletal elements in regenerating forelimbs of adult newt, *Notophthalmus viridescens*. *Am. Zool.* (abstract) (in press).

SINGER, M. (1946). The influence of number of nerve fibres, including a quantitative study of limb innervation. *J. exp. Zool.* **101**, 299–337.

SINGER, M. (1974). Neurotrophic control of limb regeneration in the newt. *Ann. N.Y. Acad. Sci.* **228**, 308–322.

SINGER, M. & CRAVEN, L. (1948). The growth and morphogenesis of the regenerating forelimb of adult *Triturus* following denervation at various stages of development. *J. exp. Zool.* **108**, 279–308.

SINGER, M. & EGLOFF, F. R. L. (1949). The effect of limited nerve quantities on regeneration. *J. exp. Zool.* **111**, 295–314.

SLACK, J. M. W. (1976). Determination of polarity in the amphibian limb. *Nature* **261**, 44–46.

SLACK, J. M. W. (1977). Determination of anteroposterior polarity in the axolotl forelimb by an interaction between limb and flank rudiments. *J. Embryol. exp. Morph.* **39**, 151–168.

SMITH, A. R. (1978). Digit regeneration in the amphibian – *Triturus cristatus*. *J. Embryol. exp. Morph.* **44**, 105–112.

STENHOUSE, F. O. & WILLIAMS, K. L. (1977). Patterning in *Dictyostelium discoideum*: the properties of three differentiated cell types (spore, stalk and basal disk) in the fruiting body. *Devl Biol.* **59**, 140–152.

STOCUM, D. L. (1979). Stages of forelimb regeneration in *Ambystoma maculatum*. *J. exp. Zool.* **209**, 395–416.

SUMMERBELL, D. (1976). A descriptive study of the rate of elongation and differentiation of the skeleton of the developing chick wing. *J. Embryol. exp. Morph.* **35**, 241–260.

SUMMERBELL, D. (1981). Evidence for growth control in the chick limb bud. *J. Embryol. exp. Morph.* **65** (*Supplement*) 129–150.

SUMMERBELL, D. & WOLPERT, L. (1973). Precision of development in chick limb morphogenesis. *Nature* **244**, 228–230.

SUMMERBELL, D., LEWIS, J. H. & WOLPERT, L. (1973). Positional information in chick limb morphogenesis. *Nature* **244**, 492–496.

TANK, P. W., CARLSON, B. M. & CONNELLY, T. G. (1976). A staging system for forelimb regeneration in the axolotl, *Ambystoma mexicanum*. *J. Morph.* **150**, 117–128.

TARIN, D. & STURDEE, A. P. (1971). Early limb development of *Xenopus laevis*. *J. Embryol. exp. Morph.* **26**, 169–179.

TAYLOR, A. C. & KOLLROS, J. J. (1946). Stages in the normal development of *Rana pipiens* larvae. *Anat. Rec.* **94**, 7–23.

THORNTON, C. S. (1968). Amphibian limb regeneration. *Adv. in Morphog.* **7**, 205–249.

WOLPERT, L. (1969). Positional information and the spatial pattern of cellular differentiation. *J. theor. Biol.* **25**, 1–47.

WOLPERT, L., HORNBRUCH, A. & CLARKE, M. R. B. (1974). Positional information and positional signalling in *Hydra*. *Am. Zool.* **14**, 647–663.

J. Embryol. exp. Morph. Vol. 65 (Supplement), pp. 169–185, 1981
Printed in Great Britain © Company of Biologists Limited 1981

Control of growth related to pattern specification in chick wing-bud mesenchyme

By JONATHAN COOKE[1] AND DENNIS SUMMERBELL

From the National Institute for Medical Research,
Mill Hill, London

SUMMARY

The distribution of raised mitotic index, and the co-incidence of this with lowered cell packing density, has been studied across the anteroposterior dimension of the terminal 500 μm of chick wing buds following various numbers of hours signalling from an anteriorly grafted extra Zone of Polarizing Activity (ZPA). The results show propagation of the situation that causes these correlated phenomena, from graft–host interface essentially right across the limb mesenchyme, frequently within 8 h. This contrasts with the much slower and more local succession of changes in position memory, for differentiation of a duplicated limb pattern, that also occurs in mesenchyme relatively close to the graft after this operation. The results are discussed, in relation to current ideas about the control of pattern during limb development.

INTRODUCTION

In the study of pattern control in the anteroposterior dimension of the chick wing bud, it has been recognized for some time that an enhanced rate of widening during outgrowth of the host rudiment is an early consequence of a successful graft of tissue from the posterior Zone of Polarizing Activity (ZPA, MacCabe, Gasseling & Saunders, 1973) to an anterior site. Such a graft results ultimately in mirror-image duplication of the developed limb pattern around its normal pre-axial border. Several authors (Camosso & Roncali, 1968; Fallon & Crosby, 1975; Calandra & MacCabe, 1978; MacCabe & Parker, 1975; Rowe & Fallon, 1981) have speculated on possible dual roles for the ZPA, including functions in the control of growth. We recently confirmed (Cooke & Summerbell, 1980) that this is correlated with a stimulation of the mean cell cycle rate within host mesenchyme, initiated as an enhanced probability of entry into the 'S' phase of the cycle by cells, and leading to markedly raised mitotic index over the time (12–24 h post-operative) when the elevated widening is morphologically noticeable. When the initial stages of this growth enhancement were studied as a pattern of 'S' phase incidence revealed by brief thymidine labelling before fixation, a surprising feature was the speed and distance of

---

[1] *Author's address:* National Institute for Medical Research, The Ridgeway, Mill Hill, London NW7 1AA, U.K.

its spread posteriorly across the host tissue, from the site of positional disparity among newly joined cells at the host–graft interface. More than the anterior half, and possibly all of the still undetermined mesenchyme of the bud appears to become involved, within hours, in an episode of enhanced cell division. Yet related work indicates that only some 25 % of the bud width, that nearest the graft junction, becomes assigned to produce the final duplicate pattern elements in such cases (Summerbell & Honig, 1981). The reliable change in pattern-forming values appears to require continuous experience of influences from the newly implanted ZPA for 15 h or more (Smith, 1980).

The simplest interpretation of the data so far suggested is perhaps that the growth rate within limb-bud mesenchyme is only loosely tied to the landscape of pattern-specifying signals (e.g. a gradient concentration profile), being largely an expression of how many polarizing regions are in the system rather than of precise position in relation to each ZPA. It is important to ascertain the precise relation between control of the rate of tissue production and that of pattern determination in this system, since it is essentially the first inter-cellular growth control linked to pattern formation *in vivo* that has become accessible to study. More precise knowledge of it should also direct our choice of models (e.g. Summerbell, 1979, 1981*a, b*) for the sequence of events in pattern determination.

An additional interest of the system derives from the established fact that, whatever its nature, machinery for pattern specification that is *independent* of tissue dimension or scale over a considerable range does exist in vertebrate embryos (Cooke, 1979, 1981*a, b*). This is utilized in control of the primary, or axial pattern in these embryos. Why then, in a secondary pattern-forming field active only a few hours later, should we find that a growth control system has been 'built in' to the machinery? One possible implication is that the spatially repetitive nature of the early patterns of cell differentiation in such organ rudiments as the limb, unlike the unique 'zones' of the primary axial pattern, may involve pre-patterning types of mechanism in addition to monotonic positional gradients (Turing, 1952; Newman & Frisch, 1979; McWilliams & Papageorgiou, 1978; Wolpert, 1969). Such mechanisms may impose constraints on the range of tissue dimensions compatible with production of normal numbers of pattern elements (Murray, 1981), so that a mechanism to adjust scale in relation to the boundaries of pattern (i.e. the position of the ZPA) has evolved.

We report here our attempt to understand further this relationship between growth and pattern by investigating the relative incidence of mitosis across limb-bud sections, 16 h after grafting a ZPA to the anterior margin in a way calculated to cause a profound pattern duplication as in the previous work (Cooke & Summerbell, 1980). In the present experiments however an impermeable, vertical barrier, dividing the bud into anterior and posterior fractions and passing into the body wall, has been interposed between the region of the graft/host junction and the rest of the responding mesenchyme either 6, 8 or 12 h after grafting and the potential onset of signalling from the new ZPA. We have studied the resulting changes in cell cycle across the anteroposterior dimension of the tissue separated from the new signalling region by the barrier. This tissue has been divided geographically into three equal anterior-most, middle and posterior proximodistal columns, for assessment of mitotic incidence relative to homologous columns of the control (left) limb. The 16 h post-graft time point was chosen because in previous experiments without barrier interposition the 16 h mitotic response was dramatic (up to a doubling of the index), so that we might expect to detect smaller effects with maximum sensitivity. Estimates of the probable length of the (determinate) sequence between 'S' initiation and subsequent mitosis in normal limb-bud cells are around 4–8 h. Therefore, by observing mitotic incidence at 16 h we are probably assessing the experience of the cells, as to signals controlling their mean cycle rate, at times between 8 and 12 h after the implantation of the anterior ZPA graft in the experimental limb of each pair.

We also present data to add to that already in the literature (Smith, 1980; Summerbell, 1973) as to the spectrum of morphological results when limbs are allowed to determine and differentiate pattern after such ZPA grafting operations with subsequent barrier interposition. This is, in effect, an assessment of the timing and spread of acquisition of new 'positional memory' for pattern formation by tissue near the newly implanted ZPA, that parallels our assessment of the timing and spread of initiation of enhanced tissue production in the host. Comparison of the results of these two types shows that conditions stimulating growth rate can spread essentially right across the limb bud in very few hours (less than 10) in a way which contrasts with the slower and relatively localized reorganization of cell position values for pattern formation in the anterior region. The implications of this for overall models of pattern control are discussed.

In addition, we add to the evidence for a geographical correlation between lowered cell packing density and circumstances stimulating the cell cycle in the wing bud (Summerbell, 1977) by presenting the data on cell packing alluded to in Cooke & Summerbell (1980). In cases of experimental pattern duplication, this inverse correlation may be an extension or exaggeration of that seen in the normal limb bud at these stages (Summerbell & Wolpert, 1972).

Neither packing density nor cell-cycle rate were studied in ectodermal limb-bud components. They may be highly relevant to understanding the system, and are the subject of future work.

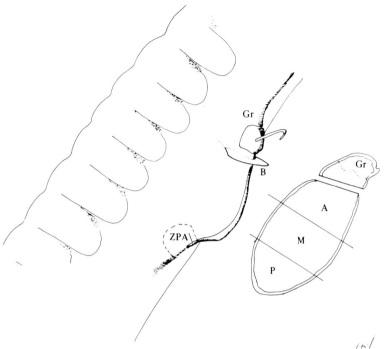

Fig. 1. The ZPA + barrier operation. A right-hand limb bud at stage 18/19 is shown, together with the outlines of the axis (somites), the site of graft implantation (Gr) with the retaining platinum pin, the tantalum foil barrier (B) and the location of the normal Zone of Polarizing Activity (ZPA), the source of grafts. A typical transverse section of the limb bud 16 h later is shown inset, with the graft and barrier sites (apparent in proximal sections of the series only) and the three dimensions of the post-barrier mesenchyme used to compose the Anterior, Middle and Posterior proximodistal columns for recording of mitotic incidence and cell packing density. Control data are from the equivalent cell populations in the left limb buds.

MATERIALS AND METHODS

Details are presented elsewhere of the embryonic operation (Summerbell, 1974) fixation, sectioning and staining, autoradiography and data collection from pre-differentiated wing buds (Cooke & Summerbell, 1980), and the assessment of differentiated skeletal patterns in advanced wing rudiments (Tickle, Summerbell & Wolpert, 1975). Figure 1 shows the version of the ZPA implantation operation used here, followed after 6, 8 or 12 h by interposition of a tantalum foil barrier. Our previous work has established that significant cell-cycle effects observed any distance from the graft/host boundary are seen only following juxtaposition of tissue having widely different position values which will lead to extra pattern formation; in our hands, the relocation of mesenchyme of posterior (ZPA) origin to an anterior site. The operations upon which the present paper is based were accordingly all ZPA grafts to positions opposite somites 15/16 at stages 18 to 19. Without later barrier placement, this leads in > 90 % of cases to enhanced width increase visible between 12 and 36 h after operation, following ultimately by differentiation of completely duplicated hand, wrist and frequently forearm structures. Embryos were treated in one of two ways after these operations.

One set of embryos was fixed and prepared for Feulgen histology at 16 h after the ZPA graft, but without prior labelling with ^{3}[H]thymidine, as only mitotic incidence was to be studied. Sets of sections at 50 μm intervals for some 500 μm behind the tip of the apical ectodermal ridge (AER) were mounted from right and left limbs to form two proximodistal series of cross sections through the mesenchymes in the parasagittal plane (i.e. parallel to the body axis). That part of each section profile lying posterior to the barrier position, or its equivalent in left control limbs, was treated as three sectors of equal anteroposterior width, that nearest the graft, the middle sector, and that furthest away (being the presumptively posterior zone of the host pattern, nearest to its own ZPA). Density in space of metaphases and anaphases was recorded in the three proximodistal columns of mesenchyme represented by the sectors of sections from each level. Mitotic incidences and ranges of cell density encountered were such that 2–300 mitotic figures and 5–7,000 cells were scanned for each column, the relative areas scanned being determined by planimetry from tracing on to paper at $\times 165$ magnification. Relative cell packing density for each column (in the plane of section concerned) was computed by random 'throwing' of a standard 'quadrat' (i.e. a frame on a television monitor at standard magnification) on to the image of mesenchyme in the appropriate third of each section, summing the nuclei recorded with an electronic colony counter and pen (150–300 per frame) and taking the mean for each column. The *relative* measure of cell density thus obtained for each column within a limb pair was used for computing their relative mitotic incidences on a per cell basis, and also as a biological parameter in its own right to correlate with those corrected mitotic incidences among the columns.

A second set of embryos was allowed to develop to 10 days of incubation after comparable operations, and the limbs prepared for analysis of their skeletal pattern in cleared whole mounts.

Certain limb pairs, after simple ZPA implantation to the right limb without subsequent barrier interposition, were subjected to geographical analysis of cell packing density correlated with the 'S' phase index recorded directly from autoradiographs. In these limb buds, which have been subjected to 1 h [^{3}H]-thymidine incorporation in the embryo immediately before fixation for the study of the 'S' phase stimulation reported elsewhere (Cooke & Summerbell, 1980), the cross sections at each level had been divided into four rather than three sectors of equal width, and these data boxes recorded separately rather than being pooled into mesenchymal columns (see Fig. 3). For the geographical layout of the boxes and thus *distribution* of enhanced labelling, see the previous paper.

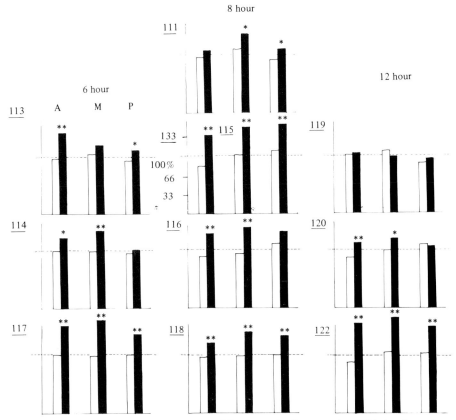

Fig. 2. Mitotic incidence on a per cell basis, 16 h after anterior ZPA implantation. Each histogram refers to a limb pair (underlined numbers), with the control (left) mesenchyme columns open and the experimental (right) columns filled in. Anterior, middle and posterior (i.e. furthest from the implant site) columns read from left to right as column pairs on the histogram. The horizontal dashed line corresponds to the mean mitotic incidence of the control columns in each limb pair, and the heights of columns register deviations from this. Asterisks show significance of *differences* between experimental and control columns at each position, at $P < 0.05*$ (usually 20 % relative incidence or more) or $P < 0.01**$ (usually 25 % relative incidence or more). Number of hours allowed for signalling by the implant before barrier interposition are marked at the top of sets of limb pairs. N.B. Only pair 119 shows a lack of stimulation, presumably because of graft failure. Only 122 and possibly 133, however, show degrees of stimulation comparable with that seen after continuous experience of an extra ZPA for 16–17 h (Cooke & Summerbell, 1980).

RESULTS

The histograms of Fig. 2 give the *relative* incidences of mitoses at grafting + 16 h, in the columns of the three mesenchymal sectors on the right and left (control) sides in limb pairs where communication between grafted ZPA and adjoining tissues had been allowed for the first 6, 8 or 12 h. The relative incidences are on a per cell basis, but are normalized to the mean incidence

Table 1. *Time required for stable change of position values in
wing-bud tissue by new ZPA signalling*

No. of hours allowed for signalling before barrier	No. of examples	Mean efficacy score
7–9	9	0
$9\frac{1}{2}$–12	9	0·5
13–15	11	0·5
16–18	9	1·4
24	3	2·0

The efficiency of given numbers of hours signalling (i.e. times from graft placement to barrier interposition) was registered by scoring 10-day-incubated, cleared wholemounts for limb skeletal pattern. Normal is: anterior → 2.3.4 → posterior. Efficacy score is → 2.3.4 = 0; → 2.2.3.4 → = 1; → 3.2.2.3.4 → = 2; → 4.3.2.2.3.4 → = 3.

It can be seen that reliable addition even of a digit 2 is not seen until at least 15 h, and that reliable complete new specification may require more than 24 h.

for the entire control limb (i.e. the level marked on the ordinate and called 100 % relative incidence). Within each pair of histogram columns, however, significance of right/left differences marked by asterisks has been assessed on the basis of absolute numbers of mitoses seen and cells scanned in the two populations. Broadly speaking a 20 % relative difference in incidence reaches significance at $P < 0·05$, while a 25 % difference reaches significance at $P < 0·01$.

Although the picture is variable (as indeed are the morphogenetic results of a series even of simple ZPA implantations) it can be seen that significant and substantial stimulation of cell cycle has frequently spread throughout the presumptive A–P tissue dimension of the buds within the initial 12 h of ZPA signalling. In at least one instance each of the 6 h, 8 h and 12 h barrier versions of the operation, stimulation has had time to reach the posterior third of the host bud in order to cause there a measurably enhanced 16 h mitotic incidence. Examination of the columns, which represent 400–500 mm depth of tissue from the bud top, suggests that the spread of the stimulating 'signal' or situation is by no means only around the periphery underneath the AER in order to reach the back, but also through the core of the undifferentiated mesenchyme. Scanning of the more proximal of the section series, from limbs showing the largest effects, confirms this view.

Data on the 'S' incidence at much earlier times after grafting do indicate that the stimulating conditions are initially concentrated in the pre-axial region near the graft (Cooke & Summerbell, 1980). The present data using the less direct mitotic indicator, some while after a restricted signalling period, show in a more positive way the dramatic rate of spread of stimulation. A concentration of the effect on the pre-axial side is hardly apparent. In the 8 h barrier limbs, however, the 60–100 % relative mitotic enhancement seen after

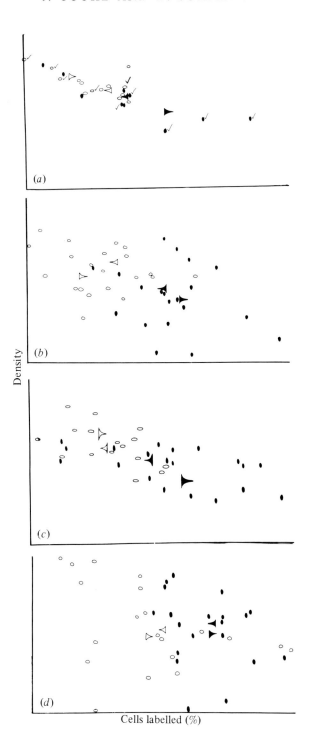

Density

Cells labelled (%)

16 h uninterrupted signalling (Cooke & Summerbell, 1980) is not reached. This indicates that the 'lead time' from 'S' phase entry up to mitosis is less than, say, 8 h (i.e. 16 minus 8). One of the 12 h barrier cases is unusual in showing no mitotic (or cell packing) effect, suggesting an unsuccessful graft, but in one of the others the overall response at around $+60\%$ relative mitosis is comparable to that after simple 16 h signalling. The cells might therefore be registering a similar situation experienced at, say 12 h in each case, making the lead time from 'S' initiation to mitosis at least 4 h (i.e. 16 minus 12). Such estimates would fit with our limited knowledge of cycle kinetics in the pre-differentiated mesenchyme.

Table 1 gives our estimate for the mean number of hours normal ZPA signalling necessary to begin the reliable organization of extra pattern elements at the pre-axial side of the future wing, as in the digit pattern 2.2.3.4 compared to the normal 2.3.4. We are in agreement with Summerbell (1973) and Smith (1980) that this is usually 12 h and may be as much as 15, while complete new specification up to 4.3.2.2.3.4 may require more than 24 h. Our method of halting signalling by passing a barrier near the ZPA, like Smith's of re-excising the ZPA plus a little extra tissue, suffers from the criticism that it may ablate the very small territory that is initially instructed to alter its fate where limb patterns like 2.2.3.4 are to be produced. If so, the results only emphasize the relatively confined territory from which tissue for the new pattern is drawn. The data of the present paper should be seen in the light of Honig's (1981) and Summerbell's (1981 *a*) finding and of the discussion in Summerbell & Honig (1981) of their independent experiments, which show that in profound and complete duplications such as we should expect to result from simple ZPA

Fig. 3. Relationships between cell packing density and entry to 'S' phase within limb pairs after simple anterior ZPA grafts. In the scatter diagrams, cell packing densities recorded within all individual data boxes of the host tissues of limb pairs (ordinate) are plotted against the corresponding labelling indices (abscissa). Highest and lowest packings observed (in the transverse plane, not per volume) were 12 800 and 6 900 cells per mm², respectively, while labelling incidence ranges from 23 to 57%.

○, Control limb (left-hand) data boxes; ●, experimental limb (right-hand) data boxes; ◄, mean of (control or experimental) posterior half-bud mesenchyme; ►, mean of (control or experimental) anterior half-bud mesenchyme. In 4a, ticks distinguish the boxes, near graft site, showing a cell cycle effect, together with the homologous control boxes.

(*a*) Pair 623 operation + 5 h. (*b*) Pair 619 operation + 9 h. (*c*) Pair 617 operation + 9 h. (*d*) Pair 634 operation + 17 h. A negative correlation between packing density and labelling index is always observed (as it is within the data boxes of control limb pairs), and is statistically significant in the cases (*a*) and (*c*) shown ($P < 0.01$). The result of ZPA grafting has been to reverse the positions of posterior and anterior halves of the mesenchyme on the labelling index axis (or, at 17 h, to equalize them) mainly by stimulation of the anterior half. Cell packing density is also decreased. For geographical distribution of the effect see Cooke & Summerbell (1980).

grafts with timing like those in the present work, some 25 % of the host tissue width anteriorly ($\sim$ 300 μm) is probably involved in founding the duplicate pattern 4.3.2.2.3.4, with a time course of at least 24 h. It can readily be appreciated from Fig. 2, Table 1, and these findings on pattern specification, that cellular responses to presence of an implanted ZPA are radically faster and more rapidly spread so far as growth control is concerned than they are in terms of altered position values (Wolpert, Hicklin & Hornbruch, 1971).

Figure 3 gives scatter diagrams of 'S' phase incidence against packing density of mesenchymal cells among the data boxes of certain individual limb-bud pairs fixed 5, 9 or 17 h after simple ZPA grafts to the right-hand limb of each pair. Control side and experimental side data boxes are distinguished, and the pre-axial (anterior) and post-axial (posterior) half mesenchymes' mean values marked out. The general inverse correlation between proportions of cells in 'S' phase during a 1 h labelling period and the mesenchymal packing density is statistically significant in the 5 h post-graft pair, and in one of the two 9 h post-graft pairs. It can be seen from the spread of positions of the right- and left-hand data and the right and left anterior and posterior means, that a decrease of packing density occurs in a coordinate way with early stimulation of the cell cycle. Up to 9 h the effects are sufficiently concentrated towards the pre-axial (graft) side of the host mesenchyme to reverse the trend seen in control limbs for the 'S' index to decline in graded fashion from back to front of the bud. In the 5 h post-graft pair, the marking of the only boxes (adjacent to the graft) which showed the dramatic effects of ZPA signalling, together with their control side partners, shows up the correlation between lowered cell packing and early cycle stimulation in a most striking way. By 17 h after operation the effects have become more generalized geographically and scarcely related to position of the graft within the system, in a way reminiscent of the mitotic response seen in parallel with 'S' enhancement at similar times (Fig. 2). As compared with the earlier limb pairs, the overall relation between packing and cell cycle is largely obscured in this example.

The scatter diagram of Fig. 4 represents the correlation between mitotic incidence per cell and cell packing density within the main series of limb pairs, those providing the geographical data on mitotic stimulation in Fig. 2. Variation in average values characterizing the mesenchymes within different embryos has been dealt with by showing percentage deviation from the average value in the control (left) limb, within each of the six proximodistal columns of each limb pair. The cluster of left limb data points is thus centred, by definition, on the origin. By itself it fails to reveal any systematic relation between the two parameters considered. Signalling by implanted anterior ZPAs for the earlier part of the 16 h period before fixation has however shifted the population of right limb data points, not only in the direction of enhanced and much more variable mitotic indices, but significantly into the lower right-hand quadrant representing diminished cell packing density. The experimental

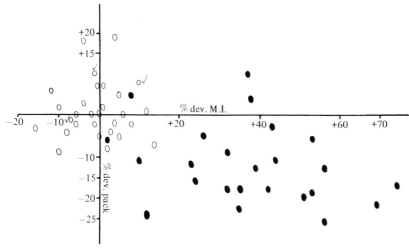

Fig. 4. Relationships between cell packing density and mitotic incidence within limb pairs 16 h after grafting. The *relative* incidences of mitosis per cell, shown in the columns of Fig. 2, are plotted as percentage deviation from their control limb means against the similar deviations for cell packing density. The origin thus represents average cell packing and mitotic incidence in the left limbs within all pairs. Control (left) data points are open circles, with the ticks marking in addition the three *right* data points of pair 119 where no stimulation was seen. Experimental data points are solid. The inverse correlation (-0.35) between packing density and mitosis within the experimental points approaches significance ($p = 0.05$), while that for the whole assemblage of data points (-0.61) is highly significant ($P < 0.005$). The presence of an extra ZPA anteriorly has diminished packing and raised mitotic index away from normal values in a highly coordinate manner. N.B. the *range* of packing values is less than for Fig. 3. This is largely because the sample for each data point is larger (up to 10 times the cell population of one of the earlier data boxes). But mitosis may be less strongly correlated with cell density than is 'S' entry, because the act of mitosis greatly restores cell numbers even though low density may have stimulated it at an earlier time.

data cluster shows an inverse correlation between packing and mitosis that borders on significance ($P < 0.05$), and when the total population of data points is considered a highly significant inverse correlation ($P < 0.005$) shows up. The three ticked data points within the control cluster are in fact those of experimental limb 119, the 12 h barrier operation which failed to result in mitotic stimulation. Sitting as they do as typical members of the control cluster, they reinforce the belief that factors diminishing cell packing and enhancing cell division rate away from normal values are (*a*) related to one another and (*b*) a function of successful communication between ZPA and surrounding tissues. The barrier was in normal position at the time of fixation in this limb.

Summerbell (1981 b) has reported elsewhere in this symposium volume that by advanced stages of wing development the sizes of the supernumerary structures in the entire or partial extra patterns of reversed polarity, caused by early ZPA grafting, are essentially normal for their stage of differentiation. Thus a completely duplicated pattern of forearm, wrist and hand structures will possess about twice the tissue mass of its single counterpart by 10 days of incubation. Observation of width changes after operations suggests that the doubling of tissue extent relative to incubation data is largely accomplished within 24 or 36 h following operation (Summerbell, 1981 a). This is reasonably close to the time taken to complete specification of limb pattern in its proximo-distal entirety after such stages ($\sim$ 44 h, Summerbell, 1974). The duplication of pattern and the width increase, after any particular ZPA graft, occur in that mesenchyme which has still to leave the 'tip' zone of 300–400 μm behind the AER during bud elongation. Honig's estimate using orthotopic quail grafts in combination with ZPA grafting (1981), and Summerbell's using the distances between two ZPA grafts necessary to produce two whole sets of structures (1981 a) agree that the tissue of anterior duplicated patterns is largely descended from a relatively local region embracing much less than half the host bud at time of operation.

Modelling of the control of pattern in the anteroposterior dimension of the bud is currently based on the idea of a monotonically graded positional signal, controlled from normal and implanted ZPA regions, possibly in the form of a diffusion gradient of morphogen (Wolpert, Lewis & Summerbell, 1975). On this basis, two extreme ways of viewing the data on early growth stimulaton accompanying pattern duplication would be as follows.

(1) A small, steep reversed landscape of positional signal is set up anteriorly, embracing tissue which plays little part in building the normal limb structures since the fate map indicates that these are mostly descended from the (normally faster growing) posterior regions of the early bud. The set of position values (Wolpert, 1969) dictating the duplicate pattern is thus present in a much smaller tissue extent than that dictating the normal pattern, whose ZPA has been active for much longer to build up a more extensive gradient landscape. The enhanced growth which then causes abnormal expansion or dilation of the extra fate map at early stages, leading to the differentiation of more normal-sized extra pattern elements, is seen largely as a consequence of small size at foundation, being due to some feedback mechanism that normalizes the sizes of tissue territories relative to the ranges of position value present in them. Early growth rate is considered to be a feedback function of the scale on which pattern determination has occurred. The 'steepness' of a positional gradient at particular times in development could in principle be utilized to set the growth rate. On this view there is no separate control signal from the ZPA

growth, simply the positional signalling system which is used in two ways (see discussion, Summerbell, 1981*a*). Such a view fails to accommodate the data, which show incursion of the growth-stimulating effect into the heart of the presumptive fate map for the original limb pattern. It is true that at very early times after ZPA implantation, and with uninterrupted signalling after 9 or 17 h, the response is concentrated in the anterior half mesenchyme (Cooke & Summerbell, 1980), but it is considerable elsewhere and frequently at the posterior border. The ZPA-plus-barrier results, which leave much stimulation but wipe out the anterior-to-posterior graded pattern otherwise seen, suggest a signal which propagates and equilibrates rapidly over the dimensions of the normal system; one which could scarcely act as the signal for *position* relative to a ZPA (and thus for limb pattern) because gradients in it would be too transient. The results of Smith (1980) and our own, showing that 12 to 15 h is the shortest time for reliable responses of cells to altered signal in terms of position value, are relevant here. They make it impossible to imagine that the system mediating cell cycle control is exclusively tied to pattern-forming values in tissue. On the simple assumptions that absolute incidence of mitotic figures in undifferentiated mesenchyme closely reflects the mean frequency of mitoses, and that all cells are potentially in cycle, the rate of expansion of tissue through most of the limb bud must be dramatically though transiently elevated following implantation of an extra ZPA at an ectopic site.

(2) The alternative extreme view is that the timecourse of cells' responses to signal, and perhaps the intercellular communication systems themselves, are different as regards control by the ZPA of position value on the one hand and growth rate on the other. Responses to growth control influences occur on a much shorter timescale, and if the signalling systems involved are in fact different, as seems likely from the data in this paper, then the growth control 'signal' is also communicated much faster. It is as if, over the 36 h when growth rate is influenced by implantation of an extra ZPA, the rate of the cycle is largely a function of the number of ZPAs contacting the bud, and only partly of position in relation to any particular ZPA. Typical width changes over this period, together with estimates of the times for determination of pattern in the proximodistal sequence, suggest that by the time duplicated regions of pattern are becoming determined, the pattern elements concerned may each be founded on a scale (i.e. using amounts of tissue) similar to that obtaining in the normal progress of development. Subsequent growth rates within the determined patterns may return to normal because the schedule of growth within established pattern parts is a fixed aspect of their determined character (Wolpert, 1978; Lewis & Wolpert, 1976) and/or because the growth-promoting activities of the ZPAs have re-established the normal amount of territory for each of them, thus returning the growth signal level to normal. If this view holds up under future experimental investigation, the growth-controlling function of ZPA tissue within the limb could be viewed as a

mechanism that ensures that standard amounts of tissue space are available in relation to the numbers of pattern elements that are to be formed. One of us (J.C.) has always experienced difficulty in supposing that biological patterns of the class of the early limb pattern could become organized solely by monotic positional signals of the gradient variety. This is because their deepest feature is their spatially periodic nature, whereby a number of dispersed centres of comparable cellular activity are laid out. The burden placed upon cellular interpretative machinery would seem too large at early stages of formation in such patterns, and the idea of a 'pre-pattern' of morphogen peaks and troughs corresponding to the future pattern of elements, and set up by reaction–diffusion kinetics or allied processes (Turing, 1952; Newman & Frisch, 1979) seems more plausible. The latter classes of process do have the constraint, unless very special accessory mechanisms were built in, that the numbers of elements formed in patterns controlled by them is almost proportional to tissue size at the time of the process. Formation of normal, complete patterns would thus require a degree of control over tissue size that goes quite beyond what is required by the primary embryonic field (e.g. Cooke, 1981a, b; Dan-Sohkawa & Sato, 1978) which controls a pattern of a different character.

The data now very strongly imply that the cellular mechanisms whereby grafted ZPA tissue controls the sequence of characters in the elements of a new, extra pattern are separate from those whereby it enhances growth, and that these are only coordinated spatially to the extent that both emanate from one site. It has been suggested that a difference in apparent rate of spread between growth stimulation and re-patterning influences could be understood in terms of a 'follow-up servo' mode of positional signalling (Wolpert et al. 1971) whereby the dynamic positional variable exists as a relatively transient set of gradient landscapes because of its diffusibility, and cell positional values for pattern formation change by slowly following these gradients. Cell cycle kinetics may be a much more immediate response to the same signal, so that effects appear to spread through more limb tissue per time. It is hard to see how such a 'single signal, dual response' system would stably set up a double, mirror-imaged limb pattern as is observed. The early spread of cycle stimulation and its geographical extent (see Fig. 2, and Cooke & Summerbell, 1980) would imply a 'flooded out' signal landscape. This would lead to obliteration of central pattern parts in a duplicate because of loss of low gradient values, during the time before determination of position values in the anteroposterior dimension is completed.

The rules of continuity and polarity followed by the sequence of element characters (digits, etc.) in relation to grafted and host ZPAs, continue to make a diffusion-controlled positional signal gradient the best concept available for this aspect of pattern. What of the propagated mesenchymal growth control? We are unable to know the causal sequence between the receipt of signal, the decrease of cell packing density and the increase of transition probability to

'S' phase. By analogy with knowledge of growth control in *in vitro* confluent cell systems (see review in Stoker, 1978), the two behavioural responses could be intimately related, with the change in cell packing possibly instrumental in stimulating the cycle. We do not yet understand enough of the mechanism of early morphogenesis in rudiments like the limb bud to know how mesenchymal packing density or shape might be controlled. It seems unlikely that cells can literally push one another apart, even though the appearance of undifferentiated bud mesenchyme from either end of the density range, in toluidine-blue-stained epon sections, shows that low-density cells simply have a more extended surface with less area of mutual contact and more intercellular 'space'. A specific chemical messenger which decreased packing indirectly through causing secretion of colloid from the cells into the space would be surprising in view of the very rapid propagation through tissue and short induction time seen for the effect. Another possibility is that shape and size of the mesenchymal population of the normal bud is constrained by positive elastic pressure and a pattern of deformability in the ectodermal sheath, with the mean cycle rate set well below the maximum possible from the nutritional situation by means of contact density. Signals controlling the growth rates of different parts of the mesenchymal cross-section could then originate as propagated changes of tensile strength within the ectoderm, leading literally to bulges underneath which the amount of space available per mesenchyme cell was increased (Summerbell, 1973). If mesenchyme is organized at all like confluent mono-layers or explants in culture, such decrease in cell density would be progressively shared and equilibrated throughout it with an appropriate evening out of the re-setting of mean cycle time (see Folkman & Moscona, 1978). A decrease in constraining forces in ectoderm caused by increased cell division there would accomplish nothing by way of explaining the rapidity of the effect, even though such an increase may finally occur as part of bud widening. If the signal acts initially to decrease the tensile strength of ectoderm, a decrease of nuclear density in the plane of the cell layer, because of a flattening of the cells, might be expected as the earliest sign of the growth-enhancing sequence of events after ZPA implantation. Future studies based on this hypothesis should address the ectoderm and apical ectodermal ridge, as well as questions of the specificity or otherwise to limb mesenchyme of the ZPA's growth-enhancing signal

REFERENCES

CALANDRA, A. J. & MACCABE, J. A. (1978). The *in vitro* maintenance of the limb bud apical ridge by cell-free preparation. *Devl Biol.* **62**, 258–269.

CAMOSSO, M. & RONCALI, L. (1968). Time sequence of the process of ectodermal ridge thickening and of mesodermal cell proliferation during apical outgrowth of the chick embryo wing bud. *Acta Embryol. Morph. exp.* **10**, 247–263.

COOKE, J. (1979). Cell number in relation to primary pattern formation in the embryo of *Xenopus laevis*. I. The cell cycle during new pattern formation in response to implanted organizers. *J. Embryol. exp. Morph.* **51**, 165–182.

COOKE, J. (1981a). The relation between scale and the completeness of pattern in vertebrate embryogenesis; model and experiments. *Symp. Amer. Soc. Zool., Seattle 1980. Amer. Zool.* (in the press).

COOKE, J. (1981b). Scale of the body pattern adjusts to available cell number in amphibian embryos. *Nature.* **290**, 775–778.

COOKE, J. & SUMMERBELL, D. (1980). Growth control early in embryonic development; the cell cycle during experimental pattern duplication in the chick wing. *Nature* **287**, 697–701.

DAN-SOHKAWA, M. & SATO, M. (1978). Studies on dwarf larvae from isolated blastomeres of the starfish, *Asterina pectinifera*. *J. Embryol. exp. Morph.* **46**, 171–185.

FALLON, J. F. & CROSBY, G. M. (1975). The relationship of the zone of polarising activity to supernumerary limb formation (twinning) in the chick wing bud. *Devl Biol.* **42**, 24–34.

FOLKMAN, J. & MOSCONA ,A. (1978). Role of cell shape in growth control. *Nature* **273**, 345–349.

HONIG, L. S. (1981). Range of a positional signal in the developing chick limb. *Nature* **291**, 72–73.

LEWIS, J. H. & WOLPERT, L. (1976). The concept of non-equivalence in development. *J. theoret. Biol.* **62**, 479–490.

MACCABE, A. B., GASSELING, M. & SAUNDERS, J. W. JR (1973). Spatiotemporal distribution of mechanisms that control outgrowth and antero-posterior polarisation of the limb bud in chick embryo. *Mech. Ageing Develop.* **2**, 1–12.

MACCABE, J. A. & PARKER, B. W. (1975). The *in vitro* maintenance of the apical ectodermal ridge of the chick wing bud: an assay for polarising activity. *Devl Biol.* **45**, 349–357.

MCWILLIAMS, H. K. & PAPAGEORGIOU, S. (1978). A model of gradient interpretation based on morphogen binding. *J. theoret. Biol.* **72**, 385–411.

MURRAY, J. D. (1981). A prepattern formation mechanism for animal coat markings. *J. theoret. Biol.* **188**, 161–199.

NEWMAN, S. A. & FRISCH, H. L. (1979). Dynamics of skeletal pattern formation in developing chick limb. *Science* **205**, 662–668.

ROWE, D. A. & FALLON, J. F. (1981). The effect of removing posterior apical ectodermal ridge of the chick wing and leg on pattern formation. *J. Embryol. exp. Morph.* **65** (*Supplement*), 309–325.

SMITH, J. C. (1980). The time required for positional signalling in the chick wing bud. *J. Embryol. exp. Morph.* **60**, 321–328.

STOCKER, M. G. P. (1978). The multiplication of cells. In *Paediatrics and Growth* (ed. D. Barltrop). The Fellowship of Postgraduate Medicine, London.

SUMMERBELL, D. (1973). *Growth and regulation in the development of the chick limb.* Ph.D. thesis, University of London.

SUMMERBELL, D. (1974) A quantitative analysis of the effect of excision of the AER from the chick limb-bud. *J. Embryol. exp. Morph.* **32**, 651–660.

SUMMERBELL, D. (1977). Regulation of deficiencies along the proximal distal axis of the chick wing bud: a quantitative analysis. *J. Embryol. exp. Morph.* **41**, 137–159.

SUMMERBELL, D. (1979). The zone of polarising activity: evidence for a role in normal chick limb morphogenesis. *J. Embryol. exp. Morph.* **50**, 217–233.

SUMMERBELL, D. (1981a). The control of growth and the development of pattern across the antero-posterior axis of the chick limb bud. *J. Embryol. exp. Morph.* **63**, 161–180.

SUMMERBELL, D. (1981b). Evidence for regulation of growth, size and pattern in the developing chick limb bud. *J. Embryol. exp. Morph.* **65** (*Supplement*), 129–150.

SUMMERBELL, D. & HONIG, L. S. (1981). The control of pattern across the antero-posterior axis of the chick limb bud by a unique signalling region. *Symp. Amer. Soc. Zool., Seattle 1980, Amer. Zool.* (in the press).

SUMMERBELL, D. & WOLPERT, L. (1972). Cell density and cell division in the early morphogenesis of the chick limb. *Nature New Biol.* **238**, 24–25.

TICKLE, C., SUMMERBELL, D. & WOLPERT, L. (1975). Positional signalling and specification of digits in chick limb morphogenesis, *Nature* **254**, 199–202.

TURING, A. (1952). The chemical basis of morphogenesis. *Phil. Trans. Roy. Soc.* B **237**, 32.

WOLPERT, L. (1969). Positional information and the spatial pattern of cellular differentiation. *J. theor. Biol.* **25**, 1–47.

WOLPERT, L. (1978). The development of the pattern of growth. In *Paediatrics and Growth* (ed. D. Barltrop). The Fellowship of Postgraduate Medicine, London.

WOLPERT, L., HICKLIN, J. & HORNBRUCH, A. (1971). Positional information and pattern regulation in regeneration of *Hydra*. In *Control Mechanisms of Growth and Differentiation*, *Symp. Soc. exp. Biol.* **25**, 391–415.

WOLPERT, L., LEWIS, J. & SUMMERBELL, D. (1975). Morphogenesis of the vertebrate limb. In *Cell Patterning*, Ciba Foundation Symposium 29.

. *Embryol. exp. Morph. Vol. 65 (Supplement), pp. 187–207, 1981*
Printed in Great Britain © Company of Biologists Limited 1981

Growth factors and pattern formation

By J. C. SMITH[1]

*From the Sidney Farber Cancer Institute and the
Department of Microbiology and Molecular Genetics,
Harvard Medical School*

SUMMARY

Growth and pattern formation occur simultaneously in many epimorphic fields and it has been suggested that specification of positional information is somehow linked to cell division. It is possible, therefore, that boundary regions responsible for the specification of positional information produce cell growth factors. In this paper I review the properties of some known growth factors, describe their effects on the cell cycle and discuss how they might act. In developing a convenient *in vitro* assay for morphogenetic factors it will be much easier to measure incorporation of [³H]thymidine into responding cells than to estimate changes in positional value.

INTRODUCTION

The development of a multicellular organism from fertilized egg to adult does not proceed by an orderly series of cell doublings. It is by differential growth and cell division (together with cytodifferentiation and pattern formation) that the features of the adult body are established (Wolpert, 1969, 1971). This is well illustrated by the development of the wrist and forearm of the chick wing. Initially, the rudiments of these two elements are very similar in length, but while the wrist remains more or less the same size the forearm grows considerably to gain its final prominence (Summerbell, 1976; Lewis, 1977). In the terminology of Lewis & Wolpert (1976) cartilage cells in the wrist and cartilage cells in the forearm are 'non-equivalent', and their different growth properties reflect their different 'positional values'.

But cell division during development should not be viewed only as a response to positional value once positional value has been specified; it is becoming clear that cell division may be an important parameter in the specification process itself. This is most obvious in the 'progress zone' model for the specification of positional information along the proximodistal axes of the chick limb bud and the regenerating amphibian limb (Summerbell, Lewis & Wolpert, 1973). In this model positional information is specified by the time a cell spends in the

[1] *Author's address:* Laboratory of Tumor Biology, Sidney Farber Cancer Institute and Department of Microbiology and Molecular Genetics, Harvard Medical School, Boston, MA 02115, U.S.A.

'progress zone' at the tip of the limb, and this time may be measured in terms of cell division cycles (Lewis, 1975; Wolpert, Tickle & Sampford, 1979).

In many systems growth and specification of positional information occur simultaneously. Along the anteroposterior axis of the chick limb bud growth and cell division are enhanced by a graft of an additional zone of polarizing activity (Tickle, Summerbell & Wolpert, 1975; Fallon & Crosby, 1977; Summerbell & Tickle, 1977; Smith & Wolpert, 1981; Summerbell, 1981), which causes mirror-image duplication of the limb about its long axis (Saunders & Gasseling, 1968). Indeed, in any epimorphic field, including the appendages and imaginal discs of developing insects and the regenerating amphibian limb (French, Bryant & Bryant, 1976) pattern formation or regulation may be causally linked to cell division (Summerbell et al. 1973). One is reminded of Holtzer's concept of the 'quantal mitosis' (Hotzler, Weintraub, Mayne & Mochan, 1972).

The question of how cell division is controlled during embryogenesis is thus of great importance to the study of pattern formation. Although Summerbell has obtained some evidence for density-dependent inhibition of growth in the chick limb bud (Summerbell & Wolpert, 1972; Summerbell, 1977; Cooke & Summerbell, 1980), most information regarding the control of cell division has been obtained from cells grown in culture. Here it has been found that a variety of factors can regulate the cell cycle: nutrients (Eagle, 1955; Ley & Tobey, 1970; Rubin, 1972), cell contact (Todaro, Lazar & Green, 1965; Dulbecco, 1970), cell shape (Folkman & Moscona, 1978), cell substrate (Gospodarowicz & Ill, 1980; Gospodarowicz, Delgado & Vlodavsky, 1980), growth factors (Sato & Ross, 1979) and inhibitors of cell growth (Balazs, 1979). However, it is not clear which, if any, of these might control cell division in developing embryos. In this review I shall describe the effects of some growth factors on cells in culture and in vivo, discuss how these growth factors might act and consider whether these, or similar, molecules might be active during development and perhaps even represent the elusive morphogens.

GROWTH FACTORS

Most cells in culture require serum (the liquid portion of clotted blood) in order to proliferate (Carrel, 1912; Eagle, 1955; Todaro et al. 1965). Cells derived from tumours or exposed in vitro to oncogenic viruses or chemicals have a reduced serum requirement (reviewed by Brooks, 1975). Serum contains many components which function in different ways to promote growth (see Stiles, Cochran & Scher, 1981). Some of these, such as trace elements, amino acids and lipids, fulfil a nutritional requirement; others may transport such low molecular weight nutrients (transferrin, for example, is an iron carrier); some, like fibronectin, mediate cell attachment. In addition, serum contains the substances we now know as growth factors – hormones which stimulate growth but are not required as nutrients; that is, they are not molecules that may be used within the cell as metabolic substrates or co-factors (Gospodarowicz & Moran, 1976). (While this definition of a growth factor is adequate for the purposes of this review, it should be pointed out that until the molecular mechanism of action of a growth factor is elucidated its metabolic function is unknown.) Of all its functions, the most important role of serum does appear to be to provide hormones or growth factors. Hayashi & Sato (1976) have succeeded in growing an established rat pituitary cell line in a defined serum-free medium containing physiological concentrations of four hormones and transferrin.

Serum is a complex mixture of a variety of substances and the concentration of mitogens is low; thus, growth factors have been isolated from serum only with the greatest difficulty (see Gospodarowicz & Moran, 1976). More success has been obtained in isolating growth factors from tissue or from cultured cells. A fibroblast growth factor (FGF) has been isolated from bovine pituitary and brain (Gospodarowicz, 1975; Gospodarowicz, Bialecki & Greenburg, 1978 *a*); an epidermal growth factor (EGF) from the male mouse submaxillary gland (Cohen, 1962); and nerve growth factor (NGF) from the same source (Cohen, 1960). Cultured adipocytes produce a potent growth factor for endothelial cells (Castellot, Karnovsky & Spiegelman, 1980).

Below, I shall describe briefly the properties of six growth factors which may be involved in development. Although none of these has been implicated in any way in the interactions occurring during pattern formation they could give an idea of the kind of molecule to look for.

Endothelial cell growth factors

There is much interest in factors that control endothelial cell growth and migration because these are the processes involved in vascularization of tissues during both development and tumour growth (Ausprunk & Folkman, 1977). Growth factors for endothelial cells cultured *in vitro* have been isolated from a variety of sources; from tumour-cell-conditioned medium (Folkman, Haudenschild & Zetter, 1979; Zetter, 1980); bovine hypothalamus (Maciag *et al.* 1979); and cultured 3T3 cells (Birdwell, Gospodarowicz & Nicolson, 1977; McAuslan, Hannan & Reilly, 1980). For developmental biologists perhaps the most interesting is a growth factor produced by differentiated 3T3 adipocytes, disvered by Castellot *et al.* (1980).

These authors have partially characterized a growth factor produced by 3T3-F442A adipocytes (Green & Kehinde, 1976). The factor is non-dialysable and it is not inactivated by heat or proteases. It is only produced by differentiated adipocytes, and of all the cell types examined only vascular endothelial cells respond to it. Adipose tissue is very highly vascularized and it is possible that the factor identified by Castellot and colleagues stimulates vascularization *in vivo*. Whether the factor also stimulates migration of endothelial cells remains an important question.

Epidermal growth factor

Epidermal growth factor was first isolated by Cohen (1962) from the sub-maxillary gland of the adult male mouse (mEGF). Extracts of the submaxillary gland injected into newborn mice induced precocious eyelid opening and incisor eruption, later shown to be due to a stimulation of epidermal growth and keratinization (Cohen, 1965). This mEGF was found to be a heat-stable single-chain polypeptide with 53 amino-acid residues. It has a pI of 4·5, three disulphide bonds (which are required for biological activity) and a molecular weight of 6045 (see Carpenter & Cohen, 1979).

A similar molecule (hEGF) has been isolated from human urine (Cohen & Carpenter, 1975; Savage & Harper, 1981). Of its 53 amino acids 37 are common to mEGF and its three disulphide bonds are in the same relative positions. Human EGF appears to be identical to urogastrone, a gastric antisecretory hormone isolated from urine (Gregory, 1975).

The biological activities of mEGF and hEGF are very similar and they possess some common antigenic sites. Both, as mentioned above, will stimulate epidermal growth and keratinization on injection into newborn mice. They will also stimulate epithelial cell proliferation in several organ culture systems, including the chick embryo cornea (Savage & Cohen, 1973). In tissue culture mEGF (hEGF was not tested) has little effect on the growth of small colonies of early passage human keratinocytes but does enhance growth of larger colonies. It also prolongs the culture lifetime of these cells (Rheinwald & Green, 1975).

EGF is mitogenic (at concentrations around 4 ng/ml; 0·7 nM) for a variety of fibroblast cells, both established lines, like 3T3 (Hollenberg & Cuatrecasas, 1973) and early passage human diploid fibroblasts (Cohen & Carpenter, 1975). Malignant transformation is sometimes associated with a loss of the growth requirement for EGF (Cherington, Smith & Pardee, 1979).

EGF is a powerful tool for studying the interactions of hormones with their target cells. It can be tagged with radioactive iodine (Carpenter & Cohen, 1976), fluorescein (Haigler, Ash, Singer & Cohen, 1978), rhodamine (Shechter et al. 1978b) and ferritin (McKanna, Haigler & Cohen, 1979) while remaining active and able to bind to its receptor. These studies have demonstrated that on binding EGF the receptors aggregate and are taken into the cell by endocytosis. It is not clear whether internalization is required for the mitogenic action of EGF; this is discussed below.

Erythropoietin

The production of erythrocytes in both normal and anaemic mammals is regulated by erythropoietin, a blood-borne hormone whose level in the circulation is under the control of the kidney. The effects of erythropoietin have been studied on explants of bone marrow and foetal liver. The hormone stimulates mitosis of proerythroblasts and promotes the differentiation of this enlarged population of erythroid precursors to erythroblasts, with the concomitant onset of globin synthesis (Chui, Djaldetti, Marks & Rifkind, 1971; Stephenson & Axelrad, 1971). Enrichment of cell populations for proerythroblasts has confirmed that these are the target cells for erythropoietin (Cantor, Morris, Marks & Rifkind, 1972) and enabled study of the expression of the globin gene (reviewed by Browder, 1980). The first detectable effect of erythropoietin is a stimulation of 4S, 5S and ribosomal RNA synthesis, but synthesis of globin mRNA does not occur until the proerythroblasts have undergone a round of DNA synthesis and mitosis. DNA synthesis is essential for the onset of globin mRNA synthesis, for if it is inhibited by hydroxyurea globin mRNA is not produced. Erythropoietin may, therefore, control a 'quantal' cell cycle (Holtzer *et al.* 1972).

Analysis of the nature of erythropoietin has been hampered because there is little starting material (plasma or urine from anaemic animals or man) available and the purification procedure is inefficient (Adamson & Brown, 1978). Isolated from sheep, erythropoietin appears to be a glycoprotein of molecular weight 46000 (Goldwasser & Kung, 1972) whereas the human form may have a molecular weight half of this (Dorado, Espada, Langton & Brandon, 1974). About 11 % of the molecule consists of sialic acid, which is necessary for its *in vivo* activity; desialated erythropoietin is removed from circulation by the liver (Morell, Irvine, Sternlieb & Scheinberg, 1968). Desialated erythropoietin is, however, active *in vitro* (Goldwasser, Kung & Eliason, 1974).

Fibroblast growth factor

Fibroblast growth factors (FGFs) have been isolated from bovine brain and pituitary by Gospodarowicz and his colleagues (Gospodarowicz, 1974, 1975; Gospodarowicz, Moran & Bialecki, 1976). There is presently some confusion about the nature of these factors. Pituitary FGF appears to be a basic molecule with a molecular weight of 13000 on SDS–polyacrylamide gels under reducing conditions. It is active at 1 ng/ml (0·1 nM) (Gospodarowicz, 1975). Purification of brain FGF led to the isolation of two biologically active polypeptides, FGF-1 and FGF-2, which seemed to originate from a common precursor (Gospodarowicz *et al.* 1978*a*). These were later shown to be produced by limited proteolysis of myelin basic protein (Westall, Lennon & Gospodarowicz, 1978). However, Bradshaw and his colleagues (Thomas *et al.* 1980) have now shown clearly that the mitogenic activity of brain FGF preparations resides in a small acidic (pI between 4·8 and 5·8) protein that represents only 1 % of the protein in preparations of brain FGF made according to the procedure of Gospodarowicz. This protein is not related to myelin basic protein.

Preparations of pituitary FGF and the 'crude' brain FGF are mitogenic for many cell types. These include 3T3 cells, foreskin fibroblasts, glial cells, kidney fibroblasts, amniotic cells, chondrocytes, myoblasts, endothelial cells, smooth muscle cells, adrenal cortex cells, granulosa cells and blastemal cells (reviewed by Gospodarowicz, Moran & Mescher, 1978 b).

Fibroblast growth factors have been implicated in the control of vertebrate limb regeneration, although the rationale for these experiments (the assumption that FGF is a breakdown product of myelin basic protein) now appears invalid. The regeneration of the amphibian limb is dependent upon innervation (Todd, 1823; Singer, 1952, 1974), and the suggestion (see Gospodarowicz & Mescher, 1980) was that peptides with FGF activity would be released from injured peripheral myelin. These FGFs might direct regeneration by acting directly on blastemal cells, by promoting endothelial cell growth and thereby vascularization (see Smith & Wolpert, 1975), and by stimulating Schwann cell proliferation and remyelination of regenerating nerves (Gospodarowicz & Mescher, 1980). To test the hypothesis that FGF regulates amphibian limb regeneration Gospodarowicz and his colleagues injected FGF into newt forelimb blastemas which had been denervated 3 days previously. The results indicated that FGF could indeed stimulate mitotic activity in the blastemas. Furthermore, repeated injections of pituitary FGF into the amputated limbs of adult frogs (which do not normally regenerate) resulted in the initial phases of blastema formation (Gospodarowicz & Mescher, 1980).

These results are consistent with the notion that a growth factor is active during vertebrate limb regeneration, but the finding of Thomas *et al.* (1980) that FGF is not a breakdown product of myelin basic protein makes it less likely that the growth factor is FGF.

Nerve growth factor

The first evidence for a nerve growth factor (NGF) was obtained in 1948 by Bueker. Implants of mouse sarcoma 180 to the body wall of chick embryos caused a 20–40 % enlargement of the dorsal root ganglions that innervated the tumour. Motor neurons were not stimulated to grow. Levi-Montalcini & Hamburger (1951, 1953) then showed that sympathetic ganglia were also enlarged and contributed nerves to the tumour. Furthermore, by transplanting sarcoma 37 or 180 to the chorioallantoic membrane, they demonstrated that the effects of the tumours were mediated by a diffusible agent. This agent became known as nerve growth factor.

The first attempts to purify NGF used sarcoma 180 as a starting material but it was then found that NGF is present in snake venom (Cohen & Levi-Montalcini, 1956) and in adult male mouse submaxillary glands (Cohen, 1960). NGF purified from the submaxillary glands of adult male mice (now the usual source) is a dimer of two identical peptide chains (each consisting of 118 amino acids, 3 disulphide bonds and with a molecular weight of 13259) linked by non-covalent forces.

NGF is not mitogenic *in vivo* or *in vitro*. The hyperplasia of sensory ganglia exposed to NGF, first observed by Bueker (1948), is probably due to enhanced neuronal survival (see Mobley *et al.* 1977); a variety of experiments have demonstrated that NGF plays a role in maintaining the survival of sympathetic and some embryonic sensory neurons (Levi-Montalcini & Booker, 1960; Levi-Montalcini & Angeleti, 1963). In addition, NGF causes an increase in cell body size and in neurite outgrowth in responsive cells; it may be chemotactic for axon growth; and it may play a role in the maintenance of the differentiated state of some neurons (reviewed by Greene & Shooter, 1980).

There appear to be three cellular receptors for NGF: on the plasma membrane, on the nucleus and at the synaptic ending (see review by Vinores & Guroff, 1980). The membrane and nuclear receptors were demonstrated by [^{125}I]NGF binding studies on a variety of cells (including dorsal root ganglia cells and the NGF-responsive clone PC 12). Synaptic receptors were first proposed as a result of experiments in which [^{125}I]NGF was injected into the anterior chamber of the eye of a mouse. It was found that there was a preferential accumulation of radioactivity in the superior cervical ganglion of the injected side, and this was interpreted to be due to retrograde axonal transport following uptake of NGF at the synaptic ending (Hendry, Stockel, Thoenen & Iverson, 1974). This retrograde transport seems to be of biological significance; if it is interrupted by axotomy neuronal degradation results, and this effect can be reversed by administration of exogenous NGF (Hendry & Campbell, 1976).

Somatomedins

Growth hormone (GH) promotes skeletal growth by stimulating epiphyseal cartilage cells to proliferate and increase matrix secretion (see Williams & Hughes, 1977). However, GH does not act directly, but through substances now known as somatomedins (Daughaday *et al.* 1972), whose levels *in vivo* are under the control of GH. This was first proposed by Salmon & Daughaday (1957) who observed that GH had no effect on explants of cartilage cultured *in vitro* but incorporation of [^{35}S]sulphate into proteoglycans was enhanced by addition of normal serum or serum from hypophysectomized rats which had been treated with GH.

Purification and characterization of substances capable of stimulating incorporation of [^{35}S]sulphate into cartilage explants demonstrated that the effects of these agents are not restricted to cartilage. Thus, substances classified as somatomedins should meet the following criteria: their concentration in serum should be regulated by GH; they should stimulate incorporation of sulphate into proteoglycans of cartilage; they should be mitogenic for fibroblasts; and they should have insulin-like effects on adipose and muscle cells (Daughaday *et al.* 1972).

Four factors or groups of factors which meet these criteria have been identified. These are: somatomedin A (Hall, 1972); somatomedin C (Van Wyk *et al.* 1974); insulin-like growth factors (IGF) (Rinderknecht & Humbel, 1976); and multiplication-stimulating activity (MSA) (Pierson & Temin, 1972; Smith &

Temin, 1974). A peptide isolated by following stimulation of [^{3}H]thymidine uptake into human glial cells was termed somatomedin B (Uthne, 1973; Westermark, Wasteson & Uthne, 1975) but it cannot be classed as a true somatomedin because it does not stimulate incorporation of [^{35}S]sulphate into cartilage explants.

Characterization of the somatomedins has been slow because of the difficulties in obtaining pure material. The liver is widely assumed to be the site of synthesis but the somatomedins are not concentrated there nor in any other organ (Chochinov & Daughaday, 1976; Daughaday, 1977; Rechler *et al.* 1979). Consequently, large amounts of blood are required. It has been found that all four factors are single-chain acid-soluble polypeptides of molecular weights 5000 to 10000. All are structurally related to insulin and compete for binding to the insulin receptor (reviewed by Blundell & Humbel, 1980).

That the somatomedins bind to the insulin receptor explains their weak insulin-like effects, but there is a second set of receptors with a high affinity for the somatomedins and a low affinity for insulin. It is through these receptors that the growth-promoting effects of the somatomedins and of insulin are exerted (see Blundell & Humbel, 1980).

Although the somatomedins are presumed to mediate the effect of GH during somatic growth there has been no direct confirmation of this because large quantities of somatomedins are difficult to obtain (Van Wyk & Underwood, 1978). However, it has recently been demonstrated directly that somatomedin C, at least, is active *in vivo*. DNA synthesis and mitosis are abolished in the lens epithelium of the frog after hypophysectomy but may be restored by injection of purified somatomedin C or human GH (Rothstein *et al.* 1980).

REGULATION OF THE CELL CYCLE BY GROWTH FACTORS

Many growth factors, like the endothelial cell growth factor, erythropoietin and NGF, have specific target cells. Others, like EGF, FGF and the somatomedins have little target specificity but may be highly specific for the portion of the cell cycle in which they act. This was first suggested by experiments in which growth factors were observed to function synergistically, suggesting that they controlled different events in the cell cycle. More sophisticated experiments have been able to dissect the sequence in which these growth factors act.

Synergism between growth factors

Synergism between growth factors was first observed by Armelin (1973), who found that hydrocortisone would enhance the growth-promoting effects of pituitary extracts. Subsequently, Gospodarowicz & Moran (1974) demonstrated that pituitary FGF would only induce optimal DNA synthesis in the presence of a corticosteroid and insulin.

More recently, Jimenez de Asua and colleagues (1977*a*, *b*, 1979, 1981) have studied the initiation of DNA synthesis in quiescent, confluent Swiss 3T3 cells. They found that prostaglandin $F_{2\alpha}$ ($PGF_{2\alpha}$) and insulin act synergistically to initiate DNA synthesis 15 h after addition of the hormones. By adding the hormones at different times they established that $PGF_{2\alpha}$ alone could initiate DNA synthesis, and addition of insulin to such $PGF_{2\alpha}$-treated cultures did not alter the lag phase but did increase the rate at which the cells entered the S phase. Hydrocortisone would inhibit growth stimulation by $PGF_{2\alpha}$, but only if it was added within 3 h. These results suggested that there are two steps in G_0/G_1 which control the entry of cells into S phase. The first is mediated by $PGF_{2\alpha}$ and inhibited by hydrocortisone, the second is regulated by insulin and controls the rate of entry of cells into S phase. Recently Jimenez de Asua *et al.* (1981) have demonstrated that EGF, like $PGF_{2\alpha}$, will stimulate the initiation of DNA synthesis but that these two factors act through different cellular or biochemical pathways.

Competence and progression

Pledger, Stiles, Antoniades & Scher (1977, 1978) have attempted to characterize the factors in serum that are required for the growth of BALB/c-3T3 cells. Serum can be separated into two sets of components that function synergistically to regulate growth. One component is the platelet-derived growth factor (PDGF), a heat-stable cationic polypeptide contained in the α-granules of platelets and released when blood clots (Kohler & Lipton, 1974; Ross, Glomset, Kariya & Harker, 1974; Antoniades, Scher & Stiles, 1979; Kaplan *et al.* 1979). The other set of components is contained within platelet-poor plasma, which is prepared from unclotted blood and therefore lacks PDGF.

Brief treatment of quiescent, density-arrested BALB/c-3T3 cells with PDGF renders the cells 'competent' to replicate their DNA. Competence persists for many hours after removal of PDGF from the medium but PDGF-treated cells make no 'progress' through G_0/G_1 towards S phase until they are exposed to the complementary set of growth factors contained in platelet-poor plasma; 12 h later the cells begin to enter S phase. Two growth-arrest points have been identified in this 12 h period. One, the V point, is located 6 h before S phase; the other, the W point, is at the G_1/S boundary. The rate of entry of cells into S phase from the W point depends upon the concentration of plasma.

The 'progression activity' of platelet-poor plasma can be replaced by EGF, somatomedin C and transferrin (Stiles *et al.* 1979*a*, *b*; Leof, Wharton, Van Wyk & Pledger, 1980). Somatomedin C is required in the latter half of the progression sequence, at the V and W points (Stiles *et al.* 1979*b*; Wharton, Van Wyk & Pledger, 1981). By this time there has already occurred in the cells an increase in receptors for somatomedin C, in response to PDGF and factors in somatomedin C-deficient plasma (Clemmons, Van Wyk & Pledger, 1980).

MECHANISM OF ACTION OF GROWTH FACTORS

Different growth factors act on different cells and at different points in the cell cycle. It is very unlikely, therefore, that they will work by similar mechanisms, and to make generalizations about the ways in which growth factors function is difficult. Below, I describe the early responses of cells to growth stimulation, the interaction of ^{125}I-labelled EGF with cells and recent evidence that at least one growth factor, PDGF, acts via a cytoplasmic 'second message'. Finally, I suggest that growth factors *in vivo* need not act while in solution; they could be active while adherent to the extracellular matrix.

Cellular responses to growth factors

Stimulation of growth in quiescent cells by serum or by purified growth factors rapidly (within a few minutes) induces a complex set of changes in the responding cells. These include stimulation of the Na–K pump (probably initiated by H^+/Na^+ exchange) (Rozengurt & Heppel, 1975; Rozengurt & Mendoza, 1980; Moolenaar, Mummery, van der Saag & de Laat, 1981); increase in 2-deoxy-D-glucose uptake (Sefton & Rubin, 1971); phosphorylation, and therefore trapping, of intracellular uridine (Rozengurt, Stein & Wigglesworth, 1977); changes in the uptake of different amino acids (Hochstadt, Quinlan, Owen & Cooper, 1979); increase in phosphate uptake (Cunningham & Pardee, 1969); and changes in intracellular cyclic nucleotide concentrations (Seifert & Rudland, 1974). These responses can occur in the absence of protein synthesis but later changes include increases in the rates of protein and RNA synthesis (see Hershko, Mamont, Shields & Tomkins, 1971; Pardee, Dubrow, Hamlin & Kletzein, 1978). These reactions have been termed as a whole the 'pleiotypic response' (Hershko *et al.* 1971).

It is not clear how these events are directed towards the onset of DNA synthesis and cell division. One suggestion is that changes in nutrient transport directly trigger and sustain DNA synthesis and cell proliferation by providing the materials needed for growth (Holley, 1972). Rubin and his colleagues (Rubin & Sanui, 1979) have suggested that Mg^{2+} ions mediate the response to extracellular stimuli by increasing rates of protein synthesis. It is certainly true that inhibition of protein or RNA synthesis after serum stimulation of quiescent cells will inhibit or delay entry into S phase (Brooks, 1977; Chadwick, Ignotz, Ignotz & Lieberman, 1980) and that inhibition of protein synthesis in cycling cells will greatly extend the G_1 period (Rossow, Riddle & Pardee, 1977). Furthermore, a variant line of Chinese Hamster cells which lacks a measurable G_1 period has been shown to have an elevated rate of protein synthesis during mitosis (Rao & Sunkara, 1980); a G_1 period could be induced by slowing their rate of protein synthesis with cycloheximide (Liskay, Kornfeld, Fullerton & Evans, 1980).

However, at least for quiescent 3T3 fibroblasts the elevated rate of protein synthesis that is induced by growth factors is not sufficient to induce DNA synthesis. Quiescent BALB/c-3T3 cells treated briefly with PDGF and then incubated in medium containing platelet-poor plasma experience an increase in protein synthesis and enter S phase after a minimum lag of 12 h. Brief treatment of cells with PDGF without subsequent incubation in plasma-containing medium or continuous incubation in medium containing plasma produce a similar increase in protein synthesis, but the treated cells do not enter S phase (Cochran, Lillquist, Scher & Stiles, 1981).

Interactions of [^{125}I]EGF with responding cells

Epidermal growth factor is frequently used to study the interactions of growth factors with their target cells because it is available in large quantities, it may be tagged in a variety of ways while remaining active and as an acidic peptide it is easy to handle. Human and mouse EGF have been radio-iodinated. Both compete for the same receptor and these receptors are present on a variety of cultured cells (see Carpenter & Cohen, 1979). Different human fibroblasts were reported to have 40000 and 100000 binding sites per cell (Hollenberg & Cuatrecasas, 1975; Carpenter, Lembach, Morrison & Cohen, 1975).

The EGF receptors are normally randomly distributed, but on binding the growth factor they aggregate and are taken into the cell by endocytosis (see review by Hopkins, 1980). Once inside the cell the EGF–receptor complexes fuse with lysosomes where both EGF and the receptor are degraded (Carpenter & Cohen, 1976; King, Hernaez & Cuatrecasas, 1980). Carpenter & Cohen (1976) demonstrated that the degradation, but not the internalization, of EGF was prevented by chloroquine, an inhibitor of lysosomal protease activity. They also observed that internalization of EGF was associated with a loss of EGF binding sites from the cell surface. This phenomenon has been termed 'down regulation' and has also been observed with insulin (Gavin *et al.* 1974) and human GH (Lesniak & Roth, 1976).

The roles of EGF–receptor internalization and degradation in the mitogenic response to EGF are unclear. It is known that EGF must be present in the medium for several hours in order to induce DNA synthesis (Carpenter & Cohen, 1975; Shechter, Hernaez & Cuatrecasas, 1978a; Aharonov, Pruss & Herschman, 1978) and that only a small fraction of the EGF receptors need be occupied in order to induce the maximal mitogenic response (Das & Fox, 1978; Shechter *et al.* 1978a). Furthermore, King *et al.* (1980) have demonstrated that degraded receptors are continually replaced by newly synthesized receptors so that the process seems to be geared towards a constant transport of EGF and its receptor into the cell (see King, Hernaez-Davis & Cuatrecasas, 1981).

The importance of the internalization and subsequent degradation of the EGF–receptor complex in the induction of the mitogenic response has been studied with inhibitors of the breakdown process. Savion, Vlodavsky & Gospodarowicz (1980) attempted to inhibit degradation of [^{125}I]EGF with leupeptin, and found that the mitogenic response to EGF was unaffected. This suggested that breakdown of the EGF–receptor complex was not required for mitogenesis. However, King *et al.* (1981) showed that primary alkylamines, which also inhibit intracellular processing of ligand–receptor complexes (but not internalization, as had previously been thought), will inhibit mitogenesis in response to EGF. Hence, the role of the degradation of the EGF–receptor complex remains unknown. Hopkins (1980) suggests that there is no compelling reason why internalization and degradation should play a part in the mitogenic response and prefers the view that signalling by the EGF–receptor complex begins on the cell surface.

PDGF induces a cytoplasmic 'second message'

The cytoplasm of cells undergoing DNA synthesis contains agents which will initiate DNA synthesis in non-replicating nuclei. This has been demonstrated by cell fusion experiments in which fusion with an S phase cell initiates DNA replication in the nuclei of G_0/G_1 cells (Rao & Johnson, 1970; Graves, 1972; Mercer & Schlegel, 1980) or of chick erythrocytes (Harris, 1965; Lipsich, Lucas & Kates, 1978) and by *in vitro* studies in which cytoplasmic extracts from growing cells promote DNA synthesis in isolated frog cell nuclei (Jazwinski, Wang & Edelman, 1976; Floros, Chang & Baserga, 1978; Das, 1980). The nature and mode of action of these cytoplasmic agents are unknown, but broadly there are two possibilities. Some may be precursors or enzymes which are produced as cells approach and enter S phase and which are directly involved in DNA replication (see Kornberg, 1980). Others may be cell-cycle regulatory molecules produced, perhaps, in response to extracellular growth factors. Such regulatory molecules would be of greater interest in the study of the control of cell growth because the critical events controlling cell division occur before the onset of DNA synthesis (see above).

Until recently, it had not been possible to distinguish between these two possibilities because exposure to growth factors could not be uncoupled from growth itself. However, the discovery that PDGF can render cells 'competent', but that competent cells will not 'progress' towards S phase unless incubated in medium containing platelet-poor plasma enabled Smith & Stiles (1981) to do just that. Cells treated briefly with PDGF, or cytoplasts derived from them, were fused to 'recipient' cells; these recipient cells were rendered competent and when incubated in medium containing platelet-poor plasma entered S phase after the typical lag of 10–12 h. The ability of PDGF-treated cells to transfer competence in this way was abolished by inhibitors of RNA synthesis but not by inhibitors of protein synthesis. These results suggest that PDGF induces the formation of a second message, perhaps an mRNA, which is responsible for the competent state. Possible translation products of this mRNA have been identified by Pledger, Hart, Locatell & Scher (1981); PDGF rapidly induces the formation of five polypeptides (molecular weights 29000,

35000, 45000, 60000 and 70000) within treated cells, the production of which is inhibited by inhibitors of RNA synthesis.

Growth factors may bind to cell substrata

The substrate on which cells are maintained can alter cellular requirements for growth factors. For example, smooth-muscle cells maintained on the extracellular matrix produced by corneal endothelial cells are able to proliferate in medium containing platelet-poor plasma (Gospodarowicz & Ill, 1980; Gospodarowicz *et al.* 1980); the same cells maintained on tissue-culture plastic require further additions to the medium, such as PDGF or FGF (see above references and Ross *et al.* 1974). Recently, Smith *et al.* (1981) have investigated the possibility that the ability of the extracellular matrix to enhance cell proliferation is due to adherent growth factors.

Tissue-culture plates, either coated with collagen (a major component of the extracellular matrix) or uncoated, were treated briefly with PDGF or FGF and then washed thoroughly. The growth of BALB/c-3T3 cells (which, like smooth-muscle cells, require PDGF or FGF for optimum growth *in vitro*) on these treated substrates or on untreated plates was compared. Cells maintained on PDGF- or FGF-treated plates grew rapidly compared with cells on untreated plates. This stimulation of growth was due to adherent growth factor, because when half a plate was treated and quiescent cells were seeded in medium containing platelet-poor plasma and [³H]thymidine, only those cells on the treated portion of the dish became labelled after autoradiography.

This observation suggests a means by which a localized and persistent *in vivo* stimulation of cell growth might be obtained, even in the absence of soluble growth factor. If morphogenetic factors are able to adhere to extracellular matrix this would provide a basis for a 'positional memory' (Smith, 1979).

GROWTH FACTORS AND PATTERN FORMATION

I have described the properties of six growth factors thought to be active during development and attempted to give a general idea of how they act. It has been suggested (see Introduction) that growth and pattern formation are causally linked, so it is possible that molecules resembling growth factors are involved in the specification of positional information. How could this be investigated?

Two broad approaches might be adopted. While there is no reason to believe that any one of the growth factors discussed here, or indeed any known growth factor, is a 'morphogen', one should not rule out its involvement in pattern formation simply because the organ which produces it in the adult animal has not yet developed or because the field under study is remote from that organ. If a growth factor is produced in the adult the potential for making it is present in every embryonic cell. Thus, one approach would be to examine the effects of known growth factors (say, EGF or somatomedin C) in developing systems. The presence of the factors could be studied by radioimmunoassay; receptors might be identified using ^{125}I-labelled growth factors (see D'Ercole & Underwood, 1980). The presence of a growth factor and its receptor does not prove that the factor is active in stimulating growth. To demonstrate this, *in vitro* experiments might be required or exogenous growth factors could be administered *in vivo*. This could be achieved by implants of growth-factor-impregnated Silastic (Heaton, 1977) or by the use of osmotic minipumps (Stiles & Smith, work in progress).

A second approach would be to study signalling regions to see if they produce growth factors – extracts could be assayed *in vitro* or *in vivo* as above – and to search for pleiotypic responses in the responding cells. This approach might prove fruitful in the chick limb where it has been demonstrated that a graft of the zone of polarizing activity (ZPA) rapidly stimulates the growth of responding cells (Cooke & Summerbell, 1980). However, it is unlikely that any novel morphogenetic growth factor could be purified from this source. Purification of 18 μg of PDGF required 500 units of human platelets (about 78 g protein) (Antoniades *et al.* 1979). Assuming the same efficiency and recovery, purification of 18 μg of ZPA growth factor would require one to ten million dozen eggs.

I am supported by a NATO postdoctoral fellowship at the Sidney Farber Cancer Institute. I thank Chuck Stiles and Fiona Watt for their helpful comments and Lynne Dillon for preparing the manuscript.

REFERENCES

ADAMSON, J. W. & BROWN, J. E. (1978). Aspects of erythroid differentiation and proliferation. In *Molecular Control of Proliferation and Differentiation* (ed. J. Papaconstantinou & W. J. Rutter), pp. 161–179. New York, San Francisco, London: Academic Press.

AHARONOV, A., PRUSS, R. M. & HERSCHMAN, H. R. (1978). Epidermal growth factor. Relationship between receptor regulation and mitogenesis in 3T3 cells. *J. biol. Chem.* **253**, 3970–3977.

ANTONIADES, H. N., SCHER, C. D. & STILES, C. D. (1979). Purification of human platelet-derived growth factor. *Proc. natn. Acad. Sci., U.S.A.* **76**, 1809–1813.

ARMELIN, H. A. (1973). Pituitary extracts and steroid hormones in the control of 3T3 cell growth. *Proc. natn. Acad. Sci., U.S.A.* **70**, 2702–2706.

AUSPRUNK, D. & FOLKMAN, J. (1977). Migration and proliferation of endothelial cells in preformed and newly formed blood vessels during tumor angiogenesis. *Microvasc. Res.* **14**, 53–65.

BALAZS, A. (1979). *Control of Cell Proliferation by Endogenous Inhibitors.* Amsterdam, New York, Oxford: Elsevier/North-Holland Biomedical Press.

BIRDWELL, C. R., GOSPODAROWICZ, D. & NICOLSON, G. I. (1977). Factors from 3T3 cells stimulate proliferation of cultured vascular endothelial cells. *Nature, Lond.* **268**, 528–531.

BLUNDELL, T. L. & HUMBEL, R. E. (1980). Hormone families: pancreatic hormones and homologous growth factors. *Nature, Lond.* **287**, 781–787.

BROOKS, R. F. (1975). Growth regulation *in vitro* and the role of serum. In *Structure and Function of Plasma Proteins*, vol. 2 (ed. A. C. Allison), pp. 239–289. New York: Plenum Press.

BROOKS, R. F. (1977). Continuous protein synthesis is required to maintain the probability of entry into S phase. *Cell* **12**, 311–317.

BROWDER, L. W. (1980). *Developmental Biology*. Philadelphia: Saunders College.

BUEKER, E. D. (1948). Implantation of tumors in the hind limb field of the embryonic chick and the developmental response of the lumbrosacral nervous system. *Anat. Rec.* **102**, 369–390.

CANTOR, L. N., MORRIS, A. J., MARKS, P. A. & RIFKIND, R. A. (1972). Purification of erythropoietin-responsive cells by immune hemolysis. *Proc. natn. Acad. Sci., U.S.A.* **69**, 1337–1341.

CARPENTER, G. & COHEN, S. (1975). Human epidermal growth factor and the proliferation of human fibroblasts. *J. cell Physiol.* **88**, 227–237.

CARPENTER, G. & COHEN, S. (1976). ^{125}I-Labeled human epidermal growth factor. Binding, internalization and degradation in human fibroblasts. *J. Cell Biol.* **71**, 159–171.

CARPENTER, G. & COHEN, S. (1979). Epidermal growth factor. *Ann. Rev. Biochem.* **48**, 193–216.

CARPENTER, G., LEMBACH, K. J., MORRISON, M. M. & COHEN, S. (1975). Characterization of the binding of ^{125}I-labeled epidermal growth factor to human fibroblasts. *J. biol. Chem.* **250**, 4297–4304.

CARREL, A. (1912). On the permanent life of tissues outside the organism. *J. exp. Med.* **15**, 516–528.

CASTELLOT, J. J., JR, KARNOVSKY, M. J. & SPIEGELMAN, B. M. (1980). Potent stimulation of vascular endothelial cell growth by differentiated 3T3 adipocytes. *Proc. natn. Acad. Sci., U.S.A.* **77**, 6007–6011.

CHADWICK, D. E., IGNOTZ, G. G., IGNOTZ, R. A. & LIEBERMAN, I. (1980). Inhibitors of RNA synthesis and passage of chick embryo fibroblasts through the G_1 period. *J. cell. Physiol.* **104**, 61–72.

CHERINGTON, P. V., SMITH, B. L. & PARDEE, A. B. (1979). Loss of epidermal growth factor requirement and malignant transformation. *Proc. natn. Acad. Sci., U.S.A.* **76**, 3937–3941.

CHOCHINOV, R. H. & DAUGHADAY, W. H. (1976). Current concepts of somatomedin and other biologically related growth factors. *Diabetes* **25**, 994–1007.

CHUI, D. H. K., DJALDETTI, M., MARKS, P. A. & RIFKIND, R. A. (1971). Erythropoietin effects on fetal mouse erythroid cells. I. Cell population and hemoglobin synthesis. *J. Cell Biol.* **51**, 585–595.

CLEMMONS, D. R., VAN WYK, J. J. & PLEDGER, W. J. (1980). Sequential addition of platelet factor and plasma to BALB/c-3T3 fibroblast cultures stimulates somatomedin C binding early in cell cycle. *Proc. natn. Acad. Sci., U.S.A.* **77**, 6644–6648.

COCHRAN, B. H., LILLQUIST, J. S., SCHER, C. D. & STILES, C. D. (1981). Post-transcriptional control of protein synthesis in Balb/c-3T3 cells by platelet-derived growth factor and platelet-poor plasma. *J. cell. Physiol.* (In the Press.)

COHEN, S. (1960). Purification of a nerve-growth promoting protein from the mouse salivary gland and its neuro-cytoxic antiserum. *Proc. natn. Acad. Sci., U.S.A.* **46**, 302–311.

COHEN, S. (1962). Isolation of a mouse submaxillary gland protein accelerating incisor eruption and eyelid opening in the new-born animal. *J. biol. Chem.* **237**, 1555–1562.

COHEN, S. (1965). The stimulation of epidermal proliferation by a specific protein (EGF). *Devl Biol.* **12**, 394–397.

COHEN, S. & CARPENTER, G. (1975). Human epidermal growth factor: isolation and chemical and biological properties *Proc. natn. Acad. Sci., U.S.A.* **72**, 1317–1321.

COHEN, S. & LEVI-MONTALCINI, R. (1956). A nerve growth-stimulating factor isolated from snake venom. *Proc. natn. Acad. Sci., U.S.A.* **42**, 571–574.

COOKE, J. & SUMMERBELL, D. (1980). Cell cycle and experimental pattern duplication in the chick wing during embryonic development. *Nature, Lond.* **287**, 697–701.

CUNNINGHAM, D. D. & PARDEE, A. B. (1969). Transport changes rapidly initiated by serum addition to 'contact inhibited' 3T3 cells. *Proc. natn. Acad. Sci., U.S.A.* **64**, 1049–1056.

DAS, M. (1980). Mitogenic hormone-induced intracellular message: assay and partial characterization of an activator of DNA replication induced by epidermal growth factor. *Proc. natn. Acad. Sci., U.S.A.* **77**, 112–116.

DAS, M. & FOX, C. F. (1978). Molecular mechanism of mitogen action: processing of receptor induced by epidermal growth factor. *Proc. natn. Acad. Sci., U.S.A.* **75**, 2644–2648.

DAUGHADAY, W. H. (1977). Hormonal regulation of growth by somatomedins and other tissue growth factors. *Clin. Endocrinol. Metab.* **6**, 117–135.

DAUGHADAY, W. H., HALL, K., RABEN, M., SALMON, W. D., VAN DEN BRANDE, J. L. & VAN WYK, J. J. (1972). Somatomedin: proposed designation for sulphation factor. *Nature, Lond.* **235**, 107.

D'ERCOLE, A. J. & UNDERWOOD, L. E. (1980). Ontogeny of somatomedin during development in the mouse. Serum concentrations, molecular forms, binding proteins, and tissue receptors. *Devl Biol.* **79**, 33–45.

DORADO, M., ESPADA, J., LANGTON, A. A. & BRANDON, N. C. (1974). Molecular weight estimation of human erythropoietin by SDS–polyacrylamide gel electrophoresis. *Biochem. Med.* **10**, 1–7.

DULBECCO, R. (1970). Topoinhibition and serum requirement of transformed and untransformed cells. *Nature, Lond.* **227**, 802–806.

EAGLE, H. (1955). Nutrition needs of mammalian cells in tissue culture. *Science* **122**, 501–504.

FALLON, J. F. & CROSBY, G. M. (1977). Polarizing zone activity in limb buds of amniotes. In *Vertebrate Limb and Somite Morphogenesis* (ed. D. A. Ede, J. R. Hichliffe & M. Balls), pp. 55–69. Cambridge and London: Cambridge University Press.

FLOROS, J., CHANG, H. & BASERGA, R. (1978). Stimulated DNA synthesis in frog nuclei by cytoplasmic extracts of temperature-sensitive mammalian cells. *Science* **201**, 651–652.

FOLKMAN, J., HAUDENSCHILD, C. & ZETTER, B. (1979). Long-term culture of capillary endothelial cells. *Proc. natn. Acad. Sci., U.S.A.* **76**, 5217–5221.

FOLKMAN, J. & MOSCONA, A. (1978). Role of cell shape in growth control. *Nature, Lond.* **273**, 345–349.

FRENCH, V., BRYANT, P. J. & BRYANT, S. V. (1976). Pattern regulation in epimorphic fields. *Science* **193**, 969–981.

GAVIN, J., ROTH, J., NEVILLE, D., deMEYTS, P. & BUELL, D. N. (1974). Insulin-dependent regulation of insulin receptor concentrations: a direct demonstration in cell culture. *Proc. natn. Acad. Sci., U.S.A.* **71**, 84–88.

GOLDWASSER, E. & KUNG, C. K.-H. (1972). The molecular weight of sheep plasma erythropoietin. *J. biol. Chem.* **247**, 5159–5160.

GOLDWASSER, E., KUNG, C. K.-H. & ELIASON, J. (1974). On the mechanism of erythropoietin-induced differentiation. XIII. The role of sialic acid in erythropoietin action. *J. biol. Chem.* **249**, 4202–4206.

GOSPODAROWICZ, D. (1974). Location of fibroblast growth factor and its effect alone and with hydrocortisone on 3T3 cell growth. *Nature, Lond.* **249**, 123–127.

GOSPODAROWICZ, D. (1975). Purification of a fibroblast growth factor from bovine pituitary. *J. biol. Chem.* **250**, 2515–2520.

GOSPODAROWICZ, D., BIALECKI, H. & GREENBURG, G. (1978*a*). Purification of the fibroblast growth factor activity from bovine brain. *J. biol. Chem.* **253**, 3736–3743.

GOSPODAROWICZ, D., DELGADO, D. & VLODAVSKY, I. (1980). Permissive effect of the extracellular matrix on cell proliferation *in vitro*. *Proc. natn. Acad. Sci., U.S.A.* **77**, 4094–4098.

GOSPODAROWICZ, D. & ILL, C. R. (1980). Do plasma and serum have different abilities to promote cell growth? *Proc. natn. Acad. Sci., U.S.A.* **77**, 2726–2730.

GOSPODAROWICZ, D. & MESCHER, A. L. (1980). Fibroblast growth factor and the control of vertebrate regeneration and repair. *Ann. N.Y. Acad. Sci.* **339**, 151–174.

GOSPODAROWICZ, D. & MORAN, J. S. (1974). Stimulation of division of sparse and confluent 3T3 cell populations by a fibroblast growth factor, dexamethasone, and insulin. *Proc. natn. Acad. Sci., U.S.A.* **71**, 4584–4588.

GOSPODAROWICZ, D. & MORAN, J. S. (1976). Growth factors in mammalian cell culture. *Ann. Rev. Biochem.* **45**, 531–558.

GOSPODAROWICZ, D., MORAN, J. S. & BIALECKI, H. (1976). Mitogenic factors from the brain and the pituitary: physiological significance. In *Growth Hormone and Related Peptides* (ed. A. Pecile & E. E. Muller), pp. 141–157. Amsterdam: Excerpta Medica.

GOSPODAROWICZ, D., MORAN, J. S. & MESCHER, A. L. (1978*b*). Cellular specificities of fibroblast growth factor and epidermal growth factor. In *Molecular Control of Proliferation and*

Differentiation (ed. J. Papaconstantinou & W. J. Rutter), pp. 161–179. New York, San Francisco, London: Academic Press.

GRAVES, J. A. M. (1972). DNA synthesis in heterokaryons formed by fusion of mammalian cells from different species. *Expl Cell Res.* **72**, 393–403.

GREEN, H. & KEHINDE, O. (1976). Spontaneous heritable changes leading to increased adipose conversion in 3T3 cells. *Cell* **7**, 105–113.

GREENE, L. A. & SHOOTER, E. M. (1980). The nerve growth factor: biochemistry, synthesis and mechanism of action. *Ann. Rev. Neurosci.* **3**, 353–402.

GREGORY, H. (1975). Isolation and structure of urogastrone and its relationship to epidermal growth factor. *Nature, Lond.* **257**, 325–327.

HAIGLER, H., ASH, J. F., SINGER, S. J. & COHEN, S. (1978). Visualization by fluorescence of the binding and internalization of epidermal growth factor in human carcinoma cells A-431. *Proc. natn. Acad. Sci., U.S.A.* **75**, 3317–3321.

HALL, K. (1972). Human somatomedin. Determination, occurrence, biological activity and purification. *Acta Endocrinol. (Copenhagen)* Suppl. **163**, 1–52.

HARRIS, H. (1965). Behaviour of differentiated nuclei in heterokaryons of animal cells from different species. *Nature, Lond.* **206**, 583–588.

HAYASHI, I. & SATO, G. (1976). Replacement of serum by hormones permits growth of cells in defined medium. *Nature, Lond.* **259**, 132–134.

HEATON, M. B. (1977). A technique for introducing localized long-lasting implants in the chick embryo. *J. Embryol. exp. Morph.* **39**, 261–266.

HENDRY, I. A. & CAMPBELL, J. (1976). Morphometric analysis of rat superior cervical ganglion after axotomy and nerve growth factor treatment. *J. Neurocytol.* **5**, 351–360.

HENDRY, I. A., STOCKEL, K., THOENEN, H. & IVERSON, L. L. (1974). The retrograde axonal transport of nerve growth factor. *Brain Res.* **68**, 103–121.

HERSHKO, A., MAMONT, P., SHIELDS, R. & TOMKINS, G. M. (1971). 'Pleiotypic response.' *Nature, New Biol.* **232**, 206–211.

HOCHSTADT, J., QUINLAN, D. C., OWEN, A. J. & COOPER, K. O. (1979). Regulation of transport upon interaction of fibroblast growth factor and epidermal growth factor with quiescent (G_0) 3T3 cells and plasma-membrane vesicles isolated from them. In *Hormones and Cell Culture* (ed. G. H. Sato & R. Ross), pp. 751–771. Cold Spring Harbor Conferences on Cell Proliferation, volume 6.

HOLLENBERG, M. D. & CUATRECASAS, P. (1973). Epidermal growth factor: receptors in human fibroblasts and modulation of action by cholera toxin. *Proc. natn. Acad. Sci., U.S.A.* **70**, 2964–2968.

HOLLENBERG, M. D. & CUATRECASAS, P. (1975). Insulin and epidermal growth factor. Human fibroblast receptors related to deoxyribonucleic acid synthesis and amino acid uptake. *J. biol. Chem.* **250**, 3845–3853.

HOLLEY, R. W. (1972). A unifying hypothesis concerning the nature of malignant growth. *Proc. natn. Acad. Sci., U.S.A.* **69**, 2840–2841.

HOLTZER, H., WEINTRAUB, H., MAYNE, R. & MOCHAN, B. (1972). The cell cycle, cell lineages, and cell differentiation. *Curr. Topics devl. Biol.* **7**, 229–256.

HOPKINS, C. R. (1980). Epidermal growth factor and mitogenesis. *Nature, Lond.* **286**, 205–206.

JAZWINSKI, S. M., WANG, J. L. & EDELMAN, G. M. (1976). Initiation of replication in chromosomal DNA induced by extracts from proliferating cells. *Proc. natn. Acad. Sci., U.S.A.* **73**, 2231–2235.

JIMENEZ DE ASUA, L., CARR, B., CLINGAN, D. & RUDLAND, P. (1977a). Specific glucocorticoid inhibition of growth promoting effects of prostaglandin $F_{2\alpha}$ on 3T3 cells. *Nature, Lond.* **265**, 450–452.

JIMENEZ DE ASUA, L., O'FARRELL, M. K., BENNETT, D., CLINGAN, D. & RUDLAND, P. (1977b). Temporal sequence of hormonal interactions during the prereplicative phase of quiescent cultured fibroblasts. *Proc. natn. Acad. Sci., U.S.A.* **74**, 3845–3849.

JIMENEZ DE ASUA, L., RICHMOND, K. M. V., O'FARRELL, M. K., OTTO, A. M., KUBLER, A. M. & RUDLAND, P. S. (1979). Growth factors and hormones interact in a series of temporal steps to regulate the rate of initiation of DNA synthesis in mouse fibroblasts. In *Hormones*

204　　　J. C. SMITH

and Cell Culture (ed. G. H. Sato & R. Ross), pp. 403–424. Cold Spring Harbor Conferences on Cell Proliferation, volume 6.

JIMENEZ DE ASUA, L., RICHMOND, K. M. V. & OTTO, A. M. (1981). Two growth factors and two hormones regulate initiation of DNA synthesis in cultured mouse cells through different pathways of events. *Proc. natn. Acad. Sci., U.S.A.* **78**, 1004–1008.

KAPLAN, D. R., CHAO, F. C., STILES, C. D., ANTONIADES, H. N. & SCHER, C. D. (1979). Platelet α-granules contain a growth factor for fibroblasts. *Blood* **53**, 1043–1052.

KING, A. C., HERNAEZ, L. J. & CUATRECASAS, P. (1980). Lysomotropic amines cause intracellular accumulation of receptors for epidermal growth factor. *Proc. natn. Acad. Sci., U.S.A.* **77**, 3283–3287.

KING, A. C., HERNAEZ-DAVIS, L. & CUATRECASAS, P. (1981). Lysosomotropic amines inhibit mitogenesis induced by growth factors. *Proc. natn. Acad. Sci., U.S.A.* **78**, 717–721.

KOHLER, N. & LIPTON, A. (1974). Platelets as a source of fibroblast growth-promoting activity. *Expl Cell Res.* **87**, 297–301.

KORNBERG, A. (1980). *DNA Replication*. San Francisco: W. H. Freeman.

LEOF, E. B., WHARTON, W., VAN WYK, J. J. & PLEDGER, W. J. (1980). Epidermal growth factor and somatomedin C control G_1 progression of competent BALB/c-3T3 cells: somatomedin C regulates commitment to DNA synthesis. *J. Cell Biol.* **87**, 5a (abstract).

LESNIAK, M. A. & ROTH, J. (1976). Regulation of receptor concentration by homologous hormone: effect of human growth hormone on its receptor in IM-9 lymphocytes. *J. biol. Chem.* **251**, 3720–3729.

LEVI-MONTALCINI, R. & ANGELETTI, P. V. (1963). Essential role of the nerve growth factor in the survival and maintenance of dissociated sensory and sympathetic embryonic nerve cells *in vitro*. *Devl Biol.* **7**, 619–628.

LEVI-MONTALCINI, R. & BOOKER, B. (1960). Destruction of the sympathetic ganglia in mammals by an antiserum to the nerve growth protein. *Proc. natn. Acad. Sci., U.S.A.* **46**, 384–391.

LEVI-MONTALCINI, R. & HAMBURGER, V. (1951). Selective growth stimulating effects of mouse sarcoma on the sensory and sympathetic nervous system of the chick embryo. *J. exp. Zool.* **116**, 321–362.

LEVI-MONTALCINI, R. & HAMBURGER, V. (1953). A diffusible agent of mouse sarcoma, producing hyperplasia of sympathetic ganglia and hyperneurotization of viscera in the chick embryo. *J. exp. Zool.* **123**, 233–288.

LEWIS, J. H. (1975). Fate maps and the pattern of cell division: a calculation for the chick wing-bud. *J. Embryol. exp. Morph.* **33**, 419–434.

LEWIS, J. (1977). Growth and determination in the developing limb. In *Vertebrate Limb and Somite Morphogenesis* (ed. D. A. Ede, J. R. Hinchliffe & M. Balls), pp. 215–228. Cambridge & London: Cambridge University Press.

LEWIS, J. H. & WOLPERT, L. (1976). The principle of non-equivalence in development. *J. theor. Biol.* **62**, 479–490.

LEY, K. D. & TOBEY, R. A. (1970). Regulation of initiation of DNA synthesis in chinese hamster cells. *J. Cell Biol.* **47**, 453–459.

LIPSICH, L. A., LUCAS, J. J. & KATES, J. R. (1978). Cell cycle dependence of the reactivation of chick erythrocyte nuclei after transplantation into mouse L929 cell cytoplasts. *J. cell. Physiol.* **97**, 199–208.

LISKAY, R. M., KORNFELD, B., FULLERTON, P. & EVANS, R. (1980). Protein synthesis and the presence or absence of a measurable G_1 in cultured Chinese Hamster cells. *J. cell. Physiol.* **104**, 461–467.

MCAUSLAN, B. R., HANNAN, G. N. & REILLY, W. (1980). Characterization of an endothelial cell proliferation factor from cultured 3T3 cells. *Expl Cell Res.* **128**, 95–101.

MACIAG, T., CERUNDOLO, J., ILSLEY, S., KELLEY, P. R. & FORAND, R. (1979). An endothelial cell growth factor from bovine hypothalamus: identification and partial characterization. *Proc. natn. Acad. Sci., U.S.A.* **76**, 5674–5678.

MCKANNA, J. A., HAIGLER, H. T. & COHEN, S. (1979). Hormone receptor topology and dynamics: morphological analysis using ferritin-labeled epidermal growth factor. *Proc. natn. Acad. Sci., U.S.A.* **76**, 5689–5693.

MERCER, W. E. & SCHLEGEL, R. A. (1980). Cell cycle re-entry of quiescent mammalian nuclei following heterokaryon formation. *Expl Cell Res.* **128**, 431–438.

MOBLEY, W. C., SERVER, A. C., ISHII, D. N., RIOPELLE, R. J. & SHOOTER, E. M. (1977). Never growth factor. *New Eng. J. Med.* **297**, 1096–1104, 1149–1158, 1211–1218.

MOOLENAAR, W. H., MUMMERY, C. L., VAN DER SAAG, P. T. & DE LAAT, S. W. (1981). Rapid ionic events and the initiation of growth in serum-stimulated neuroblastoma cells. *Cell* **23**, 789–798.

MORELL, A. G., IRVINE, R. A., STERNLIEB, I. & SCHEINBERG, H. (1968). Physical and chemical studies on ceruloplasmin. V. Metabolic studies on sialic acid-free ceruloplasmin *in vivo*. *J. biol. Chem.* **243**, 155–159.

PARDEE, A. B., DUBROW, R., HAMLIN, J. L. & KLETZIEN, R. F. (1978). Animal cell cycle. *Ann. Rev. Biochem.* **47**, 715–750.

PIERSON, R. W., JR & TEMIN, H. M. (1972). The partial purification from calf serum of a fraction with multiplication-stimulating activity for chicken fibroblasts in cell culture and with non-suppressible insulin-like activity. *J. cell. Physiol.* **79**, 319–330.

PLEDGER, W. J., HART, C. A., LOCATELL, K. L. & SCHER, C. D. (1981). Platelet-derived growth factor modulated proteins: constitutive synthesis by a transformed cell line. *Proc. natn. Acad. Sci., U.S.A.* (In the Press.)

PLEDGER, W. J., STILES, C. D., ANTONIADES, H. N. & SCHER, C. D. (1977). Induction of DNA synthesis in BALB/c-3T3 cells by serum components: reevaluation of the commitment process. *Proc. natn. Acad. Sci., U.S.A.* **74**, 4481–4485.

PLEDGER, W. J., STILES, C. D., ANTONIADES, H. N. & SCHER, C. D. (1978). An ordered sequence of events is required before BALB/c-3T3 cells become committed to DNA synthesis. *Proc. natn. Acad. Sci., U.S.A.* **75**, 2839–2843.

RAO, P. N. & JOHNSON, R. T. (1970). Mammalian cell fusion: studies on the regulation of DNA synthesis and mitosis. *Nature, Lond.* **225**, 159–164.

RAO, P. N. & SUNKARA, P. S. (1980). Correlation between the high rate of protein synthesis during mitosis and the absence of G_1 period in V79-8 cells. *Expl Cell Res.* **125**, 507–511.

RECHLER, M. M., EISEN, H. J., HIGA, O. Z., NISSLEY, S. P., MOSES, A. C., SCHILLING, E. E., FENNOY, I., BRUNI, C. B., PHILLIPS, L. S. & BAIRD, K. L. (1979). Characterization of a somatomedin (insulin-like growth factor) synthesized by fetal rat liver organ cultures. *J. biol. Chem.* **254**, 7942–7950.

RHEINWALD, J. G. & GREEN, H. (1975). Epidermal growth factor and the multiplication of cultured human epidermal keratinocytes. *Nature, Lond.* **265**, 421–424.

RINDERKNECHT, E. & HUMBEL, R. E. (1976). Amino-terminal sequences of two polypeptides from human serum with nonsuppressible insulin-like and cell-growth-promoting activities: evidence for structural homology with insulin B chain. *Proc. natn. Acad. Sci., U.S.A.* **73**, 4379 4385.

ROSS, R., GLOMSET, J., KARIYA, B. & HARKER, L. (1974). A platelet-dependent serum factor that stimulates the proliferation of arterial smooth muscle cells *in vitro*. *Proc. natn. Acad. Sci., U.S.A.* **71**, 1207–1210.

ROSSOW, P., RIDDLE, V. G. H. & PARDEE, A. B. (1979). Synthesis of a labile, serum-dependent protein in early G_1 controls animal cell growth. *Proc. natn. Acad. Sci., U.S.A.* **76**, 4446–4450.

ROTHSTEIN, H., VAN WYK, J. J., HAYDEN, J. H., GORDON, S. R. & WEINSIEDER, A. (1980). Somatomedin C: restoration in vivo of cycle traverse in G_0/G_1 blocked cells of hypophysectomized animals. *Science* **208**, 410–412.

ROZENGURT, E. & HEPPEL, L. A. (1975). Serum rapidly stimulates ouabain-sensitive ^{86}Rb$^+$ influx in quiescent 3T3 cells. *Proc. natn. Acad. Sci., U.S.A.* **72**, 4492–4495.

ROZENGURT, E. & MENDOZA, S. (1980). Monovalent ion fluxes and the control of cell proliferation in cultured fibroblasts. *Ann. N.Y. Acad. Sci.* **339**, 175–190.

ROZENGURT, E., STEIN, W. & WIGGLESWORTH, N. (1977). Uptake of nucleosides in density-inhibited cultures of 3T3 cells. *Nature, Lond.* **267**, 442–444.

RUBIN, H. (1972). Inhibition of DNA synthesis in animal cells by ethylene diamine tetraacetate and its reversal by zinc. *Proc. natn. Acad. Sci., U.S.A.* **69**, 712–716.

RUBIN, A. H. & SANUI, H. (1979). The coordinate response of cells to hormones and its mediation by the intracellular availability of magnesium. In *Hormones and Cell Culture* (ed. G. H. Sato and R. Ross), pp. 741–750. Cold Spring Harbor Conferences on Cell Proliferation, volume 6.

SALMON, W. D., JR & DAUGHADAY, W. H. (1957). A hormonally controlled serum factor which stimulated sulfate incorporation by cartilage *in vitro. J. Lab. clin. Med.* **49**, 825–836.

SATO, G. H. & ROSS, R. (eds). (1979). *Hormones and Cell Culture* (Cold Spring Harbor conferences on cell proliferation; volume 6).

SAUNDERS, J. W., JR & GASSELING, M. T. (1968). Ectodermal–mesenchymal interactions in the origin of limb symmetry. In *Epithelial–Mesenchymal Interactions* (ed. R. Fleischmajer & R. E. Binningham), pp. 78–97. Baltimore: Williams & Wilkins.

SAVAGE, C. R., JR & COHEN, S. (1973). Proliferation of corneal epithelium induced by epidermal growth factor. *Expl Eye Res.* **15**, 361–366.

SAVAGE, C. R., JR & HARPER, R. (1981). Human epidermal growth factor/urogastrone: rapid purification procedure and partial characterization. *Anal. Biochem.* **111**, 195–202.

SAVION, N., VLODAVSKY, I. & GOSPODAROWICZ, D. (1980). Role of the degradation process in the mitogenic effect of epidermal growth factor. *Proc. natn. Acad. Sci., U.S.A.* **77**, 1466–1470.

SEFTON, B. & RUBIN, H. (1971). Stimulation of glucose transport in cultures of density-inhibited chick embryo cells. *Proc. natn. Acad. Sci., U.S.A.* **68**, 3154–3157.

SEIFERT, W. & RUDLAND, P. (1974). Possible involvement of cyclic GMP in growth control of cultured mouse cells. *Nature, Lond.* **248**, 138–140.

SHECHTER, Y., HERNAEZ, L. & CUATRECASAS, P. (1978*a*). Epidermal growth factor: biological activity requires persistent occupation of high-affinity cell surface receptors. *Proc. natn. Acad. Sci., U.S.A.* **75**, 5788–5791.

SHECHTER, Y., SCHLESSINGER, J., JACOBS, S., CHANG, K. & CUATRECASAS, P. (1978*b*). Fluorescent labeling of hormone receptors in viable cells: preparation and properties of highly fluorescent derivatives of epidermal growth factor and insulin. *Proc. natn. Acad. Sci., U.S.A.* **75**, 2135–2139.

SINGER, M. (1952). The influence of the nerve in regeneration of the amphibian extremity. *Q. Rev. Biol.* **27**, 169–200.

SINGER, M. (1974). Neurotrophic control of limb regeneration in the newt. *Ann. N. Y. Acad. Sci.* **228**, 308–322.

SMITH, A. R. & WOLPERT, L. (1975). Nerves and angiogenesis in amphibian limb regeneration. *Nature, Lond.* **257**, 224–225.

SMITH, J. C. & WOLPERT, L. (1981). Pattern formation along the anteroposterior axis of the chick wing: the increase in width following a polarizing region graft and the effect of X-irradiation. *J. Embryol. exp. Morph.* **63**, 127–144.

SMITH, J. C. (1979). Evidence for a positional memory in the development of the chick wing bud. *J. Embryol. exp. Morph.* **52**, 105–113.

SMITH, J. C., LILLQUIST, J. S., GOON, D. S., SCHER, C. D. & STILES, C. D. (1981). Growth factors adherent to cell substrata are active *in situ*. (In preparation.)

SMITH, J. C. & STILES, C. D. (1981). Cytoplasmic transfer of the mitogenic response to platelet-derived growth factor. *Proc. natn. Acad. Sci., U.S.A.* (In the Press.)

SMITH, G. L. & TEMIN, H. M. (1974). Purified multiplication-stimulating activity from rat liver cell conditioned medium: comparison of biological activities with calf serum, insulin and somatomedin. *J. cell. Physiol.* **84**, 181–192.

STEPHENSON, J. R. & AXELRAD, A. A. (1971). Separation of erythropoietin-sensitive cells from hemopoietic spleen colony-forming stem cells of mouse fetal liver by unit gravity sedimentation. *Blood* **37**, 417–427.

STILES, C. D., CAPONE, G. T., SCHER, C. D., ANTONIADES, H. N., VAN WYK, J. J. & PLEDGER, W. J. (1979*a*). Dual control of cell growth by somatomedins and platelet-derived growth factor. *Proc. natn. Acad. Sci., U.S.A.* **76**, 1279–1283.

STILES, C. D., COCHRAN, B. H. & SCHER, C. D. (1981). Regulation of the mammalian cell cycle by hormones. In *The Cell Cycle*, Soc. Exp. Biology Seminar Series (ed. P. E. G. John), pp. 119–137. Cambridge and London: Cambridge University Press.

STILES, C. D., PLEDGER, W. J., VAN WYK, J. J., ANTONIADES, H. & SCHER, C. D. (1979*b*).

Hormonal control of early events in the BALB/c-3T3 cell cycle: commitment to DNA synthesis. In *Hormones and Cell Culture* (ed. G. H. Sato & R. Ross), pp. 425–439. Cold Spring Harbor Conferences on Cell Proliferation, volume 6.

SUMMERBELL, D. (1976). A descriptive study of the rate of elongation and differentiation of the developing chick wing. *J. Embryol. exp. Morph.* **35**, 241–260.

SUMMERBELL, D. (1977). Reduction in the rate of outgrowth, cell density, and cell division following removal of the apical ectodermal ridge of the chick limb bud. *J. Embryol. exp. Morph.* **40**, 1–21.

SUMMERBELL, D. (1981). The control of growth and development of pattern across the antero-posterior axis of the chick limb bud. *J. Embryol. exp. Morph.* (In the Press.)

SUMMERBELL, D., LEWIS, J. H. & WOLPERT, L. (1973). Positional information in chick limb morphogenesis. *Nature, Lond.* **244**, 492–496.

SUMMERBELL, D. & TICKLE, C. (1977). Pattern formation along the antero-posterior axis of the chick limb bud. In *Vertebrate Limb and Somite Morphogenesis* (ed. D. A. Ede, J. R. Hinchliffe & M. Balls), pp. 41–53. Cambridge and London: Cambridge University Press.

SUMMERBELL, D. & WOLPERT, L. (1972). Cell density and cell division in the early morphogenesis of the chick limb. *Nature New Biology, Lond.* **239**, 24–26.

THOMAS, K. A., RILEY, M. C., LEMMON, S. K., BAGLAN, N. C. & BRADSHAW, R. A. (1980). Brain fibroblast growth factor. Nonidentity with myelin basic protein fragments. *J. biol. Chem.* **255**, 5517–5520.

TICKLE, C., SUMMERBELL, D. & WOLPERT, L. (1975). Positional signalling and specification of digits in chick limb morphogenesis. *Nature, Lond.* **254**, 199–202.

TODARO, G. J., LAZAR, G. & GREEN, H. (1965). The initiation of cell division in a contact-inhibited line. *J. Cell comp. Physiol.* **66**, 325–333.

TODD, T. J. (1823). On the process of reproduction of the members of the aquatic salamander. *Q. J. Mic. Sci. Lit. Arts* **16**, 84–96.

UTHNE, K. (1973). Human somatomedins. Purification and some studies on their biological actions. *Acta Endocrinol. (Copenhagen)* Suppl. **175**, 1–35.

VAN WYK, J. J. & UNDERWOOD, L. E. (1978). The somatomedins and their actions. In *Biochemical Actions of Hormones*, vol. 5 (ed. G. Litwack), pp. 101–147. New York: Academic Press.

VAN WYK, J. J., UNDERWOOD, L. E., HINTZ, R. L., CLEMMONS, D. R., VIONA, S. J. & WEAVER, R. P. (1974). The somatomedins: a family of insulinlike hormones under growth hormone control. *Rec. Progr. Horm. Res.* **30**, 259–318.

VINORES, S. & GUROFF, G. (1980). Nerve growth factor: mechanism of action. *Ann. Rev. Biophys. Bioeng.* **9**, 223–257.

WESTALL, F. C., LENNON, V. A. & GOSPODAROWICZ, D. (1978). Brain-derived fibroblast growth factor: identity with a fragment of the basic protein of myelin. *Proc. natn. Acad. Sci., U.S.A.* **75**, 4675–4678.

WESTERMARK, B., WASTESON, A. & UTHNE, K. (1975). Initiation of DNA synthesis of stationary human glia-like cells by a polypeptide fraction from human plasma containing somatomedin activity. *Expl Cell Res.* **96**, 58–62.

WHARTON, W., VAN WYK, J. J. & PLEDGER, W. J. (1981). Inhibition of BALB/c-3T3 cells in late G_1: commitment to DNA synthesis controlled by somatomedin C. *J. cell. Physiol.* **107**, 31–39.

WILLIAMS, J. P. G. & HUGHES, P. C. R. (1977). Hormonal regulation of postnatal limb growth in mammals. In *Vertebrate Limb and Somite Morphogenesis* (ed. D. A. Ede, J. R. Hinchliffe & M. Balls), pp. 281–292. Cambridge and London: Cambridge University Press.

WOLPERT, L. (1969). Positional information and the spatial pattern of cellular differentiation. *J. theor. Biol.* **25**, 1–47.

WOLPERT, L. (1971). Positional information and pattern formation. *Curr. Topics devl Biol.* **6**, 183–224.

WOLPERT, L., TICKLE, C. & SAMPFORD, M. (1979). The effect of cell killing by X-irradiation on pattern formation in the chick limb. *J. Embryol. exp. Morph.* **50**, 175–198.

ZETTER, B. R. (1980). Migration of capillary endothelial cells is stimulated by tumour-derived factors. *Nature, Lond.* **285**, 41–43.

J. Embryol. exp. Morph. Vol. 65 (Supplement), pp. 209–224, 1981
Printed in Great Britain © Company of Biologists Limited 1981

Projections from sensory neurons developing at ectopic sites in insects

By HILARY ANDERSON[1]

From the European Molecular Biology Laboratory, Heidelberg

SUMMARY

This paper reviews recent experiments which attempt to gain more understanding about the recognition processes involved in the formation of neuronal connexions by studying the degree of specificity with which sensory neurons form their central connexions. This is done by generating ectopic neurons (either by transplantation or by genetic mutation) whose axons grow into novel regions of the central nervous system, and then examining their projections and synapses.

The sensory systems reviewed are: the *Antennapedia, spineless-aristapedia, proboscipedia,* and *bithorax* homeotic mutants of *Drosphila melanogaster*; the cercus-to-giant interneuron system of crickets, and the wind-sensitive hair system of locusts.

The results show that ectopic neurons form projections that are discrete and characteristic, not random and chaotic. In those cases where single classes of sensilla have been studied, they follow either their normal CNS pathways or those pathways normally used by their segmental homologues.

Ectopic sensory neurons can also form appropriate functional connexions in some cases but not in others. Possible reasons are discussed, but detailed understanding of the underlying events requires further experimentation.

INTRODUCTION

The nervous system is an outstanding example of spatial organization, from the level of its individual neurons each with their appropriate biochemical properties and characteristic morphologies, to the level of complex networks of many neurons with their precise patterns of synaptic connexions.

In insects the basic framework of the central nervous system (CNS) is laid down in the embryo to form a fully functioning nervous system at the time of hatching. Sensory neurons continue to be added during postembryonic life and must navigate through this complex framework to appropriate regions of the CNS and there form synapses with appropriate target neurons.

In the establishment of these precise patterns of sensory connexion it is thought that specific recognition of pre-established pathways and targets is very important. One way to discover more about these recognition processes is to define the degree of specificity shown by sensory neurons for growing along

[1] *Author's address:* European Molecular Biology Laboratory, Meyerhofstrasse 1, 6900 Heidelberg, Federal Republic of Germany.

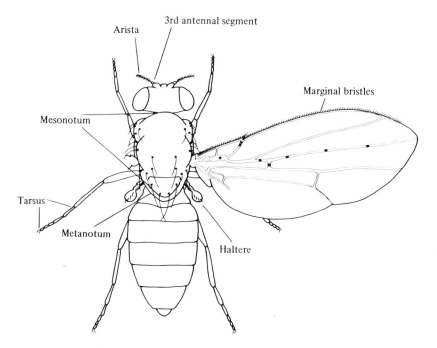

Fig. 1. Diagram of *Drosophila melanogaster* to show the location of sensilla studied in homeotic mutants. Large identifiable bristles (macrochaetes) on the mesothoracic notum are individually drawn, for further details see Ghysen (1980). The notum is also covered with many small bristles (microchaetes) which are not shown. Small bristles are shown on the haltere end-knob. Large black dots on the wing veins represent large campaniform sensilla. Small dots on the proximal wing and haltere stalk represent small campaniform sensilla.

particular pathways and for synapsing with particular central neurons. This may be done by generating ectopic neurons whose axons grow into novel regions of the CNS, and examining their resulting projections and synapses.

Recent experiments using this approach have employed extracellular and intracellular electrophysiological recording techniques, and have benefitted enormously from the development of intracellular staining techniques using horseradish peroxidase (HRP) and cobalt; the projections of groups of neurons and individual neurons can now be observed anatomically at the level of fine tertiary branches within wholemount preparations and sectioned material. In this way it has been possible to map with precision those pathways within the CNS that can be followed by different classes of sensory neurons. Three experimental systems have been studied at this new level of resolution: the homeotic mutants of *Drosphila*, the cercus-to-giant interneuron system of crickets, and the wind-sensitive hairs of locusts.

In insects, a sensory receptor (a sensillum) consists of a group of cells produced by the mitotic divisions of epidermal cells. Within the group are cells which secrete the external cuticular structures of the sensillum and one or more sensory neurons with axons extending from the periphery into the CNS. The ectopic neurons examined in these studies were generated in one of two ways. The first is by surgically grafting pieces of epidermis (either before or after the sensilla have developed) from one location to another. This method has been used in the locust *Schistocerca gregaria* and the cricket *Acheta domesticus*. An alternative method for the fruitfly *Drosophila melanogaster* is the use of homeotic mutants. In these mutants one set of body structures is replaced by another set, e.g. in the mutant *Antennapedia* mesothoracic leg structures develop in place of antennal structures. Thus sensory neurons on the body part in question are 'genetically grafted' to an ectopic site.

This review will consider in turn each of the experimental systems from the perspective of recently published work or unpublished observations.

ANTENNAPEDIA AND *SPINELESS-ARISTAPEDIA* MUTANTS OF *DROSOPHILA*

In the homeotic mutant *Antennapedia* (*Antp*), the antenna is replaced by a mesothoracic leg, whilst in *spineless-aristapedia* (*ss^a*), the distalmost segment of the antenna, the arista, and parts of the third antennal segment, are replaced by the distalmost segment of a leg, the tarsus. The location of these body parts is shown in Fig. 1. The arista contains several sensory neurons (Stocker & Lawrence, 1981) but the sensilla from which they originate have not yet been described. Scanning electron micrographs show that the normal tarsus bears contact chemoreceptors (also called taste receptors) and mechanoreceptive bristles (Green, 1979). The homeotic appendages bear several types of sensillum including the contact chemoreceptors and mechanoreceptors characteristic of the tarsus (Green, 1979) and they have a nerve which is characteristic of the mesothoracic leg nerve in number and arrangement of axons (Stocker, 1979).

Deak (1976) and Stocker (1977) have shown that stimulation with sugar solutions of the chemoreceptors on the homeotic legs regularly elicits the proboscis extension reflex, a behaviour normally elicited by stimulation of thoracic legs but not of antennae. Stimulation of a single sensory bristle is sufficient to elicit this reflex in other flies (Dethier, 1955), so a cautious conclusion is that in the mutants at least one, but not necessarily all or many, of the homeotic receptors form functional connexions within this reflex pathway.

With *Antp* flies, degeneration staining revealed that this response was not achieved by the growth of neurons from the homeotic leg to the thoracic ganglion where normal leg neurons project. In fact the homeotic neurons projected only to the antennal centres of the brain (Stocker, Edwards, Palka & Schubiger, 1976). The projection within the antennal glomeruli was somewhat different from that of a normal antenna: it was restricted to an area near the site of entry of the antennal nerve into the ipsilateral antennal glomerulus, whereas the normal antennal projection spreads to the periphery of the antennal glomeruli on both ipsilateral and contralateral sides of the brain.

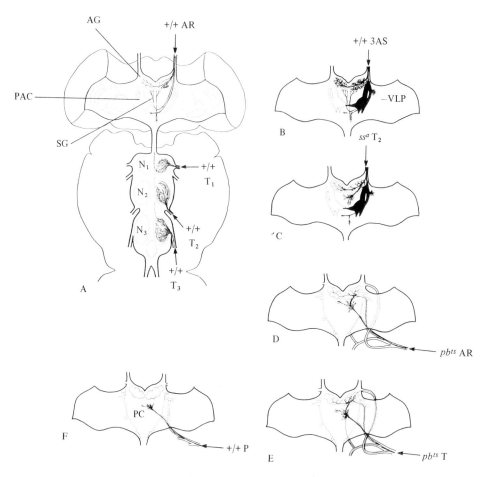

Fig. 2. Sensory projections from normal, ss^a and pb^{ts} homeotic appendages in *D. melanogaster*. (*a*) Schematic diagram of the fly brain and thoracic ganglion showing the centres to which the normal arista and the tarsi of the three thoracic legs project. (*b–f*) Schematic diagrams of the fly brain showing the projections from: (*b*) the normal third antennal segment, (*c*) the ss^a homeotic tarsus, (*d*) the pb^{ts} homeotic arista, (*e*) the pb^{ts} homeotic tarsus, (*f*) the normal proboscis. AR, arista; AG, antennal glomerulus; PAC, posterior antennal centre; SG, suboesophageal ganglion; N_1, N_2, N_3, prothoracic, mesothoracic, and metathoracic neuromeres respectively of the thoracic ganglion. T_1, T_2, T_3, prothoracic, mesothoracic and metathoracic tarsi respectively; 3AS, third antennal segment; VLP, ventrolateral protocerebrum; PC, proboscis centre of the suboesophageal ganglion (after Stocker & Lawrence, 1981; Stocker & Schorderet, 1981; Stocker, 1981).

A more detailed analysis has now been made for ss^a flies using cobalt filling to compare the projections from homeotic tarsi, normal tarsi and normal aristae (Stocker & Lawrence, 1981). The results are shown in Fig. 2. As with *Antp*, the homeotic ss^a tarsi never project to the thoracic ganglion areas used by normal tarsal neurons. They do project to normal antennal areas: ipsilateral and contralateral antennal glomeruli (but with a different pattern of terminals), ipsilateral posterior antennal centre and suboesophageal ganglion, and the ventrolateral protocerebrum. They also form a novel tract passing from the ipsilateral antennal glomerulus to the proboscis centre of the suboesophageal ganglion (Fig. 2c). Since the motor neurons innervating the proboscis are also located here (Stocker & Schorderet, 1981), it seems likely that it is this part of the homeotic projection which activates the proboscis extension reflex. However, in *Antp* flies, which also show this reflex, a projection was not detected in this region.

PROBOSCIPEDIA MUTANTS OF DROSOPHILA

In *proboscipedia* (pb^{ts}) flies, the proboscis is transformed into a tarsus if the flies are reared at 29 °C or into an arista if reared at 17 °C. The normal proboscis bears several taste bristles which can be selectively filled with cobalt. The bristles probably have a mixed mechanosensory and chemosensory function; they have the appearance of trichoid sensilla, possess terminal pores characteristic of chemoreceptors, and are innervated by three to five neurons (Stocker & Schorderet, 1981). The sensilla on the homeotic pb^{ts} tarsus and arista have not been described in detail. The projections from these appendages have been investigated by Stocker (1981), and are shown in Fig. 2.

Sensory axons from the normal proboscis project through the labial nerve to a specific region, the 'proboscis centre' of the suboesophageal ganglion (Stocker & Schorderet, 1981). Axons from homeotic tarsi and homeotic aristae that replace the proboscis may take the labial nerve and/or certain other nerves to the CNS. The central projection pattern is generally not affected by the site of entrance into the brain, although central tracts may be followed in opposite directions. Fibres reach the proboscis centre but do not go to that part of the suboesophageal ganglion normally innervated by the arista, nor to the tarsal regions of the thoracic ganglion. In this respect the neurons from homeotic tarsi and aristae behave like proboscis neurons. However, unlike proboscis neurons they also project into the antennal glomerulus where they terminate in a pattern not like that of a normal arista, but like that of a homeotic ss^a tarsus. It would be interesting to know whether these various appendages have some sensilla in common and if these form the common elements of the projections.

BITHORAX MUTANTS OF DROSOPHILA

Several studies on the *bithorax* mutants have examined the projections from identified receptors into different neuromeres of the thoracic ganglion (Ghysen, 1978; Palka, Lawrence & Hart, 1979; Ghysen, 1980; Ghysen & Janson, 1980; Palka & Schubiger, 1980; Strausfeld & Singh, 1980).

In normal flies the mesothoracic notum bears a characteristic pattern of innervated bristles, and the mesothoracic dorsal appendage – the wing – bears 4–5 large campaniform sensilla on the third vein, several rows of bristles on the

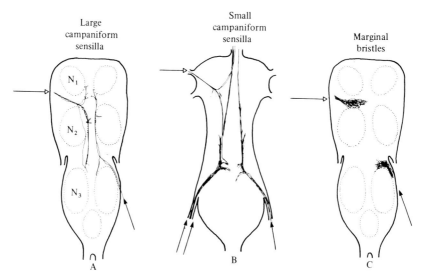

Fig. 3. Diagrams of thoracic ganglia of bx^3 pbx flies showing the ventral (a, c) and dorsal (b) sensory projections from: (a) large campaniform sensilla of the normal wing (open arrow) and homeotic wing (closed arrow); (b) small campaniform sensilla of the normal wing (open arrow), homeotic wing (closed arrow) and normal haltere (double arrow); (c) marginal bristles of the normal wing (open arrow) and homeotic wing (closed arrow). N_1, N_2, N_3, prothoracic, mesothoracic and metathoracic neuromeres respectively of the thoracic ganglion (after Palka *et al.* 1979; Ghysen & Jansen, 1980).

margin of the wing, and many small campaniform sensilla near the base of the wing (Fig. 1). The metathoracic notum is very small and devoid of bristles. The metathoracic dorsal appendage – the haltere – has no large campaniform sensilla, but does have scattered bristles on the end-knob and many small campaniform sensilla on the stalk (Fig. 1). The end-knob and stalk are considered to be homologous, respectively, to the blade and the base of the wing (Ouweneel, 1973; Morata, 1975; Morata & Garcia-Bellido, 1976). Each class of sensillum has a characteristic projection within the CNS (Ghysen, 1978; Palka *et al.* 1979), those of receptors in homologous regions being very similar (Ghysen, 1978, 1980).

The homeotic mutant *bithorax postbithorax* (bx^3 pbx) has metathoracic structures transformed into mesothoracic structures. The transformation is not always complete and may result in sensilla which appear untransformed or intermediate between wing and haltere type (Palka *et al.* 1979). The sensilla on the homeotic notum are indistinguishable from those on the normal notum (Ghysen, 1978). On the homeotic wing, the marginal bristles and large campaniform sensilla look wing-like, but the small campaniform sensilla show a range of morphologies from wing-like to haltere-like (Cole & Palka, 1980; Palka & Schubiger, 1980).

Projections from specific bristles on the normal mesothorax and homeotic mesothorax of *bx³ pbx* flies were studied by filling with HRP (Ghysen, 1978). The projections from the two pairs of scutellar bristles of the notum, from the many microchaetes covering the notum, and from the large campaniform sensilla of the third vein of the wing, all showed the same result: although the neurons entered the CNS posteriorly through a metathoracic nerve, they formed projections like those of normal mesothoracic receptors developing *in situ* and entering the CNS more anteriorly through mesothoracic nerves. Fig. 3*a* illustrates this point for the large campaniform sensilla projection. Ghysen proposed that there are specific 'trails' within the CNS which may be recognized by appropriate axons at any point of encounter and then followed in either direction (Ghysen, 1978).

Palka *et al.* (1979) did not fill specific receptors in the mutants but filled entire projections from the homeotic wings, thus including projections from the marginal bristles and small campaniform sensilla, i.e. those sensilla which may have homologues on the metathoracic haltere. The projections from the homeotic wings had some elements that were unlike anything found in normal projections of wing or haltere but also had three prominent and consistently recognizable components. One component looked just like that formed by the large campaniform sensilla of the wing. Palka *et al.* (1979) concluded, as did Ghysen from his specific fills of large campaniform sensilla, that these receptors can form a normal projection within the CNS even when they enter the CNS at an abnormal position.

The second component was reminiscent of that produced by the marginal bristles of the normal wing, although instead of being in the mesothoracic segment of the CNS, it was in the metathoracic segment (Fig. 3*c*). Specific fills of the homeotic wing bristles have since confirmed the origin of this projection (M. Schubiger, unpublished observations). Since these bristles have been considered homologous to those on the end-knob of the haltere, and they project to the same area of the metathoracic neuromere as do end-knob bristles (Ghysen, 1978), it is difficult to define to what extent this projection is 'normal' or 'abnormal'. These neurons may be able to recognize homologous areas in the mesothoracic and metathoracic neuromeres. An alternative interpretation is that the bristle neurons are not fully transformed (the degree of transformation of the externally visible cuticular structures of the sensillum and its underlying neuron need not be the same). Evidence against this interpretation is provided by the projections from flies also bearing the mutation *wingless* (*wg*) (Palka & Schubiger, 1980). This mutation results in the replacement of the wings and halteres with mirror-image duplicates of the mesonotum and metanotum respectively. The mutation is expressed variably so that *wg* flies may either be normal in appearance, have a duplicate mesonotum, a duplicate metanotum, or both. An additional effect is that in some flies with an apparently normal haltere, the haltere nerve is misrouted so that it joins the ganglion at the level of the mesothoracic neuromere at a place appropriate for the wing nerve. When the haltere is replaced by a homeotic wing in *wg/wg*; *bx³ pbx/Ubx* flies, the homeotic nerve may also be misrouted. The projections from marginal bristles on homeotic wings, when misrouted into the mesothoracic nerve, terminate in the mesothoracic neuromere just as do those from normal wings, suggesting that the homeotic transformation is complete.

The third component of the homeotic projection looked like that of the small campaniform sensilla of the haltere (Fig. 3*b*). In this case homeotic wing sensilla appeared to be forming a haltere-like projection rather than projecting to areas appropriate to normal wing. This projection has now been further traced into the suboesophageal ganglion of the brain, where the projections of small campaniform sensilla from the wing and haltere can be more readily distinguished than in the thoracic ganglion, and confirmed to be haltere-like (Ghysen, unpublished observations). One interpretation is that the small campaniform sensilla of the wing can also recognize homologous haltere pathways. It also seems plausible that these projections arise from the incompletely transformed sensilla observed in the scanning electron micrographs. Evidence supporting this view is that in *wg* flies when these fibres are rerouted into the mesothoracic neuromere, they do not form a wing-like projection (Palka & Schubiger, 1980).

CNS EFFECTS OF HOMEOTIC MUTATIONS

An important point, discussed by all the authors of papers using *Drosophila* homeotic mutants, is the possibility that the mutations also affect the CNS. This would complicate the interpretations considerably. The metathoracic neuromere of *bithorax* flies is usually enlarged but most authors agree that the internal organization of mutant thoracic ganglia appears normal when viewed in wholemount preparations, although this offers only a low level of resolution.

Better evidence has been obtained from the use of genetic mosaics (Palka *et al.* 1979; Stocker & Lawrence, 1981; Stocker, 1981). In mosaic flies only a small patch of sensilla-bearing cuticle is homozygous for the mutation, the CNS remaining heterozygous and thus phenotypically normal. Palka *et al.* (1979) found that the projections from mosaic *bithorax* appendages differed slightly from those of wholly mutant flies, but not in any important details. Stocker & Lawrence (1981) and Stocker (1981) found no significant difference between mosaic and mutant flies in their studies of the *ssᵃ* and *pbᵗˢ* mutants.

Strausfeld & Singh (1980) examined the projections of sensory antennal fibres and certain ascending and descending interneurons in normal and mutant flies. Neurons that normally terminate in the mesothoracic neuromere did not project further to the metathoracic neuromere in *bithorax* mutants, nor was there any duplication in the metathoracic neuromere of neurons which normally develop only in the mesothoracic neuromere. They suggested that the enlargement of the metathoracic neuromere in *bithorax* flies is in part due to increased sensory input from the homeotic wing, and in part due to the corresponding increase in branching of interneurons which they observed in this region.

Green (1980) backfilled thoracic leg motor neurons with HRP. The arrangement of motor neurons in the mesothoracic and metathoracic neuromeres was quite normal in bx^3 pbx/Ubx flies, suggesting no direct genetic effect on the CNS. However, flies with another genotype, abx bx^3 pbx/Ubx, sometimes showed a strikingly different result: motor neurons in the metathoracic neuromere were now arranged like those of the mesothoracic neuromere. If this change indeed arises from a direct genetic effect (and this is being further tested in mosaic flies – S. H. Green, personal communication), it provides a powerful tool for examining the determination and development of the CNS.

THE CERCUS-TO-GIANT INTERNEURON SYSTEM OF CRICKETS

Crickets possess a pair of large posterior appendages – the cerci – that bear hundreds of sensory filiform hairs. The hairs are highly sensitive to air movements and their activity triggers the animals' escape response. The sensory neurons project to the terminal abdominal ganglion where they make synapses with several giant ascending interneurons. The regeneration of the cercus-to-giant interneuron system has been studied after transplantation of the cercus to the mesothorax (Edwards & Sahota, 1967).

Cerci were transplanted to the stump of a severed mesothoracic leg where they regenerated their sensory projections to the CNS. Silver-stained sections of the mesothoracic ganglion showed the neurons entering the CNS via the leg nerve (nerve 5) and passing in the vicinity of a major bundle of neurons that included the giant interneurons. Stimulation of the grafted cercus with air puffs produced a burst of activity, recorded extracellularly, in the anterior mesothoracic connective which was characteristic of giant interneuron activity. This activity was observed in the presence of normal posterior cerci and also after isolation of the thoracic ganglia, i.e. in the absence of the normal input region in the terminal ganglion. These results suggested that cercal afferents can recognize a target neuron at a point other than the normal region of input and this can occur even when normal cercal inputs are present (Edwards & Sahota, 1967).

THE WIND-SENSITIVE HEAD-HAIR SYSTEM OF LOCUSTS

The head of the locust bears a population of wind-sensitive hairs whose inputs contribute to the initiation, maintenance, and control of flight. The hairs are grouped into five fields (F1–F5) on each side of the head (Fig. 4a). Neurons associated with hairs in each of the fields form characteristic projections within the CNS as a function of their position on the head (Tyrer, Bacon & Davies, 1979). Examples of projections from F1 and F2 on top of the head and F3 on the side of the head are shown in Figs. 4(b, c); F1 and F2 neurons form distinctive contralateral branches in the suboesophageal ganglion and prothoracic ganglion, while F3 neurons form only an ipsilateral projection in these ganglia.

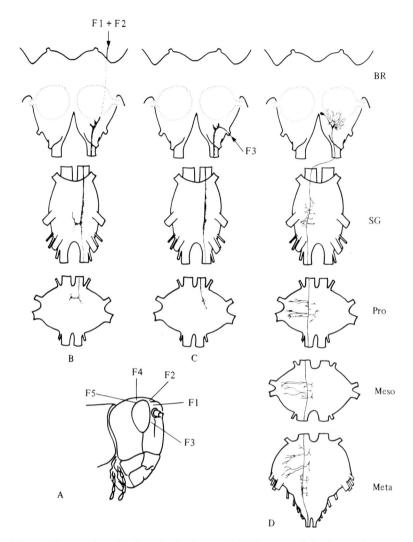

Fig. 4. Neuronal projections in the locust. (a) Diagram of the locust head to show the location of the five fields of head-hairs, F1–F5. (b) Sensory projections from F1 and F2 head-hairs. (c) Sensory projections from F3 head-hairs. (d) Projection of the TCG interneuron. BR, brain; SG, suboesophageal ganglion; Pro, prothoracic ganglion; Meso, mesothoracic ganglion; Meta, metathoracic ganglion.

Experiments in which epidermis was transplanted between different positions on the head showed that the type of anatomical projection formed was determined by the origin of the graft epidermis and not by the location of the hairs on the head at the time of their differentiation (Anderson & Bacon, 1979).

   The wind-sensitive neurons make connexions with many interneurons in the CNS. One interneuron, the Tritocerebral Commissure Giant (TCG) has been studied in some detail. Its morphology is shown in Fig. 4 (d). Extracellular

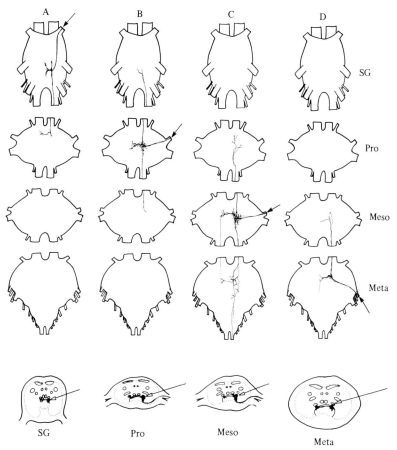

Fig. 5. Sensory projections from F1 neurons developing at 4 different ectopic locations: the posterior head (*a*), the prothorax (*b*), the mesothorax (*c*), and the metathorax (*d*). Closed arrows indicate the site of entry of the neurons into the CNS. At the foot of each column is a diagram of a transverse section through the ganglion which the neurons first enter. Within each section the neuropil is delimited by a dotted line, the major longitudinal tracts are indicated, the ventral neuropil area is cross-hatched, and the Median Ventral Tract is indicated by an open arrow. SG, suboesophageal ganglion; Pro, prothoracic ganglion; Meso, mesothoracic ganglion; Meta, metathoracic ganglion.

recording of the activity of this neuron has shown that, in the brain, it receives excitatory input from all ipsilateral fields except F 3 (Bacon & Tyrer, 1978). A more detailed intracellular study has now shown that inputs from F 3 hairs strongly inhibit the TCG (Bacon & Anderson, 1981).

Exchange of epidermis between F 2 and F 3 has shown that the polarity of the connexions formed between wind-sensitive hairs and the TCG also depends upon the origin of the epidermis and not the final location of the differentiated hairs on the head (Bacon & Anderson, 1981).

To investigate further the degree of specificity with which wind-sensitive hairs form particular projections and connexions, similar epidermal grafts were made to other body segments (the posterior head, prothorax, mesothorax, and metathorax) to cause the sensory neurons developing from the different grafts to enter the CNS at different levels (Fig. 5; Anderson, 1981 a).

The projections in the CNS were stained by cobalt backfilling. They were found to be remarkably conservative; in all cases the neurons passed directly to the ventral neuropil where they formed a discrete projection (Fig. 5). Indeed, it has not yet proved possible to misdirect these neurons into a dorsal location (unpublished observations). In those cases where the sensory neurons also formed ascending and/or descending branches to other ganglia, they always did so within a single tract, the Median Ventral Tract (Tyrer & Gregory, 1981), which runs in a similar position in each ganglion (Fig. 5). The same projection area in the ventral neuropil and the same tract are also used by sensory neurons from normal wind-sensitive hairs and other mechanoreceptive hairs of the body surface (Anderson, 1981 b).

Another feature of the projections was also conservative. F 3 neurons formed only an ipsilateral projection in all ganglia, whereas F 1 or F 2 neurons formed additional contralateral branches, although the projections of F 1 and F 2 neurons in ganglia distant from the ganglion of entry were sometimes only ipsilateral (e.g. compare the projections in the mesothoracic and prothoracic ganglia from F 1 neurons developing on the mesothorax, Fig. 5(c)).

The only variability shown by the projections was in their extent. Grafts to the posterior head that entered the suboesophageal ganglion all showed the same projection as in Fig. 5(a). There was no variation between individual preparations. Grafts to the prothorax showed some variability. Some examples projected up to the suboesophageal and down to the mesothoracic ganglion (Fig. 5b) but many formed only descending or ascending branches or were entirely local. Even more variation occurred in projections from grafts to the mesothorax and metathorax, where all combinations of local, ascending and descending components were observed. Fig. 5(c, d) show two of the more extensive projections. In contrast, the projections from surrounding host hairs were constant in extent from animal to animal. In all grafts the neurons never extended more anteriorly than the suboesophageal ganglion (Fig. 5).

Extracellular recording of the TCG failed to reveal any excitatory or inhibitory input to the TCG from the graft neurons, even though the TCG passes through these ganglia.

DISCUSSION

Choice of pathways within 'foreign' ganglia

Ectopic neurons form projections that are discrete and characteristic, not random and chaotic. The pathways taken are not merely a result of following surrounding host neurons since the host and graft projections often differ in clear and precise ways.

In those cases where the CNS route used by the normal sensilla is present, it is always taken, even if entered at an abnormal point, and it may be followed in either direction. Where the normal route is absent, but a homologous one is present, then this is taken.

These simple statements cannot yet be applied to the *Antp*, *ss*ᵃ and *pb*ᵗˢ systems where the complex organization of the brain makes it difficult to assign homologies to particular pathways and tracts. Analysis is also complicated by the fact that it is not known which sensillum types are present on all the homeotic appendages, e.g. mechanoreceptors, olfactory receptors, taste receptors, or multimodal receptors, nor which components they each contribute to the observed projections. The incomplete transformation of campaniform sensilla seen in *bithorax* flies indicates that it may not be sufficient to assume, for example, that a homeotic tarsus bears only 'tarsal' receptors.

Recognition of target neurons

Ectopic tarsal chemoreceptors in *Antp* and *ss*ᵃ flies form functional connexions within the proboscis extension reflex pathway. Whilst cobalt-filling of projections from the homeotic tarsi has provided details of the pathways taken within the brain, it has not revealed the region in which the reflex is mediated since it has not been possible to fill specifically the tarsal chemoreceptors. A behavioural comparison with homeotic *pb*ᵗˢ tarsi might be illuminating since they have only two projection areas in common – the antennal glomerulus, and the proboscis centre. Trans-neuronal filling from normal tarsal neurons in the thoracic ganglion might also help with the identification of interneurons in the normal reflex pathway. One could then ask, can ectopic sensory neurons recognize different neurons in the pathway or only the normal interneuron but at a different location?

In crickets, sensory neurons on ectopic cerci also succeed in forming functional connexions with appropriate interneurons. Again, the neural pathways underlying this response have not yet been elucidated, but since these early experiments, considerable advances have been made in understanding the details of the normal cercus-to-giant interneuron system. Two major classes of filiform hairs are now recognized: the dorsally and ventrally located T-hairs whose hair shafts vibrate transversely to the long axis of the cercus, and the medially and laterally located L-hairs whose shafts vibrate longitudinally (Palka, Levine & Schubiger, 1977). Four interneurons have been identified and their responses to T- and L-hair stimulation examined both extracellularly and intracellularly (Palka & Olberg, 1977; Matsumoto & Murphey, 1977; Levine & Murphey, 1980): each has a specific pattern of excitatory and inhibitory inputs from ipsilateral and contralateral T- and L-hairs. It seems likely that the excitatory pathways are monosynaptic, but the inhibitory pathways involve other interneurons (Levine & Murphey, 1980).

These details of interconnexion raise interesting questions about the earlier cercal transplantation study: 1. How are the regenerating afferents distributed in the mesothoracic ganglion? Do they follow a limited number of pathways? If so, are they homologous to those normally taken in the terminal ganglion? 2. What proportion of afferents regenerate connexions with these ascending interneurons? What proportion form connexions with other inappropriate interneurons? 3. Are the regenerated connexions of the appropriate sign and strength for the source of input? 4. If inhibitory responses are observed, are they mediated by interneurons which are homologous to those in the terminal ganglion? These questions are for future investigation.

Another promising system to study at this level might be the equilibrium detecting system of the cricket. At the base of each cercus is an orderly arrangement of clavate hairs that act as equilibrium detectors (Bischof, 1975). Individual clavate hairs have been backfilled and their CNS projections described in detail; they form a topographic arrangement of branches within a particular region of neuropil in the terminal ganglion, where the termination site of each afferent is related to its position and/or birthdate along and around the cercus (Murphey, Jacklet & Schuster, 1980). Furthermore, two pairs of large bilaterally symmetrical ascending interneurons receiving clavate hair input have been described (Sakaguchi & Murphey, 1980). This system is now being examined following transplantation of cerci to the mesothorax (R. K. Murphey, personal communication).

In contrast to the systems described earlier in this section, ectopic wind-sensitive hairs of the locust fail to form functional connexions with one of their appropriate interneurons, the TCG. What might account for this difference? One reason could be that only parts of an interneuron may be recognized as input regions. Those of the TCG might be restricted to its arborization in the tritocerebrum where it normally receives head-hair input. Since the ectopic hair neurons never ascend more anteriorly than the suboesophageal ganglion (Fig. 5), they would not encounter this input region. On the other hand, the definitive tests for monosynapticity of the pathway between head-hairs and TCG have not been done, so it remains possible that local interneurons might be required to mediate the input. These interneurons, or their homologues, might be absent in other ganglia, thus preventing the formation of a functional connexion. It also seems likely that sensory neurons require close proximity to an interneuron before recognition can occur. Since the TCG passes through a dorsal tract, the Dorsal Intermediate Tract, and has a dorsal arborization in all of the ventral cord ganglia (Bacon & Tyrer, 1978), while the sensory neurons have an exclusively ventral projection, the two may never come close enough for recognition. The cricket giant ascending interneurons on the other hand have some ventral branches, and run in a ventral tract, the Ventral Intermediate Tract (J. L. D.

Williams, personal communication), and may be seen in close proximity to the sensory fibres from grafted cerci (Edwards & Sahota, 1967).

The picture presented here is derived only from observation of the end-result following a perturbation. It is possible that these projections are far more extensive during their establishment, but such a dynamic view remains to be obtained.

I thank Drs Duncan Byers, Steven Green, Brian Mulloney, Dick Nassel, Margrit Schubiger and Rheinhard Stocker for valuable comments on the manuscript.

REFERENCES

ANDERSON, H. (1981 a). Development of a sensory system: the formation of central projections and connections by ectopic wind-sensitive hairs in the locust *Schistocerca gregaria* (in prep.).

ANDERSON, H. (1981 b). The organization of mechanosensory inputs within the locust central nervous system (in prep.).

ANDERSON, H. & BACON, J. (1979). Developmental determination of neuronal projection patterns from wind-sensitive hairs in the locust, *Schistocerca gregaria. Devl Biol.* **72**, 364–373.

BACON, J. & ANDERSON, H. (1981). Developmental determination of central connections from wind-sensitive hairs in the locust, *Schistocerca gregaria* (in prep.).

BACON, J. P. & TYRER, N. M. (1978). The tritocerebral commissure giant (TCG): a bimodal interneurone in the locust. *J. comp. Physiol.* **126**, 317–325.

BISCHOF, H. J. (1975). Die keulenformigen Sensillen auf den Cerci der Grille *Gryllus bimaculatus* als Schrewe receptoren. *J. comp. Physiol.* **98**, 277–288.

COLE, E. S. & PALKA, J. (1980). Sensilla on normal and homeotic wings of *Drosophila. Am. Zool.* **20**, 740.

DEAK, I. I. (1976). Demonstration of sensory neurons in the ectopic cuticle of *spineless-aristapedia*, a homeotic mutant of *Drosophila. Nature* **260**, 252–254.

DETHIER, V. G. (1955). The physiology and histology of the contact chemoreceptors of the blowfly. *Q. Rev. Biol.* **30**, 348–371.

EDWARDS, J. S. & SAHOTA, T. S. (1967). Regeneration of a sensory system: the formation of central connections by normal and transplanted cerci of the house cricket, *Acheta domesticus. J. exp. Zool.* **166**, 387–396.

GHYSEN, A. (1978). Sensory neurones recognize defined pathways in *Drosophila* central nervous system. *Nature* **274**, 869–872.

GHYSEN, A. (1980). The projection of sensory neurons in the central nervous system of *Drosophila*: choice of the appropriate pathway. *Devl Biol.* **78**, 521–541.

GHYSEN, A. & JANSON, R. (1980). Sensory pathways in *Drosophila* central nervous system. In *Development and Neurobiology of Drosophila* (eds O. Siddiqi, P. Babu, L. M. Hall & J C. Hall), pp. 247–265. New York: Plenum.

GREEN, S. H. (1979). Neuroanatomy of flies with abnormal appendages. *Rep. Div. Biol. Calif. Instit.*

GREEN, S. H. (1980). Innervation of antennae and legs in *Drosophila*: wild-type and homeotic mutants. *Soc. Neurosci. Abstr.* **6**, 677.

LEVINE, R. B. & MURPHEY, R. K. (1980). Pre- and postsynaptic inhibition of identified giant interneurons in the cricket (*Acheta domesticus*). *J. comp. Physiol.* **135**, 269–282.

MATSUMOTO, S. G. & MURPHEY, R. K. (1977). The cercus-to-giant interneuron system of crickets. IV. Patterns of connectivity between receptors and the Medial Giant Interneuron. *J. comp. Physiol.* **119**, 319–330.

MORATA, G. (1975). Analysis of gene expression during development in the homeotic mutant *Contrabithorax* of *Drosophila melanogaster. J. Embryol. exp. Morph.* **34**, 19–31.

Morata, G. & Garcia-Bellido, A. (1976). Developmental analysis of some mutants of the *bithorax* system of *Drosophila*. *Wilhelm Roux Arch. devl Biol.* **179**, 125–143.

Murphey, R. K., Jacklet, A. & Schuster, L. (1980). A topographic map of sensory cell terminal arborizations in the cricket CNS: correlation with birthday and position in a sensory array. *J. comp. Neurol.* **191**, 53–64.

Ouweneel, W. J. (1973). Replacement patterns of homoeotic wing structures in halteres. *Dros. Inf. Serv.* **50**, 102.

Palka, J., Lawrence, P. A. & Hart, H. S. (1979). Neural projection patterns from homeotic tissue of *Drosophila* studied in *bithorax* mutants and mosaics. *Devl Biol.* **69**, 549–575.

Palka, J., Levine, R. & Schubiger, M. (1977). The cercus-to-giant interneuron system of crickets. I. Some attributes of the sensory cells. *J. comp. Physiol.* **119**, 267–283.

Palka, J. & Olberg, R. (1977). The cercus-to-giant interneuron system of crickets. III. Receptive field organization. *J. comp. Physiol.* **119**, 301–317.

Palka, J. & Schubiger, M. (1980). Formation of central patterns by receptor cell axons. In *Development and Neurobiology of Drosophila* (eds O. Siddiqi, P. Babu, L. M. Hall & J. C. Hall), pp. 223–246. New York: Plenum.

Sakaguchi, D. S. & Murphey, R. K. (1980). Re-establishment of functional neuronal connections following transplantation of a sensory appendage in the equilibrium detecting system of the cricket. *Soc. Neurosci. Abs.* **6**, 678.

Stocker, R. F. (1977). Gustatory stimulation of a homeotic mutant appendage, *Antennapedia*, in *Drosophila melanogaster*. *J. comp. Physiol.* **115**, 351–361.

Stocker, R. F. (1979). Fine structural comparison of the antennal nerve in the homeotic mutant *Antennapedia* with the wild-type antennal and second leg nerves of *Drosophila melanogaster*. *J. Morphol.* **160**, 209–222.

Stocker, R. F. (1981). Homeotically displaced sensory neurons in the proboscis and antenna of *Drosophila* project into the same identified brain regions though by different pathways (in prep.).

Stocker, R. F., Edwards, J. S., Palka, J. & Schubiger, G. (1976). Projection of sensory neurons from a homeotic mutant appendage, *Antennapedia*, in *Drosophila melanogaster*. *Devl Biol.* **52**, 210–220.

Stocker, R. F. & Lawrence, P. A. (1981). Sensory projections from normal and homeotically transformed antennae in *Drosphila*. *Devl Biol.* **82**, 224–237.

Stocker, R. F. & Schorderet, M. (1981). Cobalt filling of sensory projections from internal and external mouthparts in *Drosophila*. *Cell Tiss. Res.* **216**, 513–523.

Strausfeld, N. J. & Singh, R. N. (1980). Peripheral and central nervous system projections in normal and mutant (*bithorax*) *Drosophila melanogaster*. In *Development and Neurobiology of Drosophila* (eds O. Siddiqi, P. Babu, L. M. Hall & J. C. Hall), pp. 267–291. New York: Plenum.

Tyrer, N. M., Bacon, J. P. & Davies, C. A. (1979). Sensory projections from the wind-sensitive head hairs of the locust *Schistocerca gregaria*. *Cell Tiss. Res.* **203**, 79–92.

Tyrer, N. M. & Gregory, G. E. (1981). A guide to the neuroanatomy of locust suboesophageal and thoracic ganglia. *Proc. Roy. Soc. Lond.* **B** (in press).

J. Embryol. exp. Morph. Vol. 65 (Supplement), pp. 225–241, 1981
Printed in Great Britain © Company of Biologists Limited 1981

Growth and development of pattern in the cranial neural epithelium of rat embryos during neurulation

By GILLIAN M. MORRISS-KAY[1]

From the Department of Human Anatomy, Oxford

SUMMARY

The pattern of growth and morphogenesis of the cranial neural epithelium of rat embryos during neurulation is described. Transverse sections of the midbrain/hindbrain neural epithelium at different stages (0–14 somites) show a constant area and cell number throughout neurulation, even though there is a high level of mitosis. Mitotic spindles are orientated parallel to the long axis of the embryo, so that increase in cell number occurs in this direction only. Growth is expressed only as an increase in size of the forebrain, which projects rostrad to the tip of the notochord.

In the midbrain/upper hindbrain regions, cellular organization of the neural epithelium changes from columnar to cuboidal to pseudostratified, while its shape changes from flat to biconvex to V shaped. Closure is immediately preceded by neural crest cell emigration from the lateral edges. Throughout neurulation the cranial notochord develops an increasingly convex curvature in the rostrocaudal plane. The attached neural epithelium curves with the notochord (forming the primary cranial flexure) so that as its lateral edges move dorso-medially they form a more distant concentric arc with that of the notochord, and are hence stretched during the final closure period.

The whole rat embryo culture technique was used to investigate the morphogenetic role of proteoglycans during neurulation, neural crest cell emigration and other events in the lateral edge region prior to closure, and the importance of microfilament contraction during concave curvature of the neural epithelium.

I. INTRODUCTION

The process of neurulation has been most thoroughly studied in amphibian embryos, and in the absence of much direct information on neurulation in mammalian embryos the information gained from amphibian studies is frequently assumed to be relevant here also. Studies on neural induction have not yet been carried out in mammalian embryos, but studies on morphogenesis of the neural epithelium from neural plate to closed neural tube are beginning to reveal that differences between the brains of species belonging to different vertebrate classes are evident from the onset of neurulation.

Figure 1 shows comparable transverse sections of the cranial neural

[1] *Author's address:* Department of Human Anatomy, South Parks Road, Oxford OX1 3QX, U.K.

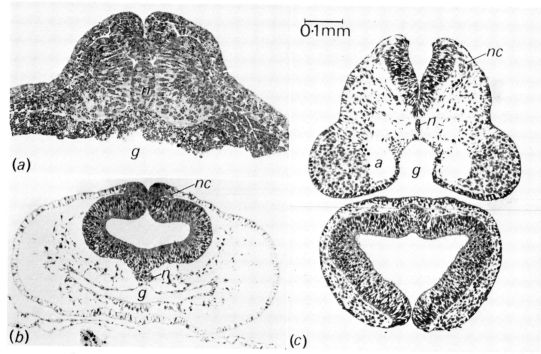

Fig. 1. Transverse sections of embryos at the time of cranial neural fold apposition. (a) *Xenopus* (stage 16), (b) chick (7-somite stage), (c) rat (13-somite stage). In the rat, the neural tube is sectioned at both forebrain (lower) and upper-hindbrain levels. *a*, 1st aortic arch; *g*, archenteron, foregut and buccal region in (a)–(c) respectively; *n*, notochord; *nc*, neural crest.

epithelium of *Xenopus*, chick and rat embryos at or just prior to the time of apposition of the neural folds. Many differences can be seen, including: neuro-epithelial cell size, appearance, numbers and organization; notochordal size and form; timing of emigration of neural crest cells; overall shape, including the existence of a cranial flexure at this stage in the mammal only. These differences reflect not only the great differences of adult structure and function, but also the lack of close phylogenetic affinities. Modern amphibians are very distant relatives of the line which evolved into the first amniotes (cotylosaurs or stem reptiles), and the pre-mammalian and pre-avian reptile lines diverged very early in the radiation of reptilian types.

This article will examine in some detail the cranial neural epithelium and associated structures of the rat embryo during neurulation. Its aim is to establish a proper descriptive basis (a) for comparison with other vertebrate (including other mammalian) embryos and (b) for experimental studies aimed at elucidating the mechanisms underlying normal and abnormal morphogenesis during development of the mammalian brain tube.

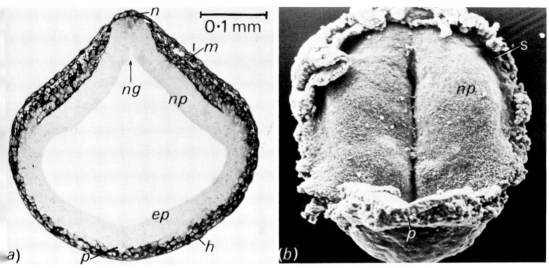

Fig. 2. Pre-somite-stage rat embryos, with flat neural plate. (*a*) One μm-thick section of embryo embedded in Spurr resin following fixation with glutaraldehyde (2·5 %) containing cetyl pyridinium chloride (100 mg/ml) and postfixation with OsO₄ containing ruthenium red (60 mg/ml). (*b*) SEM of embryo partially opened out by cutting the lateral extra-embryonic regions. *ep*, epiblast; *h*, hypoblast; *m*, primary mesenchyme; *n*, notochord; *np*, neural plate; *ng*, neural groove; *p*, primitive streak; *s*, presumptive surface ectoderm.

II. GROWTH

(*a*) *Growth of the whole embryo*

At the beginning of neurulation, the rat embryo is a small cup-shaped structure attached to a cylindrical yolk sac, both having outward-facing visceral endoderm. The amnion and chorion complete formation at this stage, and the allantois begins to develop. The neural plate is a flat, shield-shaped structure anterior to the primitive streak (Fig. 2); if the cup-shaped embryo were to be flattened out by expansion of the sides and straightening of the longitudinal axis, it would resemble a human or any other non-rodent mammalian embryo of the same stage. The embryo and its yolk sac are separated from the surrounding maternal blood sinus only by Reichert's membrane and its sparsely attached parietal endoderm cells. A continuous supply of nutrients is thus freely available to the developing embryo; proteins are broken down within the yolk-sac endoderm cells (Beck, Lloyd & Griffiths, 1967). This free availability of nutrients is reflected in considerable growth. Figure 3 shows the amount of protein in rat embryos (dissected free of their membranes) in relation to somite number throughout the period of cranial neurulation. It is not possible to relate these measurements directly to a timescale, since the rate of addition of new somites is not constant (Tam, 1981). The whole period involved was 19 h: 4 p.m. on day 9 to 11 a.m. on day 10 (day of sperm-positive vaginal smear = day 0).

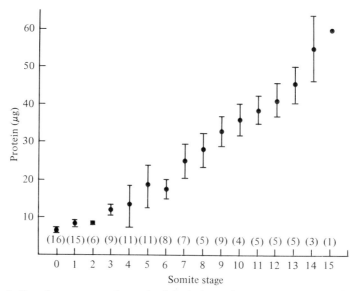

Fig. 3. Protein content, ± 1 standard deviation, of rat embryos during the period of cranial neurulation. Numbers in parentheses above the somite stage indicate the number of embryos measured. For early stages some samples contained two or three embryos: no. of samples, 0 to 7-somite stages = 3, 3, 2, 4, 6, 9, 7 and 6 respectively. Technique according to Lowry, Rosebrough, Farr & Randall (1951), using bovine serum albumen standards.

A very small amount of this increase in embryonic protein content is presumably due to incorporation of part of the yolk sac into the foregut and hindgut, which form during this period. The great majority of it, however, is real growth. There is a continuous lengthening of the body axis, and an increase in size of both head and heart.

(b) Growth of the neural epithelium

There is as yet no exact calculation on the amount of growth which occurs in the neural epithelium of amniote embryos during neurulation. However, in both the chick (Derrick, 1937) and the rat, there is a high level of mitosis in the neural epithelium, suggesting that growth of this tissue during neurulation is the rule in embryos which have a continuous supply of nutrients during this period. This is in contrast to amphibian embryos which have no extrinsic food source until hatching (stage 40 in the newt). During neurulation in *Ambystoma*, the cell number increases very little, from 113000 to 139000, from stages 13 to 19 (Gillette, 1944). In the newt *Taricha torosa*, the volume of the central nervous system does not increase from stage 13 to stage 33, i.e. during and for a considerable time after neurulation (A. G. Jacobson, 1978).

Projection drawings of transverse sections of the rat cranial neural epithelium

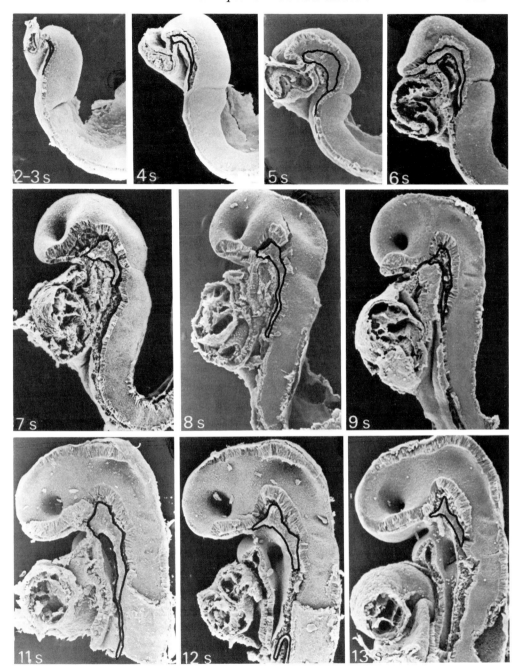

Fig. 4. SEM of cranial regions of sagittally halved embryos. Somite stages as indicated. The notochord (or, in the 9-somite specimen, its position) has been outlined in black ink to indicate its changing shape during cranial neurulation. These preparations indicate the relationship between changes in shape and position of the foregut, the growing heart and the cranial flexure, and illustrate the dramatic increase in size of the forebrain region.

at various stages (pre-somite to 14 somites) reveal that except in the developing forebrain region, the cross-sectional area remains constant throughout this period. Cell counts from scanning electron micrographs of transversely cut neural folds at midbrain or upper hindbrain levels show that the cell number in each half of the cut neuroepithelial surface is 65–70 throughout neurulation. Counts from comparable sections of plastic-embedded embryos are similar. The length of the supranotochordal neural epithelium from the preotic sulcus to the tip of the notochord also remains constant throughout cranial neurulation. This was measured on the specimens illustrated in Fig. 4, using a piece of resin-cored solder. (The measurement was 220 μm in all specimens, but due to shrinkage during preparation the actual length is likely to be slightly greater than this.)

This observed constancy of cross-sectional surface area, cross-sectional cell number, and supranotochordal neuroepithelial length is a surprising finding in view of the high level of mitosis observed. Examination of the spindle orientation in embryos cultured in medium containing colcemid (0·1 μg/ml) and in untreated embryos revealed that in the midbrain/upper hindbrain region, the orientation was almost always parallel to the long axis of the embryo, so that each mitosis contributed only to longitudinal growth; conversely, in the forebrain region the spindle orientation was predominantly in the transverse plane, allowing the great expansion which occurs in this region during neurulation (F. Tuckett, unpublished observations). Growth of the cranial neural epithelium therefore appears to be confined to the forebrain region, even though cell division occurs in all regions.

III. DEVELOPMENT OF PATTERN IN THE NEURAL EPITHELIUM

(a) Shape

The overall changes in shape seen from surface views of the cranial neural folds during neurulation have been described and illustrated elsewhere (Morriss & Solursh, 1978b; Morriss & New, 1979). They may be summarized as follows, together with the internal features illustrated in Figs. 2, 4 and 5. The neural groove is present as a midline region of cells closely apposed to the underlying notochord at pre-somite stages, when the neural plate is flat and the notochord lies within the endodermal cell layer (Fig. 2a). This close relationship of notochord and neural epithelium persists throughout cranial neurulation (not well illustrated in Fig. 4 because of shrinkage during critical-point drying). Changes in shape of the neural epithelium as a whole involve first, convex neural fold formation; then (except in the developing forebrain) flattening, followed by development of a concave curvature. At the 8- to 10-somite stage, neural crest cell emigration from the most lateral neural epithelium of the myelencephalon, then met- and mesencephalon, immediately precedes the development of a sharp mediad curvature of the consequently thinned lateral region (see section IIIb).

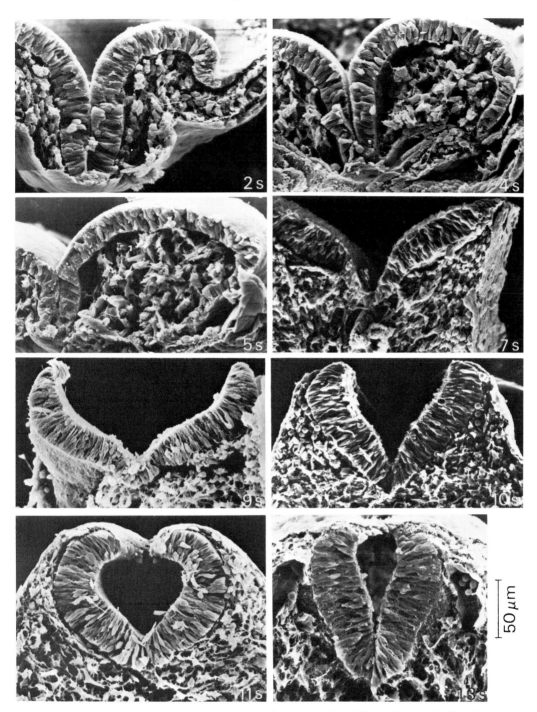

Fig. 5. SEM of embryos cut transversely across the midbrain/upper hindbrain region. Somite stages as indicated. (Four of these specimens have been illustrated in previous publications: Morriss & Solursh, 1978a, b; Morriss & New, 1979.)

Neural fold fusion begins in the upper cervical region at 7 somites, extending upwards into the hindbrain at 8 and 9 somites. At the early 10-somite stage a separate region of fusion begins, at the midbrain/forebrain junction, immediately above the developing cranial flexure. Final closure of the spindle-shaped opening between these two fusion areas is complete by 14 somites, and the apposed forebrain region (anterior neuropore) fuses soon afterwards. (This separate closure of two cranial neuropores also occurs in human embryos: see 10-somite embryo illustration, Hamilton, Boyd & Mossman (1972), p. 179. It is more pronounced in rodents, perhaps because of the more precocious development of the primary cranial flexure.)

Embryos bisected sagittally (Fig. 4) show very clearly that while the supra-notochordal neural epithelium of the midbrain/upper hindbrain region does not change in length during the 2- to 13-somite period, there is an enormous forwards extension of the anterior forebrain area, which increases greatly in size. The relative shapes and sizes of the forebrain and midbrain/hindbrain regions towards the end of cranial neurulation can be appreciated by reference to Fig. 1(c).

In summary, the observations on changes in shape and size of the forebrain and midbrain/upper hindbrain regions, together with the observations on spindle orientation during mitosis, imply that there is a net forward movement of cells towards and into the developing forebrain. Mitoses in the midbrain/hindbrain region contribute to this supply of forward-moving cells, while mitoses within the forebrain region contribute an intrinsic component to its growth.

A. G. Jacobson (1978, 1980) has assessed the degree to which the neural epithelium and notochord lengthen during neurulation in newt and chick embryos, using computer modelling. He considered that in newt embryos elongation provides a 'stretch force' which, together with microfilament-induced shrinkage of the neural plate, is responsible for neurulation. This theory is also applicable to his observations on stage-9 (9-somite) chick embryos, where elongation during a 4 h period was found to be 3–4 % in the closed region, 7 % in the open region, and 29 % in the closing region of the neural tube. Measurement of the length of the supranotochordal cranial neural epithelium on scanning electron micrographs of sagittally halved rat embryos (Fig. 4) indicates that there is *no* elongation of the cranial notochord and attached neural epithelium of the midbrain/hindbrain regions during neurulation (section IIb). However, because of the continuous development of the cranial flexure while the lateral edges of the neural epithelium are moving dorsomedially, the lateral edges come to form an arc more distant from, but concentric with, the curved notochord. The resultant elongation of the dorsal surface of the midbrain/hindbrain region may be an important contributory factor in generating the dorsomediad movement of the lateral edges of the neural folds here in later stages of neurulation.

(b) *Organization of the neural epithelium and adjacent tissues*

At the pre-somite (flat neural plate) stage, the neuroepithelial cells are columnar in form, except in the neural groove region, where those cells which are in contact with the notochord are flask-shaped; the notochord itself appears as a thickened region of endoderm at this stage (Fig. 2).

The newly formed cranial neural plate cannot be regarded as a thickened region of cells, since it is similar in appearance to the epiblast of embryos 24 h younger (Morriss, 1973) and to its own adjacent epiblast (Fig. 2). Rather, the edges of the neural plate become clearly defined as the adjacent presumptive surface ectodermal cells become thinned (0- to 5-somite stages, Fig. 4). The early neural plate is underlaid by a layer of primary mesenchyme 2–3 cells thick, i.e. the same amount as that which lies immediately adjacent to the primitive streak. During subsequent cranial neural fold development, both the number of mesenchyme cells and the volume of extracellular matrix increases (Morriss & Solursh, 1978b). Even at a very early stage, this appearance of widely separated cranial mesenchyme cells has been correlated with a neural plate/ primitive streak region differential in hyaluronate synthesis of 1·75:1 (Solursh & Morriss, 1977). Increase of mesenchyme cell number and amount of hyaluronate-rich extracellular matrix may be important factors in the generation of convex neural folds (0- to 5-somite stages). Ruthenium-red staining of this (Fig. 2) and subsequent stages indicates that the mesenchyme cells and matrix (except for those adjacent to unstained epithelium), endoderm cells, notochord and neuroepithelial basement membrane are rich in polyanionic material. The nature and possible morphogenetic role of this material has been discussed elsewhere (Morriss & Solursh, 1978a, b; Morriss-Kay & Crutch, 1981; Solursh & Morriss, 1977). It will be considered further in section iv(c). In addition to the strong staining reaction, there is slight staining of the flask-shaped supranotochordal (neural-groove-forming) neuroepithelial cells; the significance of this is not understood.

Surface views of 0- to 5-somite-stage cranial neural folds (see Morriss & Solursh, 1978b) show a broadening during this period, whereas, as reported above, there is no increase in cell number or surface area of transverse sections of the neural epithelium. The broadening in surface view is due to a slight decrease in height of the columnar neuroepithelial cells during the 0- to 5-somite stages, and to an opening out of the neural groove as the foregut forms (Fig. 5). By the 6-somite stage the neural epithelium has begun to thicken, through increase in height of its cells; embryos at this stage show a sharper neural surface–ectoderm junction than previous stages.

Increase in cell height within the now pseudostratified neural epithelium continues during the 7- to 10-somite stages. Cell elongation involves the development of narrow-necked cells with overlapping apices. Microtubules are a feature of the neck region, while in the apical region a prominent line of microfilament bundles attached to desmosome-like junctions appears. Because of the cell overlapping, these junctions are predominantly oriented parallel to the apical surface (illustrated in Morriss & New, 1979). The microfilament/junctional line can be clearly seen in light microscopic preparations at these and subsequent stages, and has given rise to the misleading term 'internal limiting membrane' of classical descriptions. The cellular organization of the major part of the cranial neural epithelium in later stages of neurulation persists after neurulation is complete (post-closure stages are illustrated in Seymour & Berry, 1975).

These apical region changes, in orientation of junctions and development of apparently contracting microfilament bundles, occur throughout the whole cranial neural epithelium in the forebrain region. At midbrain and hindbrain levels, they are absent from the most lateral cells. While the apical surface changes are occurring in the main part of the neural epithelium, the basement membrane of the lateral region loses continuity, and neural crest cell emigration begins. The timing varies according to region: it can be seen in the otic region of 8-somite embryos (Morriss & Thorogood, 1978), and in the midbrain/upper hindbrain region of 9- to 10-somite embryos (Morriss & New, 1979). The difference in timing may be correlated with a stage just prior to the mediad curvature of the lateral edges which precedes closure; closure occurs in the otic region before the midbrain/upper hindbrain region. It was previously proposed (Morriss & New, 1979) that emigration of neural crest cells from the lateral edge of the neural epithelium is a prerequisite for development of the sharp mediad curvature of this region. The hypothesis depended on the existence of a longitudinal stretch force similar to that discussed by A. G. Jacobson (1978) in the context of newt neurulation. We have shown in section III(a) that rostro-caudal elongation of the lateral edges does occur as they approach each other in the dorsal midline, as a consequence of the continuously developing cranial flexure. (The lateral neuroepithelial cells which remain *in situ* form the thin dorsal roof of the closed neural tube.)

IV. EXPERIMENTAL STUDIES USING CULTURED EMBRYOS

The 'whole rat embryo culture' technique of New (for review, see New, 1978) is particularly well suited to the study of embryos during neurulation, since during this period the embryo and yolk sac are directly exposed to the culture medium, and substances added to the medium have free access to the embryonic cells. We have cultured embryos in medium containing cytochalasin to investigate the role of microfilaments during neuroepithelial morphogenesis, and in medium containing β-D-xyloside to investigate the morphogenetic role of proteoglycans. Also summarized here are the results of a study on the effects of high oxygen levels on neuroepithelial shape and structure, since it revealed a number of cellular features correlated with neuroepithelial curvature in the later stages of midbrain/hindbrain neurulation.

Experimental details not provided here may be found in Morriss & New (1979) and Morriss-Kay & Crutch (1981).

(a) The effect of cytochalasins

M. Jacobson (1978) writes: 'It is now safe to conclude that changes in shape of neural epithelial cells during neurulation are due to intracellular microtubules which produce cell elongation and to intracellular microfilaments which produce apical constriction of the neuroectodermal cells.' Since the observations of Baker & Schroeder (1967) on the shape and position of apical microfilament bundles in the neuroepithelium of *Xenopus* embryos during neurulation, it has become accepted that microfilament contraction is directly responsible for generating neuroepithelial curvature. However, A. G. Jacobson (1978) found that isolated newt neural plates will undergo shrinkage in culture, as microfilament contraction occurs, but in the absence of underlying notochord they do not become curved to form neural tubes however long they are cultured. His observations suggest that in the newt, microfilament contraction brings about narrowing of the necks of the elongating neural plate cells, but does not itself bring about curvature of the neural epithelium.

In order to investigate the role of microfilament contraction in mammalian neurulation, we cultured rat embryos in medium containing $0.5 \mu g/ml$ cytochalasin B. (This concentration was found by transmission electron microscopy (TEM) to disrupt microfilament bundles while having a minimal effect on other aspects of cell structure.) Embryos of 9 to 11 somites were selected for these experiments, and were separated into two groups – those with and those without midbrain apposition (Figs 6*a* and *c*). At this stage, the whole cranial neural epithelium is developing an increasingly concave curvature in transverse section (see section III*a*); at the ultrastructural level microfilament bundles, associated with desmosome-like junctions, form a prominent line close to and parallel with the apical surface.

Changes in neural fold shape were evident within $\frac{1}{2}$ h of culture; the appearance after $1\frac{1}{2}$ h is illustrated by scanning electron micrographs in Fig. 6. In all embryos the neural folds splayed outwards, but the results were more pronounced in those which had not achieved midbrain apposition at the start of culture.

Similar results have been achieved using cytochalasin D (0.15 and $0.2 \mu g/ml$), and in these experiments embryos were washed and transferred to fresh medium after a 2 h exposure period. All of these embryos recovered neural fold shape and continued neurulation to form a closed brain tube, although it was of an abnormal shape, and the whole embryo was reduced in size compared with co-cultured control embryos (results to be reported in detail elsewhere).

These results suggest that microfilament contraction is essential for developing and maintaining the concave curvature of the neural epithelium. However, in the midbrain/hindbrain region, development of concave curvature and its

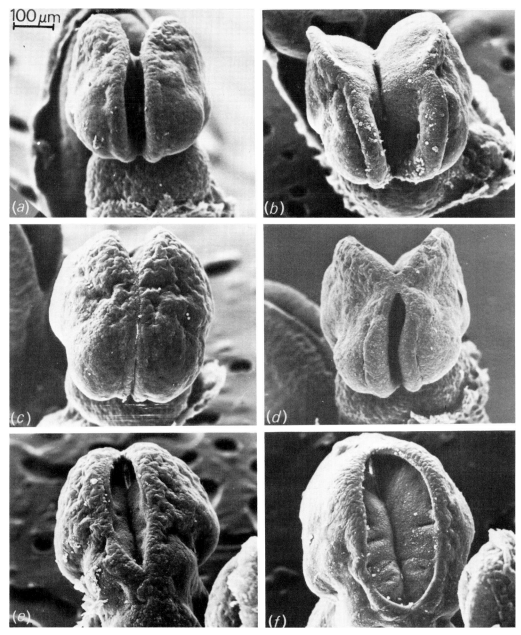

Fig. 6. SEM of heads of control embryos (left) and embryos cultured for $1\frac{1}{2}$ h in medium containing $0.5\,\mu$g/ml cytochalasin B (right). (*a*), (*b*) Frontal views of early 10-somite-stage embryos prior to apposition at the forebrain/midbrain junction; (*c*) and (*d*) 10 to 11-somite embryos with forebrain/midbrain apposition; (*e*) and (*f*) Dorsal views of 11-somite-stage embryos to show shape of the midbrain/hindbrain open area.

reversal by cytochalasins is a phenomenon of the transverse plane only; this and the rostrad flow of cells within and from this region suggest that micro-filament contraction is only of morphogenetic significance in the transverse plane here, whereas in the forebrain it is probably important in all planes.

Recovery of neural fold shape and completion of neural tube formation in washed embryos after 2 h exposure to cytochalasin shows that short-term interruption of microfilament contraction does not necessarily result in open neural tube defects, but does have an effect on the size and shape of the neural tube and other embryonic structures. In view of the known roles of micro-filaments in cell movement, cell division and other aspects of cell function, this result is not surprising. Taken as a whole, these results do not support the view that open neural tube defects in human development could be due to dietary intake of a substance that inhibits the action of microfilaments (Linville & Shepard, 1972), since these congenital abnormalities frequently occur without other abnormalities.

(b) The effects of high oxygen levels

Early work on the technique of rat embryo culture was carried out using the normal tissue culture gas phase of 5 % CO_2 in air, i.e. 20 % O_2 (New, 1978). Development of the technique involved modification of the culture medium and apparatus, resulting in great improvements in embryonic growth and morphogenesis. However, a high proportion of cultured embryos still developed with misshapen or unclosed brain tubes. This problem was overcome by using a gas phase containing only 5 % O_2 during the period of neurulation, although 20 % O_2 was found to support normal development for the subsequent 24 h (New, Coppola & Cockroft, 1976a, b).

Failure of normal cranial neural tube formation in embryos cultured with a 20 % O_2 gas phase was investigated in a morphogenetic study (Morriss & New, 1979). TEM showed that in embryos developing *in vivo*, or with a 5 % O_2 gas phase *in vitro*, during cranial neurulation, the mitochondria have few cristae and suggest anaerobic metabolism. In embryos cultured with a 20 % O_2 gas phase, the mitochondria had the appearance typical of those of cells with aerobic metabolism. Direct correlation between aerobic metabolism and abnormal cranial neurulation on the one hand, and anaerobic metabolism and normal cranial neurulation on the other, was suggested by two observations. First, during normal *in vivo* neurulation, the embryo is increasing in size and is becoming separated from the maternal oxygen source by the enveloping yolk sac (the yolk sac circulation begins to develop only towards the end of the cranial neurulation period). Secondly, a further increase in environmental O_2 levels *in vitro* (40 % O_2) resulted in both a further increase in internal mito-chondrial surface area, and an increase in the degree of abnormality of cranial neurulation.

Examination of the midbrain/hindbrain ('broad cranial neural fold') region

of the neural epithelium of high O_2-cultured embryos indicated several differences from control (5 % O_2-cultured and *in vivo*) embryos. From the 9- to 10-somite
stage onwards there was less narrowing of the V shape of the neural folds, and
less pronounced development of a concave shape in transverse section; also,
the lateral edges failed to develop a sharp mediad curvature. These differences
in shape of the neural epithelium were correlated with cellular differences
between normal and high O_2-cultured embryos: in the neural epithelium as a
whole, there were scattered pyknotic cells in normal but not in high O_2-cultured
embryos; in the lateral edge region of high O_2-cultured embryos, the apical
microfilament bundles and associated desmosome-like junctions extended to
the surface ectoderm border, and neural crest cell emigration was either late or
absent.

These results suggest that those changes of neuroepithelial cellular organization which occurred in control but not in high O_2-cultured embryos are essential
to the development of both the general concave neuroepithelial curvature and
the sharp mediad curvature of the lateral edges which occurs during closure of
the spindle-shaped midbrain/upper hindbrain neuropore (10- to 14-somite
stages).

(c) The effects of β-D-xyloside

During cranial neurulation in rat embryos, the neuroepithelial basement
membrane stains strongly with alcian blue, except in the most lateral regions.
The stainable material has been shown to consist largely of chondroitin sulphate,
with some heparan sulphate, and to be continuous with similar material in the
extracellular matrix and on the cell surfaces of the underlying mesenchyme
(Morriss & Solursh, 1978 *a*, *b*). Sulphated glycosaminoglycans (GAG) normally occur in nature in association with polypeptide as proteoglycan. Therefore, in order to investigate their role during cranial neurulation, we cultured
rat embryos in medium containing β-D-xyloside, a substance which inhibits
proteoglycan synthesis while stimulating the synthesis of free GAG chains
(Schwartz, Ho & Dorfman, 1976; Galligani, Hopwood, Schwartz & Dorfman,
1975). A 1 mM concentration was found to affect morphogenesis; there was also
a slight growth retardation.

Pre-somite (flat neural plate) or very early somite-stage embryos were used
for these cultures. Convex cranial neural fold development (up to 5 somites)
appeared to be normal in shape, although light microscopy revealed that even
at this stage the neuroepithelium was abnormally thin. Subsequent stages of
development, including flattening and thickening of the convex neural epithelium and development of a concave curvature, were late and deficient in
β-D-xyloside-cultured embryos. Neural crest cell emigration and mediad
curvature of the thinned lateral edges occurred while the neural folds were in
the form of a wide V-shaped gutter, i.e. lateral edge changes characteristic of
10- to 13-somite-stage embryos occurred while the major part of the neural

epithelium had an appearance similar to that of 6- to 7-somite-stage embryos. The forebrain region was everted instead of concave.

Alcian-blue staining at pH 1 confirmed that in β-D-xyloside-cultured embryos, sulphated GAG were present only at very low levels; ultrastructure of the neural epithelium showed breaks in the basement membrane and apical cell surfaces which were broad and flat, with poorly developed microfilament bundles.

These results are reported and discussed more fully elsewhere (Morriss-Kay & Crutch, 1981). We conclude that proteoglycans are important for both growth and pattern development during this period of development in rat embryos. Chondroitin sulphate-proteoglycan probably has a role in relation to neuro-epithelial cell shape and cell proliferation, while heparan sulphate may be involved in cell adhesion.

SUMMARY AND CONCLUSIONS

We have studied cranial neurulation in rat embryos through morphological and experimental observation. These studies lead to the conclusion that the observed changes in shape of the neural epithelium involve precise control of cell shape, cell–cell relationships and cell number at all positions. At midbrain and upper hindbrain levels, the cell number was found to remain constant throughout neurulation; mitoses must therefore result in a net forward movement of cells. During neurulation, the forebrain region extends forwards anterior to the tip of the notochord; influx of cells from lower levels is augmented by local cell division, resulting in much greater size of this region than elsewhere.

Changes in neural fold shape in the midbrain/upper hindbrain region are considered to involve cell elongation (bringing about epithelial thickening) and neural crest cell emigration (enabling the sharp mediad curvature of the lateral edges which precedes midline apposition). Although cell elongation may be stabilized by microtubules in the narrow neck region, apical microfilament contraction in the transverse plane is probably more important, while an increase in lateral cell contacts at subapical levels may also be involved.

Neural crest cell emigration from the lateral edge region is preceded by basement membrane breakdown. Apical microfilament bundles do not form in this region, and basement membrane proteoglycans are absent from early stages of neurulation. Correlation between the presence of basement membrane proteoglycans and the development of apical microfilament bundles is suggested by this observation and by experimental interference with proteoglycan synthesis. However, proteoglycans are present throughout neurulation, so while they may be involved in the mechanism of neuroepithelial thickening, the precise control of timing of this process probably lies within the epithelium itself.

Cranial flexure (convex flexure of the anteroposterior axis) begins at the start of cranial neurulation and continues throughout. It may play a role in closure of the midbrain/upper hindbrain spindle-shaped opening, since flexure involves elongation of the dorsal part of the closing neural tube. In contrast to observations in amphibian embryos and in the closure region of chick embryos, the notochord does not appear to elongate.

Closure of the spinal neural tube of mammalian embryos may well be found to be closely comparable to that of other amniotes, while all amniotes differ from amphibian embryos in that they form new axial structures continuously throughout neurulation. However, the observations and conclusions presented here indicate that in the cranial region, neurulation in a mammalian embryo is much more complex in both morphogenesis and morphogenetic mechanisms than in all non-mammalian vertebrates.

I wish to thank the M.R.C. for financial support throughout the period during which the work described here was carried out, Mr M. Barker for technical assistance, Mr A. Barclay for photographic assistance, and Mr P. Entwistle for use of the JEOL SEM. Experimental work described in sections IIa, IVa and IVc was carried out in collaboration with Miss B. Crutch. Miss F. Tuckett is an M.R.C. scholar.

REFERENCES

BAKER, P. C. & SCHROEDER, T. E. (1967). Cytoplasmic filaments and morphogenetic movement in the amphibian neural tube. *Devl Biol.* **15**, 432–450.

BECK, F., LLOYD, J. B. & GRIFFITHS, A. (1967). A histochemical and biochemical study of some aspects of placental function in the rat using maternal injection of horseradish peroxidase. *J. Anat.* **101**, 461–478.

BURNSIDE, B. (1971). Microtubules and microfilaments in newt neurulation. *Devl Biol.* **26**, 416–441.

DERRICK, G. E. (1937). An analysis of the early development of the chick by means of the mitotic index. *J. Morph.* **61**, 257–284.

GALLIGANI, L., HOPWOOD, J., SCHWARTZ, N. B. & DORFMAN, A. (1975). Stimulation of synthesis of free chondroitin sulfate chains by β-D-xylosides in cultured cells. *J. biol. Chem.* **250**, 5400–5406.

GILLETTE, R. (1944). Cell number and cell size in the ectoderm during neurulation. *J. exp. Zool.* **96**, 201–222.

HAMILTON, W. J. & MOSSMAN, H. W. (1972). *Hamilton, Boyd and Mossman's Human Embryology*, 4th ed., p. 179. Cambridge: Heffer.

JACOBSON, A. G. (1978). Some forces that shape the nervous system. *Zoon* **6**, 13–21.

JACOBSON, A. G. (1980). Computer modelling in morphogenesis. *Am. Zool.* **20**, 669–677.

JACOBSON, M. (1978). *Developmental Neurobiology*, 2nd ed. New York & London: Plenum Press.

LINVILLE, G. P. & SHEPARD, T. H. (1972). Neural tube closure defects caused by cytochalasin B. *Nature New Biology* **236**, 246–247.

LOWRY, O. H., ROSEBROUGH, N. J., FARR, A. L. & RANDALL, R. F. (1951). Protein measurement with the folin phenol reagent. *J. biol. Chem.* **193**, 265–269.

MORRISS, G. M. (1973). The ultrastructural effects of excess maternal vitamin A on the primitive streak stage rat embryo. *J. Embryol. exp. Morph.* **30**, 219–242.

MORRISS, G. M. & NEW, D. A. T. (1979). Effect of oxygen concentration on morphogenesis of cranial neural folds and neural crest in cultured rat embryos. *J. Embryol. exp. Morph.* **54**, 17–35.

MORRISS, G. M. & THOROGOOD, P. V. (1978). An approach to cranial neural crest cell migration and differentiation in mammalian embryos. In *Development in Mammals*, vol. 3 (ed. M. H. Johnson), pp. 363–412. Amsterdam: North-Holland.

MORRISS, G. M. & SOLURSH, M. (1978a). Regional differences in mesenchymal cell morphology and glycosaminoglycans in early neural-fold stage rat embryos. *J. Embryol. exp. Morph.* **46**, 37–52.

MORRISS, G. M. & SOLURSH, M. (1978b). The role of primary mesenchyme in normal and abnormal morphogenesis of mammalian neural folds. *Zoon* **6**, 33–38.

MORRISS-KAY, G. M. & CRUTCH, B. (1981). Culture of rat embryos with β-D-xyloside: evidence of a role for proteoglycans in neurulation. *J. Anat.* (in press).

NEW, D. A. T. (1978). Whole-embryo culture and the study of mammalian embryos during organogenesis. *Biol. Rev.* **53**, 81–122.

NEW, D. A. T., COPPOLA, P. T. & COCKROFT, D. L. (1976a). Improved development of head-fold rat embryos in culture resulting from low oxygen and modifications of the culture serum. *J. Reprod. Fert.* **48**, 219–222.

NEW, D. A. T., COPPOLA, P. T. & COCKROFT, D. L. (1976b). Comparison of growth *in vitro* and *in vivo* of post-implantation rat embryos. *J. Embryol. exp. Morph.* **36**, 133–144.

SCHWARTZ, N. B., HO, P-L. & DORFMAN, A. (1976). Effect of β-xylosides on synthesis of cartilage-specific proteoglycan in chondrocyte cultures. *Biochem. biophys. Res. Commun.* **71**, 851–856.

SEYMOUR, R. M. & BERRY, M. (1975). Scanning and transmission electron microscope studies of interkinetic nuclear migration in the cerebral vesicles of the rat. *J. Comp. Neurol.* **160**, 105–126.

SNOW, M. H. L. (1976). Embryo growth during the immediate postimplantation period. *Ciba Fnd. Symp.* **40**, 53–66.

SOLURSH, M. & MORRISS, G. M. (1977). Glycosaminoglycan synthesis in rat embryos during the formation of the primary mesenchyme and neural folds. *Devl Biol.* **57**, 75–86.

TAM, P. P. L. (1981). Control of somitogenesis in mouse embryos. *J. Embryol. exp. Morph.* **65** (*Supplement*), 103–128.

J. Embryol. exp. Morph. Vol. 65 (Supplement), pp. 243–267, 1981
Printed in Great Britain © Company of Biologists Limited 1981

Morphogenetic behaviour of the rat embryonic ectoderm as a renal homograft

By ANTON ŠVAJGER,[1] BOŽICA LEVAK-ŠVAJGER,[2]
LJILJANA KOSTOVIĆ-KNEŽEVIĆ[1] AND
ŽELIMIR BRADAMANTE[1]

*From the Institute of Histology and Embryology and the Institute of Biology,
Faculty of Medicine, University of Zagreb, Yugoslavia*

SUMMARY

Halves of transversely or longitudinally cut primary ectoderm of the pre-primitive streak and the early primitive streak rat embryonic shield developed after 15–30 days in renal homografts into benign teratomas composed of various adult tissues, often in perfect organ-specific associations. No clear difference exists in histological composition of grafted halves of the same embryonic ectoderm.

The primary ectoderm of the pre-primitive streak rat embryonic shield grafted under the kidney capsule for 2 days displayed an atypical morphogenetic behaviour, characterized by diffuse breaking up of the original epithelial layer into mesenchyme. Some of these cells associated into cystic or tubular epithelial structures.

The definitive ectoderm of the head-fold-stage rat embryo grown as renal homograft for 1–3 days gave rise to groups of mesenchymal cells. These migrated from the basal side of the ectoderm in a manner which mimicked either the formation of the embryonic mesoderm or the initial migration of neural crest cells. This latter morphogenetic activity was retained in the entire neural epithelium of the early somite embryo but was only seen in the caudal open portion of the neural groove at the 10- to 12-somite stage.

The efficient histogenesis in grafts of dissected primary ectoderm and the atypical morphogenetic behaviour of grafted primary and definitive rat embryonic ectoderm were discussed in the light of current concepts on mosaic and regulative development, interactive events during embryogenesis and positioning and patterning of cells by controlled morphogenetic cell displacement.

INTRODUCTION

The space between the fibrous capsule and the parenchyma of the adult kidney offers a suitable environment for growth and differentiation of whole rat egg cylinders or separated germ layers. After a period of 15–30 days following transfer of such embryonic pieces, teratomas develop whose elaborate histological composition presumably reflects the developmental capacities of embryonic cells at the moment of transplantation (Škreb & Švajger, 1975; Škreb, Švajger & Levak-Švajger, 1971, 1976). Thus the developmental capacity

[1] *Author's address*: Institute of Histology and Embryology, Faculty of Medicine, University of Zagreb, Šalata 3, P.O. Box 166, 41001 Zagreb, Yugoslavia.

[2] *Author's address*: Institute of Biology, Faculty of Medicine, University of Zagreb, Šalata 3, P.O. Box 166, 41001 Zagreb, Yugoslavia.

of the isolated rat embryonic ectoderm is greater if taken from the pre-primitive embryo streak than if removed from the head-fold stage of development (Levak-Švajger & Švajger, 1971, 1974). Towards the end of this developmental period the embryonic endoderm similarly displays regionally restricted capacities to differentiate into segments of the definitive gut when transplanted together with the adjacent mesoderm (Švajger & Levak-Švajger, 1974).

Similar results were obtained from testicular grafts of isolated germ layers of the mouse (Diwan & Stevens, 1976), thus indicating that the mechanism of definitive germ layer formation might be the same in both species. Moreover, this principle of gastrulation (origin of all three definitive embryonic germ layers from the primary ectoderm) seems to be even more general, for it has been observed also in the avian embryo, following the use of more direct methods (Nicolet, 1971).

The presumed morphogenetic movements of gastrulation in the rodent embryonic shield could be similar to the events in the chick blastoderm: a restricted area of extraordinarily rapid cell proliferation within the primary ectoderm (proliferative zone or centre) generates cells which, by expansion forces cause a considerable mass of cells to migrate through the primitive streak (Snow, 1976a, b, 1977). The primitive streak can be regarded as both the passageway for cells during gastrulation and the first, incomplete axis of bilateral symmetry of the embryo.

The temporally and spatially coordinated migration of cells through the primitive streak and the primitive node results in an ordered displacement of coherent cell sheets within the embryo. The final result of these morphogenetic cell movements is an orderly apposition of definitive germ layers to one another which makes possible inductive tissue interactions bringing about a spatial pattern of cellular differentiation and morphogenesis within the embryonic body.

When the early, pre-primitive streak embryonic ectoderm is transferred to an ectopic site such as the space under the kidney capsule, one can hardly expect that a regular primitive streak will form in this drastically altered physical environment. However, ectodermal, mesodermal and endodermal tissues regularly differentiate in these grafts and some tissues exhibit marked organ-specific cellular differentiation, of reasonably normal topography and spatial distribution (Škreb & Švajger, 1975). On the other hand, when at the head-fold stage the primitive streak and the primitive node regions are removed and only the anterior region grafted, mesodermal tissues (cartilage, bone, muscle) are regularly found in teratomas (Švajger & Levak-Švajger, 1976; Levak-Švajger & Švajger, 1979). Obviously, an adaptation of morphogenetic mechanisms to the ectopic environment should be presumed to exist in the transplanted ectoderm.

In order to shed more light on this problem the present investigation was undertaken to answer the following questions:

(a) Does the partial or complete removal of the primitive streak region in the early rat embryonic ectoderm essentially influence the degree, the diversity and the organ-specificity of its histological differentiation as a renal homograft (series I, see Materials and Methods)?

(*b*) How do the future endodermal and mesodermal cells leave the primary ectoderm transplanted under the kidney capsule (series II)?

(*c*) How do mesenchymal cells originate and where do they come from in the renal grafts of the head-fold-stage rat embryonic ectoderm (series III)?

(*d*) At which developmental stage does the grafted embryonic ectoderm lose the ability to give rise to mesenchymal cells (series IV and V)?

MATERIALS AND METHODS

Embryos

Albino rats of the inbred Fischer strain were used in the experiments. Gestation was considered to have begun early in the morning when sperm was found in the vaginal smear. Twenty-four hours later the eggs were considered to be 1 day old.

Embryos belonging to the following post-implantation stages (after Witschi, see New, 1966) were used:

Stage 11 (gestation day 8): the pre-primitive streak (pre-gastrula), two-layered embryonic shield.

Stage 12 (gestation day $8\frac{1}{2}$): the early primitive streak (early gastrula), start of mesoderm formation.

Stage 13 (gestation day 9): head-fold (late gastrula), head process, neural plate.

Stage 14 (gestation day $9\frac{1}{2}$): neural groove, somites 1–3, start of foregut formation.

Stage 15 (gestation day 10): partly closed neural tube, somites 10–12, first aortic arch.

Pregnant females were anaesthetized with ether and the entire conceptuses (embryo + extraembryonic parts) were isolated from uteri with watchmaker's forceps in sterile Tyrode's saline. All extraembryonic structures were carefully removed and the isolated embryos were subjected to further manipulation.

Isolation of the embryonic ectoderm

The embryonic ectoderm was isolated from the pre-somite embryonic shields (stages 11, 12 and 13) and the early somite embryos (stage 14) by using the standard procedure for the separation of germ layers (treatment with proteolytic enzymes at 4 °C + microdissection) described in detail by Švajger & Levak-Švajger (1975). The stage-15 embryos (10–12 somites) were cut transversely into three segments before the treatment with enzymes and mechanical manipulations. According to the design of each experimental series, the isolated ectoderm was further dissected into pieces to be used as grafts.

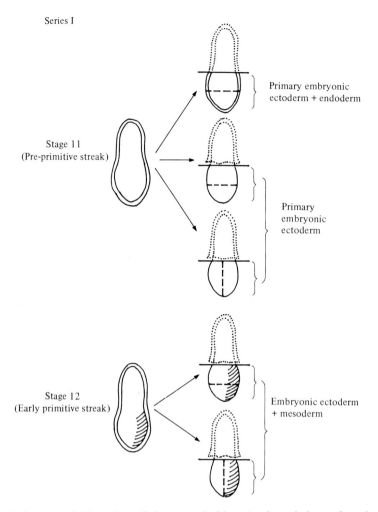

Fig. 1. Schema of dissection of the pre-primitive streak and the early primitive streak rat primary ectoderm by transverse or longitudinal cuts (Experiment series I). ···, Contours of removed extra-embryonic parts; ——, cut for removal of extra-embryonic parts; – – –, cut through the embryonic ectoderm.

Transplantation and fixation of grafts

Isolated pieces of embryonic ectoderm were transferred by means of a braking pipette under the capsule of the right kidney of an adult (3 months) male rat of the same strain. The recipient animals were killed by ether 1–30 days after transplantation. Grafts were excised with a razor blade together with a block of adherent host renal tissue. Very small grafts (fixed 1–2 days following transfer) were dissected out after a 30-min prefixation of the whole graft-bearing host kidney.

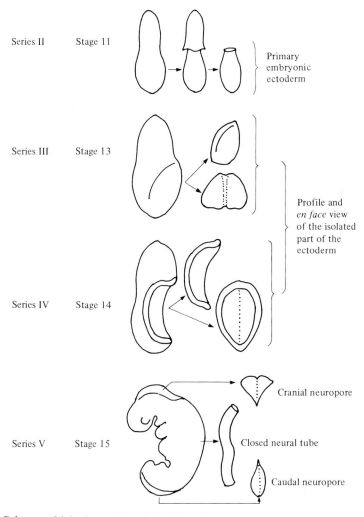

Fig. 2. Schema which shows the origin of ectodermal areas which were grafted for 1–2 days to observe the morphogenetic behaviour of ectodermal cells (Experiment series II–V).

Histological procedures

(*a*) *Paraffin sections*. Grafts were fixed in Zenker's fluid, embedded in paraffin wax, serially sectioned at 7 μm and stained with haemalum and eosin.

(*b*) *Semi-thin sections*. Grafts were fixed for 1 h in a mixture of 1% glutaraldehyde and 1% paraformaldehyde in 0·1 M phosphate buffer at 4 °C. They were washed in the same buffer and then post-fixed for 1 h in 1% osmic acid in the same buffer. Serial ethanolic dehydration was followed by embedding in Durcopan (Fluka). The 1 μm serial sections were stained with toluidine blue.

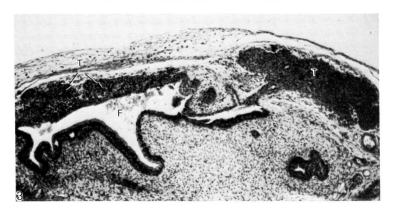

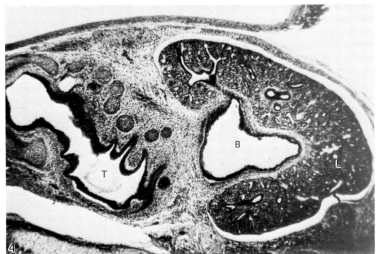

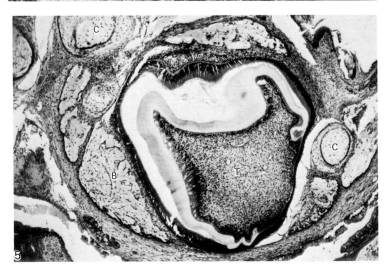

Design of experimental series

The work was divided into five series of experiments characterized by the developmental stage of embryos used, the region of the ectoderm used as graft and the period of cultivation in the host kidney. The purpose of each series was explained in the Introduction.

Series I. Halves of pre-gastrula and early gastrula embryonic ectoderm. Embryonic ectoderm belonging to developmental stages 11 (pre-primitive streak) and 12 (early primitive streak) was divided approximately into two halves by either a transverse or a longitudinal cut (Fig. 1). Longitudinal cuts were made haphazardly, with no respect to the expected (stage 11) or actual (stage 12) position of the primitive streak. Each half was transplanted separately under the kidney capsule. In 23 of the transversely cut stage-11 embryonic shields the primary endoderm was not removed from the ectoderm. The halves of the stage-12 ectoderm were transplanted together with the mesodermal wings emerging from the primitive streak. The grafts were fixed after 15–30 days and examined histologically for the presence and distribution of mature tissues (Total number: 88 embryos = 176 grafts).

Series II. Whole pre-gastrula embryonic ectoderm. The primary ectoderm of the pre-primitive-streak (stage-11) embryo was transplanted and fixed after 2 days (23 grafts, Fig. 2).

Series III. Incomplete late gastrula embryonic ectoderm. The posterior part and the tip of the stage-13 egg cylinder (areas containing the primitive streak and the primitive node respectively) were cut off prior to the germ layer separation procedure, and the rest of the embryonic ectoderm was transplanted for 1–3 days (152 grafts, Fig. 2).

Series IV. Incomplete early-somite-stage ectoderm. As in the previous series, the primitive streak and the primitive node regions were removed from the stage-14 embryo. The isolated and grafted ectoderm roughly corresponded to the neural fold. It very probably contained presumptive areas of both the neural epithelium and a part of the surface ectoderm (future epidermis) which are not yet sharply demarcated at this stage (Waterman, 1976; ectoderm classes II–V of Verwoerd & van Oostrom, 1979; 13 grafts, Fig. 2).

Series V. Neuroectoderm of the stage-15 embryo. Prior to treatment with enzymes the whole embryo was transversely cut into three segments which contained: (*a*) the cranial neuropore (from which the rostral part with the optic

Fig. 3. Origin of thymus (T) from the foregut (F) epithelium. Experiment series I. H.E. × 100.

Fig. 4. Respiratory tract differentiation and morphogenesis in the teratoma. T, trachea; B, bronchial bifurcation; L, lung lobe. Experiment series I. H.E. × 50.

Fig. 5. A complex structure reminiscent of the foetal mandible, developed in the teratoma. T, tooth germ; B, membrane bone; C, hyaline cartilage. Experiment series I. H.E. × 70.

vesicles was removed), (*b*) the closed neural tube, and (*c*) the caudal neuropore (from which the tail bud with remnants of the primitive streak was removed). After treatment with enzymes the closed neural tube was cleanly separated from the overlying surface ectoderm. The thick neuroepithelium of the cranial and caudal neuropore was easily isolated from its continuous surface ectoderm along the distinct demarcation line (Waterman, 1976). Each of the three segments of the neuroectoderm was separately transplanted and fixed after 2 days (32 grafts, Fig. 2).

RESULTS

Series I (Figs. 3–5). Halves of the pre-primitive streak and the early primitive streak embryonic ectoderm transplanted to the host kidney for 15–30 days developed into solid tumours of various sizes, comprising a multitude of well-differentiated tissues, often arranged in a clearly recognizable organ-specific association. The chaotic arrangement of adult tissues conformed to the definition of benign or mature embryo-derived teratoma (Damjanov & Solter, 1974; Solter, Damjanov & Koprowski, 1975).

Tissues found in grafts were derivatives of all three definitive germ layers:

Ectodermal tissues: skin (epidermis, hairs, sebaceous and mammary glands), neural tissues (brain, neural retina, choroid plexus, ganglia) and other ectodermal derivatives (lentoids, oral cavity with teeth, and salivary glands).

Endodermal tissues: foregut-derived epithelia (glands, thymus, thyroid, parathyroid, oesophagus, stomach, respiratory tube, lungs), mid- and hindgut-derived epithelia (small and large intestine, urogenital sinus, prostatic gland).

Mesodermal tissues: white and brown adipose tissues, cartilage, membrane and enchondral bone, smooth, skeletal and heart muscle).

Well-expressed organ-typical differentiation and combinations of tissues were regularly observed. Endoderm-derived epithelia displayed a wide range of segment-specific differentiations (stratified squamous epithelium of the oesophagus and forestomach, typical surface and glandular epithelium of the glandular stomach, small and large intestine, pseudostratified ciliated epithelium with a continuous transition into the epithelial lining of the lobe-shaped lung (Fig. 4). Glandular derivatives of the primitive gut always showed a typical regional origin: thymus and thyroid originated from the foregut epithelium (Fig. 3) while the prostatic gland appeared in direct continuity with the urogenital sinus, which was closely topographically related to the large intestine. The pattern of cellular differentiation in mesodermal tissues also displayed distinct organ-specificity. The pseudostratified ciliated epithelium of the respiratory tube was thus associated with pieces of non-ossifying hyaline cartilage, while the stomach and intestinal epithelium was surrounded by two layers of smooth muscle. Small ganglia were often found in close proximity of the muscular intestinal wall or even between the two muscular layers (intramural ganglia). Among especially peculiar tissue combinations, found only in single grafts were: a tooth germ surrounded by membrane bone and pieces of hyaline (Meckel's?) cartilage, mimicking the whole foetal mandible, and the optic cup enclosing a mass of lentoid cells. Interestingly, in the later case lentoid cells did not originate from the surface ectoderm, but from cylindrical epithelium which most probably belonged to the anterior edge of the optic cup. The whole tissue complex was thus strongly reminiscent of the Wolffian lens regeneration in amphibians.

Several difficulties, most of them of a technical nature (lower rate of successful grafts of halves when compared with grafts of whole embryonic shields, random topographical distribution of cuts through the embryonic ectoderm) make it impossible to perform any exact numerical analysis of differences in the histological composition of grafts, especially of pairs of grafts (halves of the same embryonic shield). Therefore only the following general statements can be made:

(*a*) Each graft, regardless of its origin, contained adult tissues derivatives of all three definitive germ layers.

(*b*) Neural tissue was present in all grafts, and therefore in both grafts of a pair.

(*c*) The tissue composition of particular grafts varied a great deal, but without any distinct prevalence of particular tissues in particular categories of grafts.

(*d*) Even tissues which probably originate from restricted areas of the primary ectoderm *in situ* (heart, respiratory tube, stomach, intestine, prostatic gland), could be found in grafts of both halves of the same embryonic ectoderm.

Series II (Figs. 6–9). The primary ectoderm of the pre-primitive streak rat embryonic shield displayed, as a renal homograft, remarkable deviations of morphogenetic behaviour in comparison with its normal development *in situ*. By 2 days after transplantation the original, compact epithelial organization of the primary ectoderm was hardly recognizable (Fig. 6). The grafts still consisted of undifferentiated cells, but only a small part of them retained the epithelial configuration. In general, three types of cell associations could be discerned (Figs. 7, 8, 9): (*a*) sheets of tightly packed epithelial cells (remnants of the original ectoderm?), (*b*) sheets of cells which still retained the two-dimensional pattern of the epithelium, but the contact between neighbouring cells was loosened (these epithelial cells sometimes formed cystic or tubular structures with irregular outlines, Fig. 8), and (*c*) mesenchyme-like masses of loosely dispersed, irregularly outlined cells (Fig. 9). These cells originated from various portions of the grafted primary ectoderm, with apparently no respect to the topographical position of the primitive streak *in situ*. It is impossible to associate any of the observed cell assemblies with a final tissue and/or organ derivative.

Series III (Figs. 10–19). The head-fold-stage rat embryonic ectoderm continued to develop, as renal a homograft, into both the high pseudostratified columnar epithelium of the neural tube (neuroepithelium) and the simple, lower epithelium of the surface ectoderm or the future epidermis (classes IV–VI and I–III respectively of Verwoerd & van Oostrom, 1979). Immediately after grafting the ectoderm was either extended as a flat sheet or folded with its basal

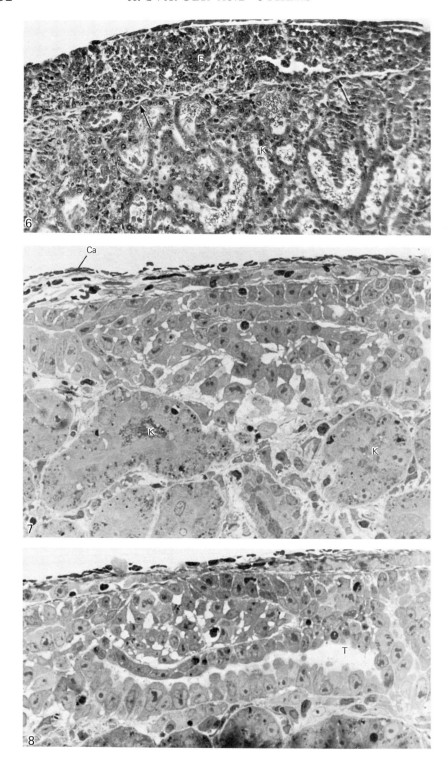

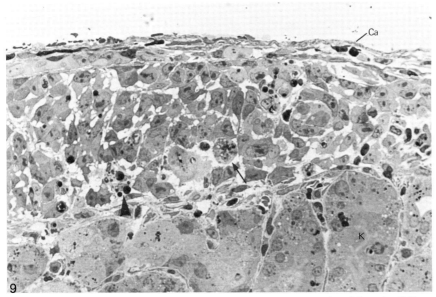

Fig. 9. Detail of the primary ectoderm 2 days after transplantation. Note the diffuse breaking up of the ectoderm into mesenchyme. Mitotic figures (arrow) and dead cells (arrowhead) are also seen. K, host kidney tubules; Ca, renal capsule. Experiment series II. Semi-thin section. × 600.

side outwards. Two or three days later it kept this original shape or gave rise, at least partly, to cystic, tubular or rosette-like structures (Figs. 10–12, 17, 18). Each of these epithelial forms seems to produce groups of cells, roughly similar to mesenchyme. These cells are always found on the basal side of the epithelial sheet, regardless of whether it was apposed to the parenchyma or the capsule of the host kidney, or to the periphery of neighbouring epithelial cysts or tubules.

Two ways could be distinguished by which this 'mesenchyme neoformation' seems to take place: (*a*) breaking up or dissociation of a portion of the epithelium into an amorphous group of loosely dispersed cells (as in the previous experimental series), and (*b*) protrusion of groups of more closely packed cells (often in the form of tongue-like projections) beyond the basal boundary of the epithelium. The first way seems to occur more generally, starting from all the above listed epithelial forms of the grafted ectoderm (Figs. 11, 12, 17, 19). The

Fig. 6. General appearance of the primitive ectoderm (E) 2 days after transplantation. K, host kidney. Arrows point to the sharp boundary between graft and host tissues. Experiment series II. H.E. × 300.

Fig. 7. Detail of the primary ectoderm 2 days after transplantation. K, host kidney; Ca, renal capsule. Note mesenchyme and epithelial structures. Experiment series II. Semi-thin section. × 600.

Fig. 8. Detail of the primary ectoderm 2 days after transplantation. Note the irregularly outlined epithelial tubule (T). Experiment series II. Semi-thin section. × 550.

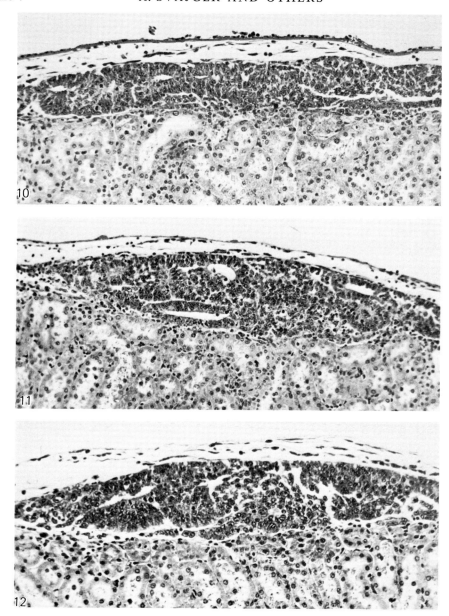

Figs. 10–12. General appearance of the head-fold stage embryonic ectoderm 2 days after transplantation. Note the mixed epithelial and mesenchymal organization of the explant. Experiment series III. H.E. × 170.

second way is strongly reminiscent of the initial migration of neural crest cells *in situ* (Morriss & Thorogood, 1978). It regularly occurs within small areas on the basal surface of a thick ectoderm, which is most probably equivalent to the neural plate (Figs. 13–16, 18, 19). Interestingly enough, this way of cell migration could sometimes be observed in grafts just at the boundary between the thick

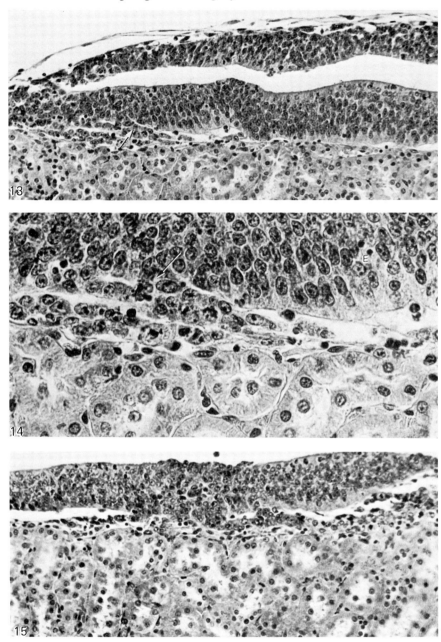

Fig. 13. Protrusion of a tongue-like projection (arrow) from the basal side of the head-fold stage embryonic ectoderm, 2 days after transplantation. Experiment series III. H.E. × 200.

Fig. 14. A higher magnification micrograph of a portion of Fig. 13. Note the outgrowth of cells (arrow). E, ectoderm. H.E. × 520.

Fig. 15. Outgrowth of closely packed cells from the basal side of the head-fold stage embryonic ectoderm, 2 days after transplantation. Experiment series III. H.E. × 200.

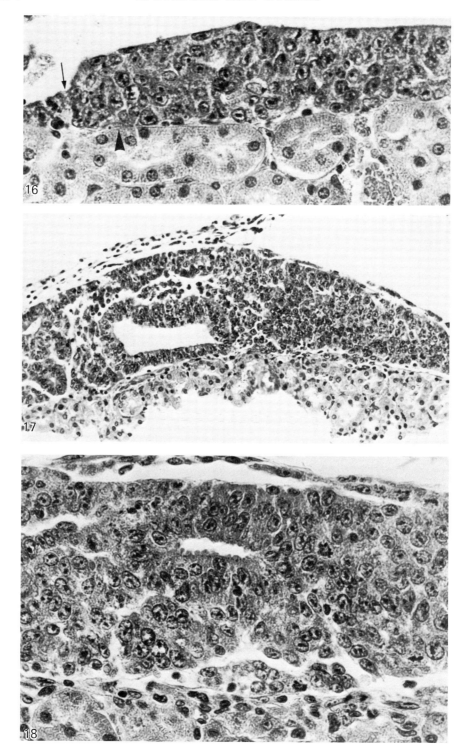

and the thin ectoderm, thus providing an almost exact copy of neural crest development *in situ* (Fig. 16).

Roughly estimated, the mitotic activity of grafted ectodermal cells did not differ essentially from its counterpart *in situ*. Mitoses could be observed predominantly within the innermost (ependymal) layer of the thick ectoderm. Dead cells were also a common finding. In some grafts massive cell death was observed in both the grafted ectodermal layer and the newly formed mesenchyme cells.

Series IV. Two days after transplantation the ectoderm of the early-somite-stage rat embryo still gave rise to new mesenchymal cells although to a considerably reduced extent and predominantly in a form reminiscent of neural crest formation *in situ*.

Series V. Two days after transplantation grafts of the closed neural tube and of the cranial, open neural groove formed relatively thick, disk-shaped and sharply outlined tumours which consisted exclusively of differentiating neural tissue with outgrowing axons and glial cells.

On the other hand, grafts of the caudal open neural groove were thin and the degree of neural tissue differentiation within them was less advanced. The outline of these grafts showed local irregularities with outgrowth of small groups of cells.

DISCUSSION

General remarks on ectopic development of early mammalian tissues

The whole experimental procedure applied in this or in similar experiments, involves a number of atypical influences upon the isolated embryonic tissue, the effects of which are poorly understood or completely unknown. These are: (*a*) temporary interruption of blood circulation, (*b*) influence of environmental constituents (saline, serum, enzymes), (*c*) surgical trauma, (*d*) loss of connexion with extraembryonic membranes, (*e*) altered physical (spatial) conditions after transplantation, (*f*) reduced supply of oxygen and nutrients before full vascularization of grafts, and (*g*) interactive influences from the host tissues. The basement lamina, which already exists between germ layers of the early embryonic shield (Adamson & Ayers, 1979; Pierce, 1966) is dissolved during the treatment with enzymes. Very probably, proteolytic enzymes can also considerably affect the composition and properties of the cell surface. The problem is best revealed by quoting Waymouth (1974): 'It is doubtful whether carefully controlled enzyme treatments are less traumatic than cutting, pressing, or

---

Fig. 16. Outgrowth of basal cells (arrowhead) at the boundary (arrow) between the thick (neural) and thin (surface, epidermal) head-fold stage ectoderm. Experiment series III. H.E. × 520.

Fig. 17. Mixed epithelial (tubular) and mesenchymal organization of the head-fold stage ectodermal explant, 2 days after transfer. Experiment series III. H.E. × 200.

Fig. 18. A rosette-like structure and groups of closely packed cells in a head-fold stage ectodermal explant, 2 days after transfer. Experiment series III. H.E. × 550.

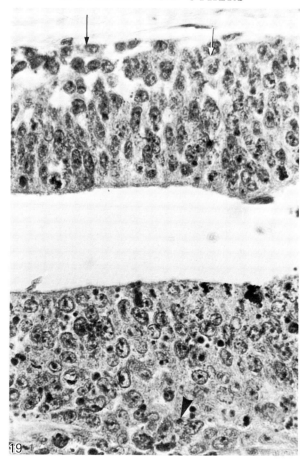

Fig. 19. Breaking up of the head-fold stage ectoderm into mesenchyme close to the host renal capsule (arrows) and protrusion of densely packed cell groups close to the kidney parenchyma (arrowhead). Experiment series III. H.E. × 520.

otherwise mechanically reducing tissues to manageable size for explantation. Any honest practitioner of cell and tissue culture will acknowledge that his art is one of survival of fittest cells in conditions that are never quite ideal'. Snow (1976a) pointed to the possibility that developmental capabilities of the isolated embryonic tissues may be modified by surgical trauma. In the present study mechanical distortions observed in some grafts fixed 2 days following transfer, have most probably arisen during the histological procedure. Morphogenetic features within the grafts did not show any particular relationship to these defects.

Prior to isolation the free and the basal surfaces of the ectoderm are in contact with the amniotic fluid and the basal lamina respectively. After transplantation under the kidney capsule the graft's new environment consists of loose connective tissue and, at the capsular side, of one or more layers of peculiar squamous cells (Bulger, 1973). We do not know whether this atypical microenvironment exerts any significant influence upon the graft. As pointed out in a previous paper (Levak-Švajger & Švajger, 1974), the great diversity of differentiation and organ-specific tissue associations in teratomas is unlikely to have been non-specifically induced by the host tissue. In the same study the presence of tissue derivatives of particular germ layers varied regularly in relation to the original germ-layer composition of the graft. It is therefore most probable that the final composition of the grafts reflects their initial developmental capacities.

Differentiation of parts of the dissected primary ectoderm (Experiment series I)

The unusually shaped rat and mouse egg cylinder with its inverted germ layers presents considerable difficulties in orientation and manipulation. A preliminary testing of developmental capacities of isolated parts of the rat embryonic ectoderm (Švajger & Levak-Švajger, 1976) showed that the borderlines between the frontal and lateral ectoderm, arbitrarily chosen in the mouse egg cylinder by Poelmann (1980) do not sharply delineate ectodermal regions with neural and surface ectodermal (epidermal) differentiative capacities. Although a regionalization with respect to mitotic activity has been demonstrated within the mouse epiblast (Snow, 1977), a detailed fate map of presumptive areas, as existing for the chick blastoderm (Rosenquist, 1966), can hardly be imagined in the embryonic shield of rodents. Moreover, the exact position of the primitive streak is difficult to foresee or even to record at its early stages.

With all this in mind the random transverse or longitudinal cutting of the early egg cylinder into two approximately equal parts seems to be a very simple and unpromising experimental design. After transverse cutting, the part of the cylinder adjacent to the extraembryonic membranes contained regions corresponding to the anterior portion of the neural plate and the posterior end of the primitive streak. The other part (with the tip of the cylinder) contained regions corresponding to the posterior portion of the neural plate and the anterior end of the primitive streak. The random longitudinal cutting resulted most probably in one half containing the primitive streak region and the other without it. One might think that in both experimental designs the two halves of the same embryonic ectoderm differed in the (partial) presence or absence of the primitive streak and the presumptive organ-forming areas. However, mature tissues, often in normal organ-specific associations, developed in all grafts, regardless of the variations in the initial developmental stage and the direction of the cut. These results suggest that within the primary ectoderm areas with different developmental potentialities are not yet sharply demarcated or, at least, that the revealing of regionally restricted prospective areas is highly complicated in the inverted egg cylinder. The other essential conclusion is that cells with endodermal and mesodermal destinations can leave the primary ectoderm in regions other than the usually positioned primitive streak.

An obvious question is, how are the specific epitheliomesenchymal inter-actions established in the absence of coordinated displacements of future endodermal and mesodermal cells through the primitive streak? This might, however, not be surprising if one remembers that even at later developmental stages factors other than strong local specificity are involved in these interactions (Lawson, 1974), and that non-cellular substrata may substitute for the local mesenchyme in supporting differentiation of digestive tract epithelia in chickens (Sumiya, 1976).

Atypical morphogenetic behaviour of the primary ectoderm
(Experiment series II)

During normal development *in situ* the formation of the primary mesenchyme from the primary ectoderm can be defined as the movement of individual cells in a migrating cell stream using the primitive streak as the passageway (Solursh & Revel, 1978). The atypical behaviour of the primary ectoderm as an explant, observed in this study, can best be defined as the 'breaking up of epithelial layers to produce mesenchyme' (Balinsky, 1975). The general appearance of this process is very similar to the partial dissociation of the epithelial somite into sclerotome (secondary mesenchyme, Hay, 1968). One might speculate that the ectoderm possesses an intrinsic tendency to produce new cell layers. After transfer to an ectopic site the onset of this activity takes place in remarkably disturbed environmental conditions. The basal lamina, whose collagen probably 'acts as a railroad track to guide the migration of the primitive streak mesenchyme' (Hay, 1973), is dissolved by enzymes, and the epithelial ectodermal sheet is tightly trapped within the subcapsular space of the host kidney. Very probably both these cirumstances may account for the deviation from the normal mechanism of primary mesenchyme production. In this context, it is interesting to note that in early mouse embryos cultivated *in vitro*, in conditions which allowed the expansion of embryonic and extraembryonic cavities, mesoderm formation occurred through a primitive-streak-like structure (Wiley & Pedersen, 1977; Wiley, Spindle & Pedersen, 1978; Libbus & Hsu, 1980), whereas the abortive mesoderm formation in cystic embryoid bodies proceeds in a modified way. Unfortunately the available data do not permit a clear comparison with the mechanism observed in the present study (Stevens, 1960; Martin, 1977; Martin, Wiley & Damjanov, 1977).

Cystic and tubular structures in explants were usually lined by epithelial cells whose irregular shape and size, as well as the occasionally loosened intercellular contacts, gave the impression that they have arisen by aggregation of mesenchymal cells. However, the 2-day interval between transplantation and fixation of grafts was obviously too long for recording all the gradual changes within the explant. At this early stage the simple morphology of these epithelial structures does not permit any prediction about the future differentiation into neural, epidermal, intestinal or mesodermal epithelia. One may note that the development of atypical tubular epithelia of mesodermal origin was also observed in T-mutant mouse embryos and interpreted as an aberrant kind of somite and notochord construction (Spiegelman, 1976).

Even more than in the previous series, it is difficult to reconcile this chaotically disturbed morphogenetic cell behaviour with the observed normal tissue differentiation and elaborate organ-typical association of older grafts. It is commonly accepted that the positioning and patterning of cells by controlled morphogenetic cell movements is a prerequisite for the construction of the basic structure and interrelationships of tissues. According to Curtis (1978) cell patterning can arise either by positioning of pre-differentiated cells or by differentiation of cells that have already taken up their final position. In the present case one might discuss the following two possibilities:

(*a*) The primary ectoderm is a heterogeneous population of small groups of pre-determined cells which already bear discrete cell surface properties necessary for future organ-specific cell–cell recognition. These cells and their immediate progeny could therefore recognize each other, associate (aggregate) and differentiate in a tissue- and organ-specific pattern regardless of the way in which they have left their original position within the primary ectoderm. In other words, the pre-determined cells may overcome the loss of their initial correct positioning and coordinated movement and find their 'required' final positions by mechanisms similar to those involved in tissue-specific sorting-out of cells from mixed aggregates *in vitro*. This principle of mosaic or polyclonal development, or 'development by means of compartments' (Garcia-Bellido, Lawrence & Morata, 1979) is consistent with some data on the early regionalization of the pre-gastrulation ectoderm in various classes of vertebrates. These include the findings of electrophoretically distinct subpopulations of cells within the undifferentiated amphibian ectoderm (Ave, Kawakami & Sameshima, 1968); of selective sorting-out of cells in aggregates prepared from unincubated chick blastoderms (Zalik & Sanders, 1974); and of regionalized mitotic activity in the mouse primitive ectoderm (Snow, 1977).

It is interesting that a mechanism analogous to that observed in this study, i.e. the formation of gut epithelium directly from the mesenchyme, operates in the tail region during the normal development ('direkt gebildetes Entoderm', Peter, 1941).

(*b*) The primary ectoderm is a homogeneous population of undetermined, pluripotent cells, which move away from their original positions and reach another place where 'first the cells are assigned positional information and then they interpret that information according to their genetic program' (Wolpert, 1978). This regulative type of development, or differentiation in response to environmental factors, has been demonstrated in various systems in vertebrates. Among the most impressive examples are: tissue or organ-specific differentiation (or metaplasia) of epithelia of the avian and mammalian embryonic membranes in response to specific or non-specific environmental stimuli (Moscona, 1959; Moscona & Carneckas, 1962; Kato & Hayashi, 1963; Yasugi & Mizuno, 1974;

Payne & Payne, 1961), contribution of Schwann sheath cells in the regeneration of the salamander limb (Wallace, 1972), differentiation of cartilage from differentiated muscle cells (Nathanson, Hilfer & Searls, 1978; Nathanson & Hay, 1980), and the conversion of cell type or transdifferentiation occurring during the Wolffian lens regeneration in amphibians (Yamada, 1977). In addition the clonal contribution of single teratocarcinoma cells to normally differentiated tissues in chimaeric mice (Illmensee & Stevens, 1979), strongly suggests epigenetic influence on their development.

All these data, however, concern the plasticity of some cells in response to unusual environmental conditions, rather than the repertoire of potencies which are realized during undisturbed development *in situ*. In other words, the occasional expression of pluripotentiality or aberrant developmental tendencies by various differentiated cells does not necessarily rule out the existence of covert populations of pre-determined cells within the primary ectoderm. Any attempt to explain the atypical behaviour of these cells in renal explants in terms of either the pre-determination or positional information, is limited by major gaps in our knowledge about what is actually going on in embryonic cells as they pass along their developmental pathway. Even clear-cut experimental results might provide only suggestive rather than definitive data if we keep in mind that 'we have practically no idea of what is really going on in cells of the blastoderm when they move, invaginate, induce or are induced, interact, become determined and begin to differentiate' (Leikola, 1976), and that 'we have no idea how positional signalling is accomplished or how cells record and remember their positional value' (Wolpert, 1978).

Atypical morphogenetic behaviour of the definitive ectoderm (Experiment series III–V)

The head-fold stage immediately follows primary induction and precedes neurulation, somitogenesis and primitive gut formation. Despite the remarkably advanced state of determination of the ectoderm, it still displays some of the atypical morphogenetic properties of the primary ectoderm: it breaks up into mesenchymal cells and forms cystic, tubular or rosette-like structures. New cells always originate from the basal side of the ectoderm, whether they are adjacent to the parenchyma or the capsule of the host kidney. This fact apparently implies an absence or lack of specificity of inductive influences by host tissues. The localized dissociation of the definitive ectoderm into a mesenchyme-like tissue might most probably be regarded as a residual capacity to form primary mesenchyme (embryonic mesoderm) even in the absence of the primitive streak. Determination of the ectoderm is therefore not yet fully stabilized and it would not be appropriate to designate the observed mesenchyme neoformation by terms such as cellular metaplasia, switch in differentiation, cell-type conversion or transdifferentiation (see Yamada, 1977).

The other form of mesenchyme production at this stage is reminiscent of neural crest formation and is most probably equivalent to it. Both primary mesenchyme formation via the primitive streak and ectomesenchyme formation via the neural crest involved local conversion of an epithelial into a mesenchymal tissue organization, probably by the same cellular mechanism. Proper identification of the neural crest-like cells depends upon whether their origin is from the neuroectoderm and results in a more condensed organization of the mesenchymal derivative (Morriss & Thorogood, 1978). A similar, but not identical organization of the newly formed mesenchyme was observed in the arrested primitive streak region of the T-mutant mouse embryo (Spiegelman & Bennett, 1974). A preliminary histological examination of the same type of explant but fixed after 15 or more days, has revealed the presence of skeletal muscle in addition to hyaline cartilage and membrane bone. These data suggest that the newly formed mesenchyme in grafts of definitive ectoderm might be analogous to both the mesoderm of primitive streak origin and the mesectoderm of neural crest origin. However, before a detailed analysis little can be said about the real developmental capacities of the neural crest-like cells in the present experimental conditions, especially as during normal development *in situ* cells of neural crest origin differ in their dependence on post-migratory tissue interactions for differentiation into various typical derivatives (Hall & Tremaine, 1979).

The restriction of the mesenchyme-forming capacity of the ectoderm transplanted at a later developmental stage corresponds to the definitive stabilization of neuroectodermal and epidermal components in the course of neural tube closure. The caudal portion of the neural tube is the last one to lose this capacity. This is compatible with developmental features *in situ*, where the posterior end of the neural tube gives rise to the mesenchyme of the tail bud (Jolly & Férester-Tadié, 1936).

It may be noted that an atypical, indistinct boundary between the closed neural tube neuroepithelium and the surrounding mesenchyme was also observed in T-mutant mouse embryos (Spiegelman, 1976).

CONCLUDING REMARKS

In the attempt to make more or less decisive conclusions about the observed phenomena we cannot avoid 'a problem that constantly faces the biologist, namely, that his concepts are limited both by the data on which he attempts to build them and by the design of the human mind' (Hay & Meier, 1978).

The main issue of this study is the finding that the primary ectoderm of the pre-gastrulation rat embryo can give rise to properly differentiated tissues and elaborate organ-specific tissue associations without the involvement of co-ordinated morphogenetic movements of cells through the primitive streak. In the ectopic environment it dissociates into mesenchyme-like cells which then apparently associate in a way reminiscent of the sorting-out and tissue-specific aggregation of embryonic cells from mixed suspensions. One may suppose that the morphogenetic aspects of gastrulation are essential for the determination of polarity and for the complete organization of pattern in the embryonic body, but not for the establishment of histotypical and organotypical cell combinations.

The various aspects and implications of altered morphogenetic cell movements in ectodermal explants can hardly be unequivocally explained within the framework of commonly held views which direct our thinking about development. As an example, how can the elaborate tissue pattern of the intestinal wall be achieved from an amorphous cell mass, when a definite organization into distinct cell layers is apparently required for the subtle 'directive' and 'permissive' interactive events (Saxén, 1977) during embryogenesis *in situ*? But surprisingly, the same 'abnormal' mechanism is at work in the tail bud of the embryo which develops *in utero*! Also, when we think about the mosaic or regulative nature of a particular developmental event, we should bear in mind that even experimentally well argued concepts, such as that on the divergence of stable cell lineages within the mouse blastocyst (Adamson & Gardner, 1979), are valid only in terms of prospective significance, and tell us little of prospective capacity (metaplastic changes of the extraembryonic endodermal epithelia!). Features like lens regeneration by transdifferentiation might support this view.

Beliefs are being expressed that 'there may be an universal mechanism whereby the translation of genetic information into spatial patterns of differentiation is achieved' (Wolpert, 1969). In fact, hardly any developmental mechanism is fully understood yet, and therefore any search for universal mechanisms appears hopeless. In order to avoid giving new names to our ignorance we had better to join C. H. Waddington (1957) in his belief that 'It seems impossible to hope that we shall ever discover any single basic mechanism of pattern formation or morphogenesis, as we may still hope to find, for instance in the mechanism of protein synthesis and its control by genes, the fundamental mechanism for substantive differentiation. In discussing pattern formation and morphogenesis, therefore, one can hardly hope to do more than provide a number of illustrations of the general nature of the processes which are at work'.

The authors wish to thank Miss Đurđica Cesar and Mrs Radmila Delaš for skilled technical assistance as well as Mrs Ivanka Kuraži for photographic work. This investigation was supported by a grant given by the Selfmanaging Community of Interest (SIZ-V) of S.R. Croatia.

REFERENCES

ADAMSON, E. D. & AYERS, S. E. (1979). The localization and synthesis of some collagen types in developing mouse embryo. *Cell* **16**, 953–965.
ADAMSON, E. D. & GARDNER, R. L. (1979). Control of early development. *Brit. med. Bull.* **35**, 113–119.
AVE, K., KAWAKAMI, I. & SAMESHIMA, M. (1968). Studies on the heterogeneity of cell populations in amphibian presumptive epidermis, with reference to primary induction. *Devl Biol.* **17**, 617–626.
BALINSKY, B. I. (1975). *An Introduction to Embryology*. Philadelphia: Saunders Company.

BULGER, R. E. (1973). Rat renal capsule: presence of layers of unique squamous cells. *Anat. Rec.* **177**, 393–408.

CURTIS, A. S. G. (1978). Cell–cell recognition: positioning and patterning systems. In *Cell–Cell Recognition*. Symp. Soc. Exper. Biol. No. 32, pp. 51–82. Cambridge: Cambridge University Press.

DAMJANOV, I. & SOLTER, D. (1974). Experimental teratoma. *Curr. Topics Path.* **59**, 69–130.

DIWAN, S. B. & STEVENS, L. C. (1976). Development of teratomas from the ectoderm of mouse egg cylinders. *J. natn Cancer Inst.* **57**, 937–939.

GARCIA-BELLIDO, A., LAWRENCE, P. A. & MORATA, G. (1979). Compartments in animal development. *Sci. Am.* **241**, 102–110.

HALL, B. K. & TREMAINE, R. (1979). Ability of neural crest cells from the embryonic chick to differentiate into cartilage before their migration from the neural tube. *Anat. Rec.* **194**, 469–476.

HAY, E. D. (1968). Organization and fine structure of epithelium and mesenchyme in the developing chick embryo. In *Epithelial-Mesenchymal Interactions* (ed. R. Fleischmajer & R. E. Billingham), pp. 31–35. Baltimore: Williams & Wilkins.

HAY, E. D. (1973). Origin and role of collagen in the embryo. *Am. Zool.* **13**, 1085–1107.

HAY, E. D. & MEIER, S. (1978). Tissue interaction in development. Concept of Embryonic induction. Inductive mechanisms in orofacial morphogenesis. In *Textbook of Oral Biology* (eds. J. Shaw, E. Sweeney, C. Cappuccino & S. Meller), pp. 3–23. Philadelphia: Saunders.

ILLMENSEE, K. & STEVENS, L. C. (1979). Teratomas and chimeras. *Sci. Am.* **240**, 121–132.

JOLLY, J. & FÉRESTER-TADIÉ, M. (1936). Recherches sur l'oeuf du rat et de la souris. *Arch. Anat. microsc.* **32**, 323–390.

KATO, Y. & HAYASHI, Y. (1963). The inductive transformation of the chorionic epithelium into skin derivatives. *Expl Cell Res.* **31**, 599–602.

LAWSON, K. A. (1974). Mesenchyme specificity in rodent salivary gland development: the response of salivary epithelium to lung mesenchyme *in vitro*. *J. Embryol. exp. Morph.* **32**, 469–493.

LEIKOLA, A. (1976). Hensen's node – the 'organizer' of the amniote embryo. *Experienta* **32**, 269–277.

LEVAK-ŠVAJGER, B. & ŠVAJGER, A. (1971). Differentiation of endodermal tissues in homo-grafts of primitive ectoderm from two-layered rat embryonic shields. *Experientia* **27**, 683–684.

LEVAK-ŠVAJGER, B. & ŠVAJGER, A. (1974). Investigation on the origin of the definitive endo-derm in the rat embryo. *J. Embryol. exp. Morph.* **32**, 445–459.

LEVAK-ŠVAJGER, B. & ŠVAJGER, A. (1979). Course of development of isolated rat embryonic ectoderm as renal homograft. *Experientia* **35**, 258–259.

LIBBUS, B. L. & HSU, YU-CHIH (1980). Sequential development and tissue organization in whole mouse embryos cultured from blastocyst to early somite stage. *Anat. Rec.* **197**, 317–329.

MARTIN, G. R. (1977). The differentiation of teratocarcinoma stem cells *in vitro*: parallels to normal embryogenesis. In *Cell Interactions in Development* (ed. M. Karkinen-Jääskel-äinen, L. Saxén & L. Weiss), pp. 59–75. London: Academic Press.

MARTIN, G. R., WILEY, L. M. & DAMJANOV, I. (1977). The development of cystic embryoid bodies *in vitro* from clonal teratocarcinoma stem cells. *Devl Biol.* **61**, 230–244.

MORRISS, G. & THOROGOOD, P. V. (1978). An approach to cranial neural crest migration and differentiation in mammalian embryos. In *Development in Mammals* (ed. M. H. Johnson), vol. 3, pp. 363–411. Amsterdam: North Holland.

MOSCONA, A. (1959). Squamous metaplasia and keratinization of chorionic epithelium of the chick embryo in eggs and in culture. *Devl Biol.* **1**, 1–23.

MOSCONA, A. & CARNECKAS, Z. I. (1962). Etiology of keratogenic metaplasia in the chorio-allantoic membrane. *Science* **129**, 1743–1744.

NATHANSON, M. A. & HAY, E. D. (1980). Analysis of cartilage differentiation from skeletal muscle grown on bone matrix. I. Ultrastructural aspects. *Devl Biol.* **78**, 301–331.

NATHANSON, M. A., HILFER, S. R. & SEARLS, R. L. (1978). Formation of cartilage by non-chondrogenic cell types. *Devl Biol.* **64**, 99–117.

NEW, D. A. T. (1966). *Culture of Vertebrate Embryos.* London: Logos Press.

NICOLET, G. (1971). Avian gastrulation. *Adv. Morphogen.* **9**, 231–262.

PAYNE, J. M. & PAYNE, S. (1961). Placental grafts in rats. *J. Embryol. exp. Morph.* **9**, 106–116.

PETER, K. (1941). Die Genese des Entoderms bei den Wirbeltieren. *Ergebn. Anat. EntwGesch.* **33**, 285–369.

PIERCE, G. B. (1966). The development of basement membranes of the mouse embryo. *Devl Biol.* **13**, 231–249.

POELMANN, R. E. (1980). Differential mitosis and degeneration patterns in relation to the alterations in the shape of the embryonic ectoderm of early post-implantation mouse embryos. *J. Embryol. exp. Morph.* **55**, 33–51.

ROSENQUIST, G. C. (1966). A radioautographic study of labelled grafts in the chick blastoderm. Development from primitive streak stages to stage 12. *Contr. Embryol. Carnegie Inst.* **38**, 71–110.

SAXÉN, L. (1977). Directive versus permissive induction: a working hypothesis. In *Cell and Tissue Interactions* (ed. J. W. Lash & M. M. Burger), pp. 1–9. New York: Raven Press.

ŠKREB, N. & ŠVAJGER, A. (1975). Experimental teratomas in rats. In *Teratomas and Differentiation* (ed. M. Sherman & D. Solter), pp. 83–97. New York: Academic Press.

ŠKREB, N., ŠVAJGER, A. & LEVAK-ŠVAJGER, B. (1971). Growth and differentiation of rat egg-cylinders under the kidney capsule. *J. Embryol. exp. Morph.* **25**, 47–56.

ŠKREB, N., ŠVAJGER, A. & LEVAK-ŠVAJGER, B. (1976). Developmental potentialities of the germ layers in mammals. In *Embryogenesis in Mammals. Ciba Found. Symp. 40* (new series), pp. 27–45. Amsterdam: Elsevier.

SNOW, M. H. L. (1976a). Embryo growth during the immediate postimplantation period. In *Embryogenesis in Mammals. Ciba Found. Symp. 40* (new series), pp. 53–70. Amsterdam: Elsevier.

SNOW, M. H. L. (1976b). Proliferative centres in embryonic development. In *Development in Mammals* (ed. M. H. Johnson), vol. 3, pp. 337–362. Amsterdam: North Holland.

SNOW, M. H. L. (1977). Gastrulation in the mouse: growth and regionalization of the epiblast. *J. Embryol. exp. Morph.* **42**, 293–303.

SOLTER, D., DAMJANOV, I. & KOPROWSKI, H. (1975). Embryo-derived teratoma: a model system in developmental and tumor biology. In *The Early Development of Mammals* (ed. M. Balls & A. E. Wild), pp. 243–264. London: Cambridge University Press.

SOLURSH, M. & REVEL, J. P. (1978). A scanning electron microscope study of cell shape and cell appendages in the primitive streak region of the rat and chick embryo. *Differentiation* **11**, 185–190.

SPIEGELMAN, M. (1976). Electron microscopy of cell associations in T-locus mutants. In *Embryogenesis in Mammals. Ciba Found. Symp. 40* (new series), pp. 199–226. Amsterdam: Elsevier.

SPIEGELMAN, M. & BENNETT, D. (1974). Fine structural study of cell migration·in the early mesoderm of normal and mutant mouse (T-locus: t^9/t^9). *J. Embryol. exp. Morph.* **32**, 723–738.

STEVENS, D. C. (1960). Embryonic potency of embryoid bodies derived from a transplantable testicular teratoma of the mouse. *Devl Biol.* **2**, 285–297.

SUMIYA, M. (1976). Differentiation of the digestive tract epithelium of the chick embryo cultured *in vitro* enveloped in a fragment of the vitelline membrane in the absence of mesenchyme. *Wilhelm Roux Arch. devl Biol.* **179**, 1–17.

ŠVAJGER, A. & LEVAK-ŠVAJGER, B. (1974). Regional developmental capacities of the rat embryonic endoderm at the head-fold stage. *J. Embryol. exp. Morph.* **32**, 461–467.

ŠVAJGER, A. & LEVAK-ŠVAJGER, B. (1975). Technique of separation of germ layers in rat embryonic shields. *Wilhelm Roux Arch. devl Biol.* **178**, 303–308.

ŠVAJGER, A. & LEVAK-ŠVAJGER, B. (1976). Differentiation in renal homografts of isolated parts of rat embryonic ectoderm. *Experientia* **32**, 378–379.

VERWOERD, C. D. A. & VAN OOSTROM, C. G. (1979). Cephalic neural crest and placodes. *Adv. Anat. Embryol. Cell Biol.* **58**, 1–75.

WADDINGTON, C. H. (1957). *Principles of Embryology.* London: George Allen & Unwin.

WALLACE, H. (1972). The components of regrowing nerves which support the regeneration of irradiated salamander limbs. *J. Embryol. exp. Morph.* **28**, 419–435.

WATERMAN, R. E. (1976). Topographical changes along the neural fold associated with neurulation in the hamster and mouse. *Am. J. Anat.* **146**, 151–171.

WAYMOUTH, C. (1974). To disaggregate or not to disaggregate. Injury and cell disaggregation, transient or permanent. *In vitro* **10**, 97–111.

WILEY, L. M. & PEDERSEN, R. A. (1977). Morphology of mouse egg cylinder development *in vitro*: a light and electron microscopic study. *J. exp. Zool.* **200**, 389–402.

WILEY, L. M., SPINDLE, A. I. & PEDERSEN, R. A. (1978). Morphology of isolated mouse inner cell mass developing *in vitro*. *Devl Biol.* **63**, 1–10.

WOLPERT, L. (1969). Positional information and the spatial pattern of cellular differentiation. *J. Theor. Biol.* **25**, 1–47.

WOLPERT, L. (1978). Pattern formation in biological development. *Sci. Am.* **239**, 154–164.

YAMADA, T. (1977). Control mechanisms in cell-type conversion in newt lens regeneration. *Monographs in Developmental Biology, Vol. 13.* Basel: S. Karger.

YASUGI, S. & MIZUNO, T. (1974). Heterotypic differentiation of chick allantoic endoderm under the influence of various mesenchymes of the digestive tract. *Wilhelm Roux Arch. devl. Biol.* **174**, 107–116.

ZALIK, S. E. & SANDERS, E. J. (1974). Selective cellular activities in the unincubated chick blastoderm. *Differentiation* **2**, 25–28.

J. Embryol. exp. Morph. Vol. 65 (Supplement), pp. 269–287, 1981
Printed in Great Britain © Company of Biologists Limited 1981

Autonomous development of parts isolated from primitive-streak-stage mouse embryos. Is development clonal?

By MICHAEL H. L. SNOW[1]

From the MRC Mammalian Development Unit,
University College London

SUMMARY

The relationship between growth rate and regionalization of amphibian, bird and mammalian embryos is briefly reviewed. In contrast to the others, mammals start gastrulation with few cells but accelerate cell proliferation coincidentally. Experiments are described which demonstrate (1) autonomous development of pieces isolated surgically from such mouse embryos, and (2) an absence of regeneration or regulation. Since such embryos regulate completely after chemically induced random cell death it is postulated that these results reflect developmental determination and a resulting mosaicism that suggests development may have a clonal basis. Maps are drawn, allocating positions to various tissues in the embryo.

INTRODUCTION

Among those classes of vertebrates that are generally oviparous early embryonic development, i.e. cleavage stages, is characterized by very rapid cell proliferation; cell cycle times of 1 h or less are common. No doubt such rapid proliferation of cell is to some extent facilitated by the large store of prefabricated substances e.g. RNA and DNA precursors, proteins and yolk, laid down in the oocyte prior to ovulation and which the developing embryo can readily utilize without recourse to its own genome and biosynthetic pathways. In the wholly viviparous eutherian mammals it would seem that the protected environment of the oviduct and uterus has enabled them to dispense with most of the material normally stored in the oocyte and to be able to rely to a much greater extent upon the functioning of the embryonic genome. Perhaps as a consequence of this, cleavage is characteristically slow with cell cycle times ranging from 10 to over 24 h. Synthesis of new RNA is detectable at the two cell stage.

However even the mammalian embryo, or at least the mouse, is not completely independent of stored maternally derived material essential to its development and lethal maternal effect mutants are known in the mouse (for reviews of

[1] *Author's address:* MRC Mammalian Development Unit, Wolfson House (UCL), 4 Stephenson Way, London, NW1 2HE, UK.

these aspects of mammalian development see Rossant & Papaioannou, 1977; van Blerkom & Manes, 1977; Chapman, West & Adler, 1977; McLaren, 1979).

Apart from the obvious, considerable, differences in cleavage rate the progress through the early morphogenetic phases of development also show widely different rates. In amphibian embryos gastrulation is initiated some 12–20 h after fertilization when an embryo may have around 5000 cells (Sze, 1953; Graham & Morgan, 1966; Cooke, 1973, 1979). In birds the embryo is often at the early primitive streak stage at the time of laying (about 18–26 h after fertilization) and may have as many as 80000 cells (Emanuelsson, 1965; Spratt, 1963). In mammals the fastest cleavers would require a week or two to achieve such cell numbers if their early cell division rate was continued, and the slower ones nearly a month. Instead mammals commence gastrulation with comparatively few cells and seem to accelerate their cell division rate coincidentally. According to Snow (1976, 1977) the mouse embryo contains only 500–600 cells when the primitive streak forms, but 24 h later has about 12–15 thousand. If this overall proliferation rate is maintained then at the neural plate stage, a convenient end point for the process of gastrulation, the mouse embryo at 8 days will contain between 70 and 100 thousand cells. Other mammalian species including man show similar growth profiles (Snow, 1981). Vertebrate embryos of neural fold stages seem therefore to be of comparable cell number although the course by which they got there may differ. Thus amphibian and bird embryos start by accumulating large numbers of cells then slow down and rearrange the population by various cell and tissue movements into an embryo. Indeed, in *Xenopus* gastrulation will continue in the absence of cell division (Cooke, 1973) and in chicks is not fundamentally impaired by mitotic inhibitors (Overton, 1958). Mammals on the other hand seem to have acquired the ability to combine rapid proliferation with morphogenesis, and mitotic inhibitors are damaging (see Snow, 1976; Snow & Tam, 1979). It is perhaps understandable that the period of rapid cell proliferation in mammals follows implantation, when the nutritional status of the embryo is considerably improved over its free living cleavage stages.

A combination of accessibility and size has made the amphibian and bird embryos ideal objects for experimental embryology. One of the subjects that has interested biologists concerns cell lineage and the processes involved in the regionalization of embryos into tracts of tissue whose developmental course can be predicted (see Toivonen, 1979). Thus in amphibian (*Triturus*) embryos Vogt (1925, 1929) was able to mark small groups of cells on the surface of the early gastrula with vital dyes and follow the movement of those cells and thus monitor their subsequent development. His 'fate maps' for urodele amphibians have been modified only slightly in the intervening years (Landström & Lovtrup, 1979; Lovtrup, 1966); although that for anurans may be significantly different (Keller, 1975, 1976; Youn, Keller & Malacinski, 1980). Experiments in which tissue is transplanted from one part of the embryo to another demonstrate that developmental fate and developmental potency are not necessarily the same. If cells of known fate in the blastula are transplanted to another site in the embryo they are generally able to participate in the normal morphogenesis of their new site (see review by Barth, 1953). However cells of the gastrula may not be so labile in their development (Spemann, 1938) and it is now clear that the process of commitment to a certain developmental course is progressive and may extend over a considerable period of morphogenesis (Forman & Slack, 1980; Slack & Forman, 1980).

In birds a rather similar overall picture is found. Marking of cell tracts with carmine or carbon particles revealed the prospective fates of areas in the chick blastoderm which, at the mid- and late-primitive-streak stages (late gastrula) were confirmed by experiments in which the appropriate embryonic regions were isolated surgically. Such embryonic parts showed considerable autonomic development. At these late stages the fate map thus becomes equivalent to a determination map (see reviews by Rudnick, 1944; Waddington, 1952; Bellairs, 1971; Hara, 1978). Nevertheless the unincubated blastoderm of the chick, duck or quail (pre- or early-primitive-streak stage) can be transected and each piece regulates to form a complete embryo (Lutz, 1949, 1955; Lutz & Lutz-Ostertag, 1963, 1964; Eyal-Giladi & Spratt, 1965; Rogulska, 1968; Eyal-Giladi, Kochav & Menashi, 1976), even though vital dye markings permit the drawing of a fate map for early-primitive-streak stages (see Waddington, 1952).

Such fate maps or determination maps are a prerequisite for further investigation of the cellular events associated with the progressive differentiation of particular organ systems and for studying interactions between tissues associated with normal embryogenesis. In mammalian embryos several studies indicate that complete regulations can occur and normal individuals result from single cells of the embryos at least up to the 8-cell stage (Moore, Adams & Rowson, 1968; Kelly, 1975). Identical twins can be made by separating blastomeres at the 2-cell stage in mice and sheep and quadruplets have been made in sheep from 4-cell embryos (Willadsen, 1979, and personal communication). Single cells or groups of cells from the inner cell mass of $3\frac{1}{2}$-day embryos have the potential to contribute to many tissues in the late fetus if grafted into a host blastocyst (Gardner, 1978). However, by this stage of development inner-cell-mass cells have lost the ability to form trophectoderm and are beginning to show signs of further restriction in their developmental potency although there is no evidence that determination of cell lineages in the fetus has yet occurred. From studies on aggregation chimeras it seems unlikely that even a fate map of normal prospective development could be drawn for these early stages of the mammalian embryo. In an aggregate of two 8-cell mouse embryos, the cells of the respective embryos remain as discrete populations and there is very little cell mingling by the time of implantation (Garner & McLaren, 1974). In the chimeric fetuses developed from such aggregates, the cell populations are very thoroughly mixed and there is very little evidence that would indicate regionalisation of the inner cell mass even to the extent that anterior–posterior or dorsoventral axes may be present. Recently, 31 aggregate chimeras, age between $6\frac{1}{2}$ and $7\frac{1}{2}$ days *post coitum* (i.e. 3–4 days after implantation and at early- to mid-primitive-streak stages) have been analysed (Matta & Snow, unpublished). These show that the cell mingling is already very extensive, but provide no indication as to how it is brought about. But for Gardner's data, the simple explanation of the cell mixing in aggregate chimeras would be that cells are determined and a process of sorting out occurs to ensure that cells of like developmental potential become located in the same region of the embryo. Clearly this cannot be so and if cell movement is a normal occurrence in single embryo inner cell masses, then no fate map could be drawn for these early stages. Anatomically the developing epiblast components of the inner cell mass become an epithelium with basement membrane and with an apical bar network at the luminal surface of the cells at about 6 days *post-coitum* and it seems reasonable to conclude that cell mingling should precede that event, and equally that regionalisation should either accompany that tissue transformation or follow it.

In the mouse a primitive streak forms at about $6\frac{1}{2}$–7 days *post-coitum*, at which time the anterior posterior axis becomes apparent. The anatomy of subsequent development, the patterns and distribution of mitosis and the short-term vital marking of rabbit embryos (Daniel & Olson, 1966), all suggest that despite the variable geometry of the mammalian embryo, it is probably very similar in layout to the bird embryo and it is tempting to assume that the developmental status and capacity of its parts will be similar (review Snow, 1978).

The development of an *in vitro* culture system allowing the normal development of early and mid-primitive-streak mouse embryos (Tam & Snow, 1980) has provided an opportunity to carry out some basic embryological experiments on the regionalization of the mouse embryo. Initially attempts were made to graft fairly large pieces of donor embryo into a host of similar stage in order to provide potential duplications of e.g. primitive streak, Hensen's node, etc. However the presence of a continually expanding yolk sac (exocoelom) invariably forced the grafts apart. Efforts to prevent yolk-sac expansion by inserting as many as three small glass tubes were frustrated by the embryo managing to seal up the tubes with the growing allantois interiorly or by a covering of primary endoderm over the outside projection. The useful observation from these experiments was that a great deal of damage could be done to the yolk sac without apparently disturbing embryonic development at all, even to the extent that head-fold formation and somitogenesis would occur without the yolk sac at all. Subsequently I have focussed upon the development of isolated pieces and of embryos from which small pieces have been removed.

The culture system used is that described by Tam & Snow (1980). In order to prevent the pieces or the operated embryos sticking to the substrate where the cells then grow as a monolayer destroying morphology, the plastic dishes were

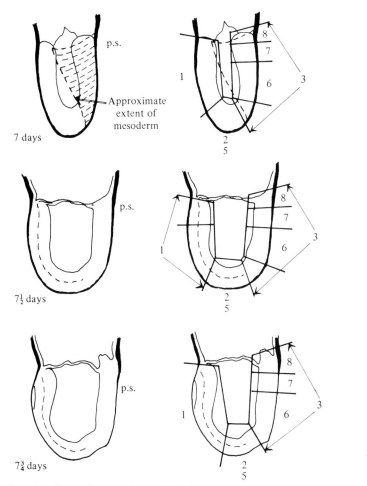

Fig. 1. Outline drawings of the embryos showing the position of the cuts used in this study. Age of embryo is indicated on left. ps = primitic streak.

coated with 1 % agar in Dulbeccos modified Eagles medium. All surgery was done by hand with tungsten needles with the aid of a dissecting microscope. Cultures were usually analyzed after 24 h of development. Because of the limitations of the culture system and the small size of the mouse embryos only early-, mid- and late-primitive-streak stages have been attempted. Figure 1 illustrates the stages of embryos used and marks in the lines along which the majority of cuts are made. The cuts transect all tissues in the embryo, i.e. they pass through the primary endoderm, mesoderm (when present) and epiblast/ectoderm. Pieces therefore contain all of these tissues. Figure 2 shows exploded diagrams of a late-streak embryo to illustrate the important pieces and their relative sizes. The pieces in the primitive streak have been removed singly or in

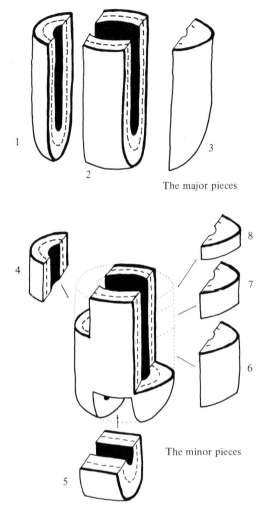

Fig. 2. Exploded diagrams of the embryos showing the position and relative size of the pieces. Pieces 4,5,7, and 8 contain some 150–200 cells.

various combinations, and a few other types of excision have been made. These will be described at the appropriate time.

Tables 1, 2, and 3 summarize the data. A piece is described as growing simply if it increases in size, and as undergoing morphogenesis if normal embryonic structures can be seen in it with a dissecting microscope. In the latter case growth of the pieces, as measured by protein content, is comparable to that in the intact embryo (Table 4). A brief record of the structures formed by the pieces (Table 2) and by the deficient embryos (Table 3) is given, and some of the pieces and their products are illustrated in Figs 3–7.

The development of the major pieces and of the embryos lacking pieces clearly demonstrates a mosaicism of developmental potential in the various

Table 1. *A summary of the number of pieces and deficient embryos cultured*

Piece	*n*	No. growing	No. differentiating
1	68	66	51
2	47	47	24*
3	57	56	48
1/2	51	51	45
1/3	8	8	8
2/3	65	65	61
4	8	8	0
5	34	34	0
6	18	18	18
7	20	18	18
8	14	14	14
6/7/8	25	23	17
7/8	24	22	22
'Sag'	26	26	22
'lat'	19	19	2
Total	484	475	350 (71·4 %)
Embryos			
−4	8	8	8
−5	40	38	37
−6	18	18	18
−7	20	20	20
−8	10	10	10
−6/7/8	25	25	25
−6/7	18	18	18
−7/8	24	23	22
−lat	20	20	20
Total	183	180	178 (98 %)

* Of 20 pieces isolated from their yolk sac only 3 showed evidence for differentiation.

regions in the primitive-streak-stage embryo. A comparison of the development of adjacent pieces shows no evidence for duplication of any structure lying at the cut boundary and thus strongly suggests that these pieces have no capacity for regeneration or regulation of either an epimorphic or morphallactic type. In the few cultures carried through 48 h of development these conclusions remain the same. Although development slows down a little during the second day in culture, no additional structures appear in any piece that would suggest regeneration.

From Tables 2 and 3 it can be seen that the capacity to form some structures change with age and also seems to shift from one piece to an adjacent piece. For instance, Piece 1 from a 7-day embryo will not form head structures. This is believed to be because the piece lacks mesoderm at the time of excision and in the culture period has no means of acquiring any; hence proper neural induction cannot occur. However it should be noted that such a piece from a rat embryo will make mesodermal structures in the teratoma it produces when grafted to an

Table 2. *Summary of the development of isolated pieces*

	Piece	n	NP*	HF*	FG	H	NT	S	TB	HG	PGC	A
≥7½ days p.c.	1	49	2†	+	+	+	−	−	−	−	−	−
	2	17	−	−	−	4	+	+	−	−	−	−
	3	38	−	−	−	−	−	−	+	+	+	+
	6	8	−	−	−	−	−	−	+	5	ND	−
	7	13	−	−	−	−	−	−	−	3	+	−
	8	10	−	−	−	−	−	−	−	−	6	+
	6/7/8	10	−	−	−	−	−	−	+	+	+	+
	7/8	5	−	−	−	−	−	−	−	1	+	+
	Sag	18	+	+	+	+	+	+	+	+	ND	+
	lat	2	−	−	−	−	−	2	−	−	−	−
7 days p.c.	1	2	+	−	−	−	−	−	−	−	−	−
	2	7	−	−	−	+	+	−	−	−	−	−
	3	10	−	−	−	−	−	−	+	+	+	+
	6	10	−	−	−	−	−	−	+	3	ND	−
	7	5	−	−	−	−	−	−	−	−	+	−
	8	4	−	−	−	−	−	−	−	−	3	+
	6/7/8	7	−	−	−	−	−	−	+	+	+	+
	7/8	17	−	−	−	−	−	−	−	4	+	+
	Sag	4	+	+	+	+	+	+	+	+	ND	+

* NP and HF are exclusive stages in cranial neural plate development.
† These embryos arrested at NP stage.
ND = not done.

Table 3. *Summary of the development of deficient embryos*

	Embryo	n	NP	HF	FG	H	NT	S	TB	HG	PGC	A
≥7½ days p.c.	−4	7	−	+*	+	+	+	+	+	+	+	+
	−5	32	−	+	+	4	+	+	+	+	+	+
	−6	8	−	+	+	+	+	+	−	3	+	+
	−7	13	−	+	+	+	+	+	+	+	−	+
	−8	10	−	+	+	+	+	+	+	+	+	−
	−6/7/8	12	−	+	+	+	+	+	−	−	−	−
	−6/7	4	−	+	+	+	+	+	−	1	−	+
	−7/8	5	−	+	+	+	+	+	+	−	−	−
	−lat	20	−	+	+	+	+	+†	+	+	+	+
7 days p.c.	−5	5	−	+	+	−	+	2	+	+	+	+
	−6	10	−	+	+	1	+	1	−	2	+	+
	−7	6	−	+	+	+	+	+	+	+	−	+
	−8	4	−	+	+	+	+	+	+	+	+	−
	−6/7/8	13	−	+	+	6	+	1	2	−	−	−
	−6/7	14	−	+	+	5	+	−	3	−	−	+
	−7/8	17	−	+	+	+	+	+	+	+	−	−

* Lacked fore brain; mid hind brain seemed normal.
† Generally one side only, but see text.

Table 4. *Protein content of intact embryos and the major pieces*
before and after our culture

$7\frac{1}{2}$ days	n	Protein (μg)
Whole egg cylinder	10	9·6
Yolk sac only	6	7·1
Embryo		2·5
Piece 1	11	1·4
Piece 2 + yolk sac	6	7·3
Piece 3	6	1·2
		9·9
$8\frac{1}{2}$ day embryo only	10	18·8
yolk sac	18	13·1
Piece 1	15	8·1
Piece 2	18	8·4
Piece 3	16	2·9
		19·4

ectopic site (Svajger, Levak-Svajger, Kostovic-Knezevic & Bradamante, 1981). Other organ-forming capacities that 'move' are for heart and somites. In younger embryos heart-forming capacity does not reside in Piece 1 but is located in Piece 2 and can be severely interfered with by removal of Piece 6 (Table 3). This can be interpreted as showing that heart mesoderm emerges from the anterior end of the primitive streak at about 7 days *p.c.* and subsequently comes to occupy a more and more anterior position until it is located in the Piece 1 from a $7\frac{1}{2}$-day embryo. Whether this shift in position is achieved by active migration of heart mesoderm cells or by differential growth of the embryo which changes shape around them is not known (see Poelman, 1980, for discussion of gastrulation movements in mouse embryos).

Somite-forming ability also changes in time and space (Tables 5 and 6). Piece 1 and 3 alone never make somites but Piece 3 seems to possess somite-generating ability if attached to Piece 2 at 7 days *p.c.* when Piece 2 alone will not make somites. Piece 2 from an embryo of $7\frac{1}{2}$ days or older will make as many somites as an intact embryo (Fig. 6), and removal of Piece 5 from the distal tip of the egg cylinder (a piece which roughly corresponds to Hensens node) reduces somite number by almost 2. The lateral piece referred to in Table 6 (lat) is a piece removed from the right or left side of the egg cylinder at $7\frac{1}{2}$ days *p.c.* or older. It is similar in shape and size to Piece 1 but does not extend significantly into the domain of that piece; it also leaves the central AP axis intact. Removal of the lateral piece interferes with somite formation on the operated side only. In only 4 of the 19 embryos which made somites following removal of the lateral piece were somites made on the operated side. They were the most posterior somites; the somites 1–4 were completely missing in two embryos and very small

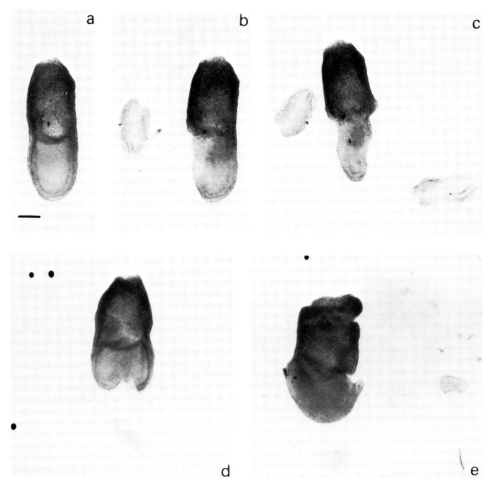

Fig. 3. (*a*) A 7½-day egg cylinder, the lower half is the embryo. (*b*) Piece 1 removed, (*c*) Piece 1 and 3 removed, (*d*) Piece 5 removed, and (*e*) Piece 7 removed from a slightly older embryo. Bar = 100 μm.

in the other two but when somites 5, 6, 7 or 8 were formed they were of normal dimension in comparison to the unoperated side. Sagittal halves (sag, Tables 1, 2 and 6) made half embryos with normal somite numbers; there is no lateral regulation or regeneration.

Three prospective tissues do not move but are of considerable interest in terms of embryonic development. Pieces 6, 7 and 8 contain prospective tail-bud, primordial germ cells (PGCs) and allantois respectively. The location of these regions is the same at 7 and 7½ days *p.c.* indicating an absence of movement of cells through this part of the primitive streak during this period. It is difficult to explain these observations on the basis of epigenetic specification of these developmental potentials since wound healing following removal from a 7-day embryo does not result in replacement of the tissue. This is adequately shown

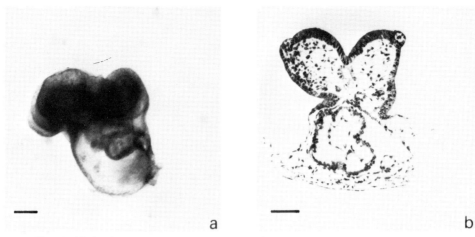

Fig. 4. (*a*) An isolated head fold and heart formed from a Piece 1, (*b*) A frontal
section of such a piece. Bar = 100 μm.

Table 5. *The somite-forming ability of various pieces and deficient embryos*

Piece	Age (days)	*n*	No. with somite	(%)	Somite No.
1/2	7	8	0		–
	7½	23	18	(78)	4·2 ± 0·4
2	7	7	0		
	7½	17	17	(100)	5·1 ± 0·5
2/3	7	4	4	(100)	5·0 ± 0·4
	7½	39	30	(77)	4·8 ± 0·3
Embryo					
− 5	7	6	2	(33)	2 and 2
	7½	32	25	(78)	4·0 ± 0·3†
− 6*	7	23	2	(9)	2 and 3
	7½	20	20	(100)	5·4 ± 0·6
Control	all	46	42	(91)	5·8 ± 0·3

* includes operations removing 6/7/8.
† Students *t*-test, significantly different from control $P < 0.05$.

with respect to the allantois but is more easily quantitated by reference to the
germ cells. Primordial germ cells in the mouse are alkaline phosphatase (AP)
positive and in normal circumstances can only be identified at about 8½ days *p.c.*
when they emerge from an AP-positive epiblast at the posterior end of the
primitive streak at the base of the allantois (Oždženski,1967; Tam & Snow,
1981). Table 7 gives the number of AP-positive cells (presumed PGC's) found in
isolated pieces or deficient embryos after 24 or 36 h culture. All the PGCs missing
from the embryos as a result of removing Pieces 7 and 8 can be accounted for
in the isolated pieces. Clearly there is no ability to replace the potential lost by
the surgical removal of these parts.

The development of the pieces which contain germ cells shows a varied

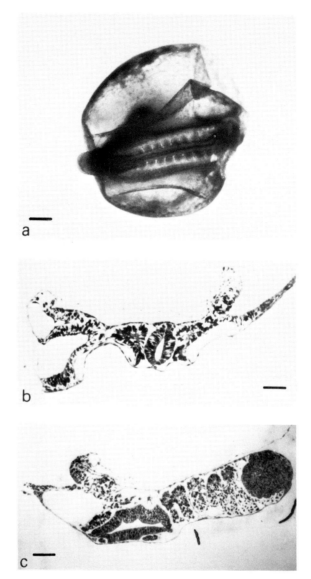

Fig. 5. (a) A cultured Piece 2, attached to its yolk sac, (b) A transverse and (c) a longitudinal section of such a piece. Bar × 100 μm.

morphology. Piece 7/8 will grow as an allantois and germ cells become distributed throughout the structure (Fig. 7a) with no apparent tendency to congregate together at any point. However, if Piece 6 is included then in those developing a hind gut invagination (Table 2) all the PGCs are localized around the endoderm lining that invagination (Fig. 7b). This indicates an affinity between endoderm and PGCs which is reminiscent of the chemotaxis claimed for chick germ cells (Dubois & Croisille, 1970; Rogulska, Oźdźeński & Komar, 1971); but the

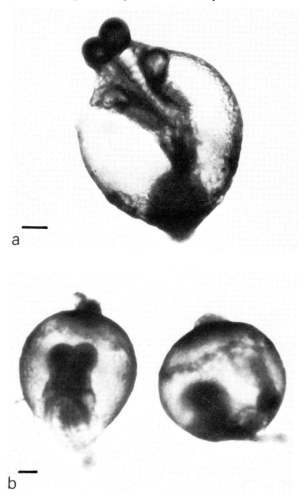

Fig. 6. (*a*) An embryo from which Piece 4 was removed at $7\frac{1}{2}$ days. The prominent mid brain has herniated through the incision and no fore brain has developed. (*b*) Two embryos from which Piece 7 was removed. Although the tail herniates through the incision the major defect in these embryos is a lack of germ cells. Bar = 100 μm.

exact means of route-finding in mammalian germ cells remains mysterious (McLaren, 1981).

The presence of these tissues in the primitive steak raises questions about the role of regression movements in somite formation. In chicks a considerable shortening of the primitive streak occurs during early somitogenesis, believed to be brought about by the anterior end of the streak retracing its steps (regressing) and 'organizing' the mesoderal tissues into somites as it does so, (see Spratt, 1947 and Bellairs, 1963, 1971, 1979). Measurements of primitive-streak length in the two strains of mice used in my studies show that the maximum length is achieved at $7\frac{1}{2}$ days and during the ensuing 24 h (when the embryo will generate

Table 6. *Somite formation in sagittally halved embryos and embryos lacking one lateral side*

		No. with somites		Somite No.	
	n	Both sides	One side	Cut side	Intact side
Control Q	42	42	–	–	5·8 ± 0·3
Sagittal	22	18	–	5·3 ± 0·4	–
Control MF1	61	61	–	–	6·7 ± 0·2
Lat-	20	4	15	1·3 ± 0·6*	6·8 ± 0·6
lat piece only	18	–	2	3 and 3	–

* see text.

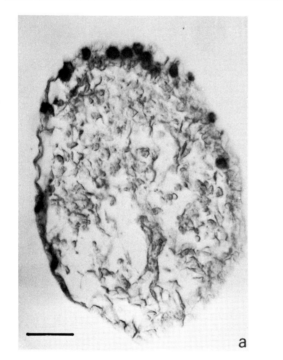 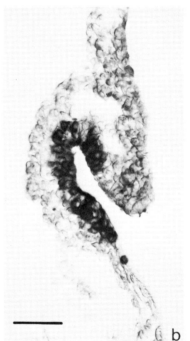

Fig. 7. (*a*) An allantois containing primordial germ cells (PGC's; darkly stained) developed from Piece 7/8, (*b*) PGC's clustered around the hind gut invagination in a cultured Piece 6/7. Bar = 50 μm.

5–7 pairs of somites) shortens only slightly (Table 8). Indeed the shortening would seem to be accounted for entirely by the loss from the posterior half of the streak of the allantois and PGC precursors, and it is to be concluded that regression movements play no part in somitogenesis in the mouse.

From the foregoing data it is possible to create a map of areas whose development can be predicted (Fig. 8). In the light of past controversies about fate maps versus determination maps I prefer to regard this as an allocation map, until

Table 7. *Formation of primordial germ cells in isolated pieces or deficient embryos*

	No. PGCs (n)†	
Piece	7 days *p.c.*	7½ days *p.c.*
1/2	0 (7)	0 (14)
6	ND	5 (3)
7	ND	67 (1)
8	27·5 (4)	15 (9)
6/7/8	ND	58 (11)
6/7	ND	ND
7/8	122 (3)	58 (15)
Embryo		
Intact	99 (6)	75 (22)
−6	73 (7)	66 (11)
−7	21 (5)	20 (7)
−8	ND	ND
−6/7/8	0 (7)	0 (18)
−6/7	14 (14)	1·5 (4)
−7/8	ND	2 (14)

* Embryos cultured from 7 days were MF1 strain, those from 7½ days strain were Q strain.
† Figure in parentheses is the number of observations.

such time as tissues have been grafted from site to site and the degree of determination of the region has been directly tested. Suffice it to say that the behaviour of the pieces removed from the primitive streak, at several developmental ages and in various combinations shows no evidence for lability in developmental potential and no abnormal arrangements of tissues have been seen. This seems to indicate a rather determined state. However it is in contrast to the picture that emerges from the study of teratomas where the usual layout of pattern is severely disturbed although good organotypic structures can be found. It must remain speculation whether or not Piece 1, for instance, could develop an ability to make additional mesodermal and/or endodermal components since cultures cannot at present be prolonged to times equivalent to those in teratoma studies.

At first sight the autonomous development of embryonic parts seems a contradiction to the recent Mitomycin C experiments of Snow & Tam (1979) in which cell number in the 7–7½-day mouse embryo can be reduced by random killing of cells to about 10 % of normal value, without disturbing gastrulation and the resulting formation of a neural-plate-stage embryo with a few pairs of somites (see also Snow, Tam & McLaren, 1981; Tam, 1981, this volume). In those experiments a very rapid reprogramming of development by the surviving cells seems necessary. One mechanism whereby this could be achieved requires that the surviving cells be developmentally labile and that the regionalization of the embryo of reduced size is by a reassignment of 'positional value' according to the site a cell now occupies in the embryonic field (Wolpert, 1969, 1971).

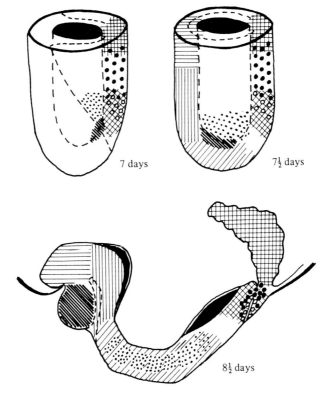

Fig. 8. A map of the embryo allocating positions of the tissues at various ages.

Table 8. *Changes in primitive-streak length during early somitogenesis*

Age (days *p.c.*)	MF1		Q	
	n	Length (*μ*m)	*n*	Length (*μ*m)
$7\frac{1}{2}$	20	358 ± 32	19	316 ± 14
8	19	271 ± 15	10	285 ± 16
$8\frac{1}{2}$	18	238 ± 13	20	273 ± 17

Alternatively, at the time cells are killed by Mitomycin C the regions of the embryo may already be determined but within each region there are a number of cells that survive the Mitomycin C damage. These cells then proliferate to refill the region or compartment for which they are determined. In the light of the data on isolated pieces a mechanism of the latter kind seems more likely, although it must be borne in mind that the cell killing by Mitomycin C may create circumstances in which lability is induced in the survivors and embryonic regulation with reduced cell numbers is an epigenetic phenomenon.

Should the 'clonal' survival of some cells in each determined tissue be correct it is difficult to avoid a comparison with the developmental 'compartments' identified in *Drosophila* imaginal discs (Garcia-Bellido 1975; Simpson, 1981, this volume). Nevertheless there is at present no observation on the excised pieces, or the deficient embryos that would indicate that 'compartments', in the sense that they are defined in *Drosophila*, are also found in mouse embryos.

I gratefully acknowledge the assistance of Patrick Tam for scoring germ cells, and Jacquie Mace for measuring primitive streaks.

REFERENCES

BARTH, L. G. (1953). *Embryology*, 2nd ed. New York: Holt, Rinehart & Winston.

BELLAIRS, R. (1963). The development of somites in the chick embryo. *J. Embryol. exp. Morph.* **11**, 697–714.

BELLAIRS, R. (1971). *Developmental Processes in Higher Vertebrates*. London: Logos Press.

BELLAIRS, R. (1979). The mechanism of somite segmentation in the chick embryo. *J. Embryol. exp. Morph.* **51**, 227–243.

CHAPMAN, V. M., WEST, J. D. & ADLER, D. A. (1977). Genetics of early mammalian embryogenesis. In *Concepts in Mammalian Embryogenesis* (ed. M. I. Sherman), pp. 95–135. MIT press.

COOKE, J. (1973). Properties of the primary organisation field in the embryo of *Xenopus laevis*. IV. Pattern formation and regulation following early inhibition of mitosis. *J. Embryol. exp. Morph.* **30**, 49–62.

COOKE, J. (1979). Cell number in relation to primary pattern formation in the embryo of *Xenopus laevis*. I. The cell cycle during new pattern formation in response to implanted organizers. *J. Embryol. exp. Morph.* **51**, 165–182.

DANIEL, J. C. & OLSON, J. D. (1966). Cell movement, proliferation and death in the formation of the embryonic axis of the rabbit. *Anat. Rec.* **156**, 123–128.

DUBOIS, R. & CROISILLE, Y. (1970). Germ-cell line and sexual differentiation in birds. *Phil. Trans. Roy. Soc. Lond.* B **259**, 73–89.

EMANUELSSON, H. (1965). Cell multiplication in the chick blastoderm up to the time of laying. *Expl Cell Res.* **39**, 386–399.

EYAL-GILADI, H. & SPRATT, N. T. (1965). The embryo-forming potencies of the young chick blastoderm. *J. Embryol. exp. Morph.* **13**, 267–274.

EYAL-GILADI, H., KOCHAV, S. & MENASHI, M. K. (1976). On the origin of primordial germ cells in the chick embryo. *Differentiation* **6**, 13–16.

FORMAN, D. & SLACK, J. M. W. (1980). Determination and cellular commitment in the embryonic amphibian mesoderm. *Nature* **286**, 492–494.

GARCIA-BELLIDO, A. (1975). Genetic control of wing disc development in *Drosophila*. In '*Cell patterning*', *Ciba Foundation Symposium* **29**, 161–182.

GARDNER, R. L. (1978). The relationship between cell lineage and differentiation in the early mouse embryo. In *Genetic Mosaics and Cell Differentiation* (ed. W. J. Gehring), pp. 205–242. Berlin-New York: Springer-Verlag.

GARNER, W. & MCLAREN, A. (1974). Cell distribution in chimaeric mouse embryos before implantation. *J. Embryol. exp. Morph.* **32**, 495–503.

GRAHAM, C. F. & MORGAN, R. W. (1966). Changes in the cell cycle during early amphibian development. *Devl Biol.* **14**, 439–460.

HARA, K. (1978). 'Spemann's Organizer' in birds. In *Organizer – A Milestone of a Half-century from Spemann* (ed. O. Nakamura & S. Toivonen), pp. 221–265. North Holland, Amsterdam: Elsevier.

KELLER, R. E. (1975). Dye mapping of the gastrula and neurula of *Xenopus laevis*. I. Prospective areas and morphogenetic movements of the superficial layer. *Devl Biol.* **42**, 222–241.

KELLER, R. E. (1976). Dye mapping of the gastrula and neurula of *Xenopus laevis*. II. Prospective areas and morphogenetic movements of the deep layer. *Devl Biol.* **51**, 118–137.

KELLY, S. J. (1975). Potency of early cleavage blastomeres of the mouse. In *The Early Development of Mammals. Brit. Soc. Devl. Biol. Symp.* 2 (ed. M. Balls & A. E. Wild), pp. 97–105. London: Cambridge University Press.

LANDSTRÖM, U. & LØVTRUP, S. (1979). Fate maps and cell differentiation in the amphibian embryo – an experimental study. *J. Embryol. exp. Morph.* **54**, 113–130.

LØVTRUP, S. (1966). Morphogenesis in the amphibian embryo. Cell type distribution, germ layer, and fate maps. *Acta Zool.* (Stockholm) **47**, 209–276.

LUTZ, H. (1949). Sur la production expérimentale de la poly-embryonie et de la monstrosité double chez les oiseaux. *Archs Anat. microsc. morph. exp.* **38**, 79–144.

LUTZ, H. (1955). Contribution expérimentale à l'etude de la formation de l'endoblaste chez les oiseaux. *J. Embryol. exp. Morph.* **3**, 59–76.

LUTZ, H. & LUTZ-OSTERTAG, Y. (1963). Sur l'orientation des embryons jumeax obtenus par fissuration parallèle à l'axe présumé du blastoderme son incubé de l'oeuf de caille (*Coturnix coturnix japonica*) *C.r. hebd. Séanc. Acad. Sci., Paris* **256**, 3752–3754.

LUTZ, H. & LUTZ-OSTERTAG, Y. (1964). Orientation des embryons obtenus par fissuration perpendiculaire à l'axe présumé de l'oeuf de caille (*Coturnix coturnix japonica*). *C. R. Soc. Biol.* **158**, 1504–1506.

MCLAREN, A. (1976). *Mammalian Chimaeras.* Cambridge: Cambridge University Press.

MCLAREN, A. (1979). The impact of pre-fertilization events on post-fertilization development in mammals. In *Maternal Effects in Development* (ed. D. R. Newth & M. Balls), *Brit. Soc. Devl. Biol. Symp.* **4**, pp. 287–320. Cambridge University Press.

MCLAREN, A. (1981). *Germ Cells and Soma: a New Look at an Old Problem.* New Haven and London: Yale University Press. (In the Press.)

MOORE, N. W., ADAMS, C. E. & ROWSON, L. E. A. (1968). Developmental potential of single blastomeres of the rabbit egg. *J. Reprod. Fert.* **17**, 527–531.

OVERTON, J. (1958). Effects of colchicine on the early chick blastoderm. *J. exp. Zool.* **139**, 329–348.

OŻDŻEŃSKI, W. (1967). Observations on the origin of primordial germ cells in the mouse. *Zoologica Pol.* **17**, 367–379.

POELMANN, R. E. (1980). Differential mitosis and degeneration patterns in relation to the alterations in the shape of the embryonic ectoderm of early post-implantation mouse embryos. *J. Embryol. exp. Morph.* **55**, 35–51.

REINIUS, S. (1965). Morphology of the mouse embryo, from the time of implantation to mesoderm formation. *Zeit. für Zellforsch.* **68**, 711–723.

ROGULSKA, T. (1968). Primordial germ cells in normal and transected duck blastoderms. *J. Embryol. exp. Morph.* **20**, 247–260.

ROGULSKA, T., OŻDŻEŃSKI, W. & KOMAR, A. (1971). Behaviour of mouse primordial germ cells in the chick embryo. *J. Embryol. exp. Morph.* **25**, 155–164.

ROSSANT, J. & PAPAIOANNOU, V. E. (1977). The biology of embryogenesis. In *Concepts in Mammalian Embryogenesis* (ed. M. I. Sherman), pp. 1–36. MIT Press.

RUDNICK, D. (1944). Early history and mechanics of the chick blastoderm. *Q. Rev. Biol.* **19**, 187–212.

SIMPSON, P. (1981). Growth and cell competition in *Drosophila. J. Embryol. exp. Morph.* **65**, (*Suppl*), 77–88.

SLACK, J. M. W. & FORMAN, D. (1980). An interaction between dorsal and ventral regions of the marginal zone in early amphibian embryos. *J. Embryol. exp. Morph.* **56**, 283–299.

SNOW, M. H. L. (1976). Embryo growth during the immediate post-implantation period. In *Embryogenesis in mammals. Ciba Foundation Symposium* **40**, pp. 53–70. Amsterdam: Elsevier/North-Holland.

SNOW, M. H. L. (1977). Gastrulation in the mouse: growth and regionalisation of the epiblast. *J. Embryol. exp. Morph.* **42**, 293–303.

SNOW, M. H. L. (1978). Proliferative centres in embryonic development. In *Development in Mammals*, vol. 3 (ed. M. H. Johnson), pp. 337–362.

SNOW, M. H. L. (1981). Growth and its control in early mammalian development. *Brit. Med. Bull.* **37** (in the Press).

SNOW, M. H. L. & TAM, P. P. L. (1979). Is compensatory growth a complicating factor in mouse teratology? *Nature* **279**, 555–557.

SNOW, M. H. L., TAM, P. P. L. & McLAREN, A. (1981). On the control and regulation of size and morphogenesis in mammalian embryos. In *Levels of Genetic Control and Development*. 39th *Symp. Am. Soc. Devl. Biol.* (ed. S. Subtelny). (In the Press.)

SPEMANN, H. (1938). *Embryonic Development and Induction*. New York: Hafner.

SPRATT, N. T. (1947). Regression and shortening of the primitive streak in the explanted chick blastoderm. *J. exp. Zool.* **104**, 69–100.

SPRATT, N. T. (1963). Role of the substratum, supracellular continuity and differential growth in morphogenetic cell movements. *Devl Biol.* **7**, 51–63.

ŠVAJGER, A., LEVAK-ŠVAJGER, B., KOSTOVIĆ-KNEŽEVIĆ, L. & BRADAMANTE, Ž. (1981). Morphogenetic behaviour of the rat embryonic ectoderm as renal homo-graft. *J. Embryol. exp. Morph.* **65**, (*Suppl*), 243–267.

SZE, L. C. (1953). Respiration of the parts of the *Rana pipiens* gastrula. *Physiol. Zool.* **26**, 212–223.

TAM, P. P. L. (1981). The control of somitogenesis in mouse embryos. *J. Embryol. exp. Morph.* **65**, (*Suppl*), 103–128.

TAM, P. P. L. & SNOW, M. H. L. (1980). The *in vitro* culture of primitive-streak-stage mouse embryos. *J. Embryol. exp. Morph.* **59**, 131–143.

TAM, P. P. L. & SNOW, M. H. L. (1981). Proliferation and migration of primordial germ cells during compensatory growth in mouse embryos. *J. Embryol. exp. Morph.* **64**, 133–147.

TOIVONEN, S. (1979). Regionalization of the embryo. In *Organizer – A Milestone of a Half-century from Spemann* (ed. O. Nakamura & S. Toivonen), pp. 119–178. Amsterdam: Elsevier/North Holland.

VAN BLERKOM, J. & MANES, C. (1977). The molecular biology of the preimplantation embryo. In *Concepts in Mammalian Embryogenesis* (ed. M. I. Sherman), pp. 37–94. MIT Press.

VOGT, W. (1925). Gestaltungsanalyse am Amphibienkeim mit örtlicher Vital-färbung. I. Methodik und Wirkungsweise der orthichen Vital-färbung mit Agar als Farbtrager. *Wilhelm Roux. Arch. EntwickMech. Org.* **106**, 542–610.

VOGT, W. (1929). Gestalangsanalyse am Amphibienkeim mit örtlicher Vital-färbung II. Gastrulation und Mesoderm bildung bei Urodelen und Anuren. *Wilhelm Roux. Arch. EntwickMech. Org.* **120**, 384–706.

WADDINGTON, C. H. (1952). *The Epigenetics of Birds*. Cambridge University Press.

WILLADSEN, S. M. (1979). A method for culture of micromanipulated sheep embryos and its use to produce monozygotic twins. *Nature* **277**, 298–300.

WOLPERT, L. (1969). Positional information and the spatial pattern of cellular differentiation. *J. theor. Biol.* **25**, 1–48.

WOLPERT, L. (1971). Positional information and pattern formation. *Curr. Top. Devl. Biol.* **6**, 183–224.

YOUN, B. W., KELLER, R. E. & MALACINSKI, G. M. (1980). An atlas of notochord and somite morphogenesis in several anuran and urodelan amphibians. *J. Embryol. exp. Morph.* **59**, 223–247.

J. Embryol. exp. Morph. Vol. 65 (*Supplement*), pp. 289–307, 1981
Printed in Great Britain © Company of Biologists Limited 1981

The role of morphogenetic cell death during abnormal limb-bud outgrowth in mice heterozygous for the dominant mutation *Hemimelia-extra toe* (*Hmˣ*)

By T. B. KNUDSEN and D. M. KOCHHAR[1]

From the Department of Anatomy, Thomas Jefferson University
Philadelphia

SUMMARY

A dominant mutation in the mouse, *Hemimelia-extra toe* (*Hmˣ*), induces congenital limb malformations in heterozygotes. Typical expression includes axial shortening of the radius, tibia and talus ('hemimelia'), with supernumerary metacarpals, metatarsals, and digits ('polydactyly'). Pathogenesis was investigated during developmental stages 16 through 22 (11th through 15th days of gestation). Full expression was apparent during stage 20 when the limb pattern was comprised of pre-cartilaginous anlagen. Formation of a pre-axial protrusion on the autopod during stage 17 or 18 was the earliest gross abnormality, and foreshadowed the development of supernumerary digits. Microscopically, there was an alteration in the pattern of physiologic cellular degeneration (PCD) programmed to occur within the zeugopod and autopod. The 'opaque patch' (mesodermal necrotic zone normally occurring between tibial and fibular anlagen) was overextended pre-axially causing resorption of the tibial pre-cartilage. Additionally, PCD normally occurring within the basal cell layer of the apical ectodermal ridge (AER) and the 'foyer primaire préaxial' was not expressed in the mutant autopod. This occurred in association with outgrowth of the protrusion. The pre-axial portion of the AER remained in an abnormally thickened, viable, proliferative state, and did not undergo scheduled degression. This may have been the basis for prolonged induction of pre-axial outgrowth. Paucity of mesenchymal cell filopodial processes extended along the basal lamina, as well as a rarefaction of the filamentous material normally associated with the mesodermal face of the basal lamina, was detected at the pre-axial AER–mesenchymal interface on stage 18. A potential involvement of epithelial–mesenchymal interactions in the induction of epithelial PCD is discussed.

INTRODUCTION

Normal skeletal pattern in the limb can be disrupted by genomic mutations, but the precise mechanisms for pathogenesis are not understood (Hinchliffe & Ede, 1967; Johnson, 1969; Cairns, 1977; Rooze, 1977). Abnormal limb pattern may also be induced in genetically normal embryos by the administration of various anti-proliferative agents during a 'critical phase' of morphogenesis

[1] *Author's address:* Department of Anatomy, Jefferson Medical College, 1020 Locust Street, Philadelphia, PA 19107, U.S.A.

(Scott, Ritter & Wilson, 1977, 1980; Kochhar & Agnish, 1977; Kochhar, Penner & McDay, 1978; Wolpert, Tickle & Sampford, 1979; Knudsen & Kochhar, 1981). The pattern of abnormalities induced by such agents is stage-dependent, and susceptibility is lost once the pre-cartilage pattern emerges. Hence, the 'critical phase' encompasses the sequence of events from initial limb-bud outgrowth to pre-cartilage condensation. In genetic mutants which exhibit malformations of limb pattern, cellular abnormalities have invariably been found during this same period of development. Pertinent examples include the polydactylous *talpid²* (Cairns, 1977) and *talpid³* (Hinchliffe & Ede, 1967) chick mutants, as well as *Dominant hemimelia* (Rooze, 1977) in the mouse. A genetic alteration in the pattern of cell death normally programmed to occur during the 'critical phase' of morphogenesis has been found in the limb buds of these mutants.

Another dominant mutation in the mouse, *Hemimelia-extra toe* (*Hmˣ*), induces specific pattern abnormalities localized to the limbs of heterozygotes. In some respects, these limb malformations are similar to those found in the *talpids*, *Dominant Hemimelia*, teratogenesis induced by anti-proliferative agents, and some human conditions (Freund, 1936). This report describes the malformations exhibited by *Hmˣ*-heterozygotes and some morphological features associated with pathogenesis during the 'critical phase' of pattern formation.

MATERIALS AND METHODS

Animals

Hemimelia-extra toe arose in the B10.D2–New/SnJ mouse strain in 1968 at the Jackson Laboratory (Bar Harbor, Maine). The colony maintained for our studies was started from seven mice received from Dr K. S. Brown, N.I.D.R., N.I.H. (Bethesda, Maryland). In our laboratory the colony was not strictly inbred, and did not produce skeletal malformations other than those associated with *Hemimelia-extra toe*. All animals were maintained on Purina Laboratory Chow *ad libitum* and kept in the light from 6 a.m. to 6 p.m. A group of seven females were caged with one male from 9 a.m. to 1 p.m. or from 6 p.m. to 9 a.m., and the presence of a vaginal plug immediately afterwards was regarded as the sign of successful mating, with this day designated as the first day of gestation. One of the parents was always heterozygous for *Hemimelia-extra toe*. Some pregnant mothers were killed by cervical dislocation on the eleventh to fifteenth days of gestation; others were allowed to deliver to obtain neonates. Further staging of gestational age during the embryonic period was adapted from the system described by Theiler (1972) which is based upon observations of developing hybrid mouse embryos at one-half-day intervals (Table 1). Embryonic stages 16 to 20 (11th–13th gestational days) were assessed by determining the number of somite pairs and other external morphological features. Fetal stage 22 (15th day) was assessed by crown–rump length and degree of interdigitation.

Table 1. *Stage designation of midgestation mouse embryos*

Day of gestation	Stage*	Somite no.†
10·0	14	13–20
10·5	15	21–29
11·0	16	30–34
11·5	17	35–39
12·0	18	40–44
12·5	19	45–49
13·0	20	50–

* Theiler (1972) designated these stages in hybrid mice where they corresponded to days 9–12 of gestation (plug day = day 0).

† The most caudal somite related to the post-axial border of the hindlimb bud was taken as somite 28.

Nomenclature

The original designation for *Hemimelia-extra toe* was *Hx* (see recent report by Kalter, 1980). However, because there is evidence for allelism on chromosome V between *Hemimelia-extra toe* and *Hammer-toe* (based upon the failure of these traits to cross over in over 700 matings), new symbols have been proposed by Dr K. S. Brown (personal communication) and will be adhered to for this study: *Hammer-toe* (*Hm^T*) and *Hemimelia-extra toe* (*Hm^x*). Thus, the normal phenotype bears a +/+ genotype, and the mutant phenotype bears a *Hm^x*/+ genotype.

Whole-mount preparations of intact skeletons

Whole normal and mutant neonates were fixed overnight in 95 % ethanol and processed for staining of their osseous skeleton by the rapid Alizarin red S dye method (Kochhar, 1973). Whole stage-22 foetuses were fixed overnight in 95 % ethanol and processed for staining of cartilage/bone by the combined alcian blue/Alizarin red S technique of Inouye (1976) with the modification that, after staining, foetuses were not macerated with potassium hydroxide but instead were dehydrated in ethanol and cleared with methyl salicylate.

Light and transmission electron microscopy

Embryo or foetuses were dissected free of their extraembryonic membranes in ice-cold Tyrode's saline and staged. Limbs from conceptuses of stages 16 to 22 were excised, fixed in cacodylated-buffered Karnovsky's fixative (Karnovsky, 1965), postfixed in 1 % osmium tetroxide in cacodylate buffer, dehydrated in ethanol, cleared in propylene oxide, then embedded in Araldite 502 (Electron Microscopy Sciences). One micron (μm) or serial 3 μm sections were cut in the anteroposterior plane, stained at 65 °C for four minutes with basic Azure II (Matheson, Coleman, and Bell), 0·25 % in 0·5 % sodium borate, and examined with a Nikon Biophot microscope. For electron microscopy, thin sections (silver)

Table 2. *Monohybrid inheritance of the* Hm^x *allele**

Parental genotype	No. of successful pregnancies	No. of offspring	Average litter size[†]	Mutant (%)	Normal (%)	Implantation sites[‡] resorbed (%)	P[§]
$+/+ \times +/+$	26	183	7·0	0	100	0	1·00
$Hm^x/+ \times +/+$	107	661	6·2	47	53	9·7	0·02–0·05
$Hm^x/+ \times Hm^x/+$	43	265	6·2	35	65	5·3	0·70–0·50

* Eleventh gestational day to first day post-partum.
† Resorbed implantation sites excluded.
‡ During eleventh to fifteenth gestational day.
§ *P*, the probability that deviation from the expected ratio occurred by chance alone, was determined by chi-square analysis with a level of significance of $P < 0.05$. The expected ratios were based on simple autosomal dominant, homozygous lethal inheritance of the Hm^x allele.

were cut on an LKB-Huxley ultramicrotome and examined with an RCA-3G electron microscope.

Autoradiography

Stage-19 embryos, freed of Reichert's membrane and parietal yolk sac, were grown in whole embryo culture by techniques described elsewhere (Kochhar, 1975). After a five-hour incubation in the presence of [methyl-^{3}H]thymidine (2 μCi/ml, New England Nuclear), embryos were freed of the visceral yolk sac and amnion, then rinsed several times in ice-cold Tyrode's saline containing 100 μg/ml unlabelled thymidine. Limb buds were excised, fixed for microscopy as described above, and serially sectioned at 3 μm. Unstained sections were coated in the dark with a 50 % aqueous dilution of nuclear track emulsion NTB-2 (Kodak) at 40 °C, air dried, refrigerated under desiccant for four weeks in light-tight boxes, developed for four minutes in D-19 photographic developer (Kodak), fixed with Kodak Rapid Fixer, washed, and then stained lightly for two minutes with Azure II.

Assessment of cellular degeneration

Dead or dying cells appeared under light microscopy as extravascular, heterogeneous, and densely azurophilic granules. Their ultrastructural features correlated well with descriptions of cellular degeneration provided by Kerr, Wyllie & Currie (1972); Schweichel & Merker (1973); and Kochhar *et al.* (1978). Zones of cellular degeneration were further confirmed, at the gross level, by incubating living embryos (dissected free of their extraembryonic membranes) at 37 °C for 30 min in Ringer's saline solution containing the supravital dye Nile Blue A (1:20000, Allied Chemical), followed by a rinse with Ringer's saline (without Nile Blue A) at 4 °C. Similar methods have been used successfully by Saunders, Gasseling & Saunders (1962), and Scott *et al.* (1977, 1980) for assessment of normal and abnormal patterns of cell death during embryonic limb development.

RESULTS

Inheritance of the Hm^x allele

Monohybrid ratios of normal and mutant phenotypes, average litter size, and the percentage of implantation sites found resorbed are summarized in Table 2. In 107 matings between a normal and a mutant parent, no litters were composed entirely of mutants. In matings where either one or both parents were mutants, resorptions were more frequent and the average litter size was 14 % less than in those matings where both parents were normal. The F1 monohybrid ratios obtained from a total of 176 matings were compared for goodness of fit (Chi-square test) to those expected if *Hemimelia-extra toe* was indeed an autosomal dominant, homozygous embryolethal mutation. *P*, the probability that observed ratios deviated from the expected by chance alone, did not exist below the level of significance ($P \leqslant 0.05$) in any of the three mating combinations, but was very low for normal-mutant matings ($0.05 < P < 0.02$).

Gross morphology of the neonatal osseous skeleton

Alizarin red S-stained preparations of neonatal osseous skeletons, along with inspection of adults, demonstrated a range of gene penetration that was generally, but not strictly, related to parentage (the more penetrant in the parent, the more penetrant in the offspring). Hindlimbs in any individual were always more severely affected than the forelimbs, and commonly one side was more affected than the other. The heterozygote syndrome of hindlimb malformations is presented in Fig. 1. Forelimbs were similarly malformed, but to a lesser degree. The most severe deformities, taken as complete gene penetration, included bilateral absence of the tibia and distal half of the radius, with polydactyly in all four paws. The fibula and ulna were normal in size but bowed or stumpy secondary to reduction of the companion element. Polydactyly included six to seven metatarsals or metacarpals with seven to eight digits per paw. Supernumerary elements were always located pre-axially, and were of normal morphology (though ossification of the middle phalanx was commonly delayed). The humerus, femur or limb girdles were never affected, and the only other skeletal malformation observed was an occasional shortened tail.

A continuum of mutant phenotypes occurred, presumably due to varying degrees of incomplete gene penetration. For convenience of description, three levels of severity were noted: intermediate, mild/intermediate, and mild (Fig. 1). Intermediate penetration included absence of the distal half of the tibia, absence or reduction of the talus, with pre-axial polydactyly in all four paws. Mild/intermediate penetration included absence or reduction of the talus with pre-axial polydactyly in all paws. Mild penetration included absence of part or all of the talus, metacarpals and metatarsals I, II, III, and the proximal phalanx of

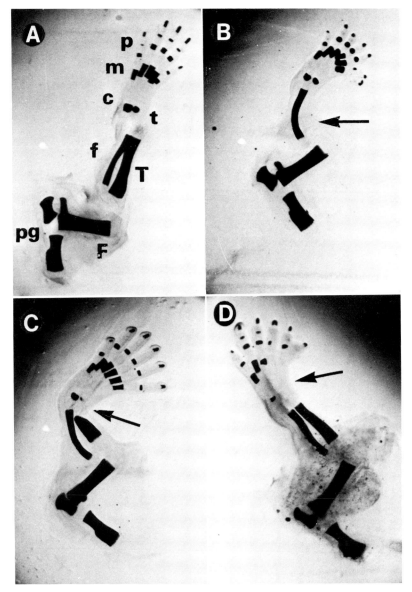

Fig. 1. Neonatal hindlimbs stained with Alizarin red S then cleared to demonstrate ossified elements. Whole mounts, $19.5 \times$. (A) $+/+$ phenotype. Pelvic girdle, *pg*; femur, *F*; tibia, *T*; fibula, *f*; talus, *t*; calcaneus, *c*; metatarsals, *m*; phalanges, *p*. (B–D) $Hm^x/+$ phenotype. Arrow denotes 'hemimelia'. (B) Complete penetration; (C) incomplete (intermediate) penetration; (D) incomplete (mild) penetration.

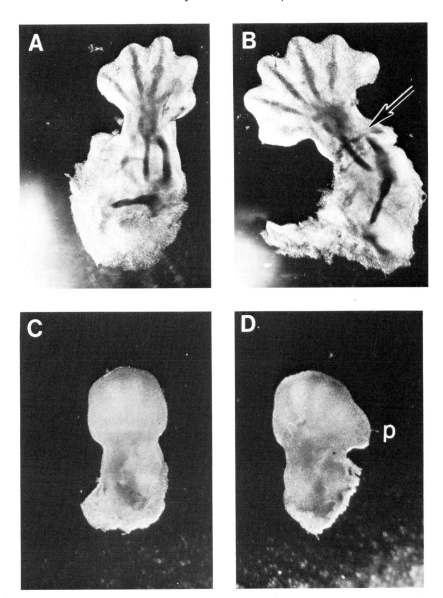

Fig. 2. Hindlimbs from +/+ (A, C) and *Hm^r*/+ (B, D) conceptuses. Whole mounts, 36×. (A, B) Combined alcian blue/Alizarin red S staining on stage 22 of gestation. Distal shortening of the tibial cartilage (arrow) with supernumerary digits occurred in the mutant. (C, D) Stage 20 of gestation. A definitive protrusion (*p*) was present pre-axially on the autopod of mutant hindlimb buds.

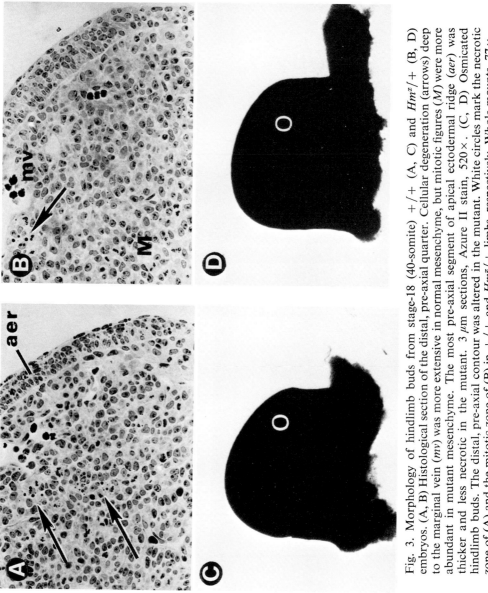

Fig. 3. Morphology of hindlimb buds from stage-18 (40-somite) $+/+$ (A, C) and $Hm^x/+$ (B, D) embryos. (A, B) Histological section of the distal, pre-axial quarter. Cellular degeneration (arrows) deep to the marginal vein (mv) was more extensive in normal mesenchyme, but mitotic figures (M) were more abundant in mutant mesenchyme. The most pre-axial segment of apical ectodermal ridge (aer) was thicker and less necrotic in the mutant. 3 μm sections, Azure II stain, 520×. (C, D) Osmicated hindlimb buds. The distal, pre-axial contour was altered in the mutant. White circles mark the necrotic zone of (A) and the mitotic zone of (B) in $+/+$ and $Hm^x/+$ limbs, respectively. Whole mounts, 77×.

pre-axial digits. In mild penetration, pre-axial polydactyly occurred in the hindpaws only, but pre-axial ectrodactyly occurred in the forepaws. Several neonates that exhibited mild penetration died shortly after birth. When autopsied, they had no milk in their stomach, an air-expanded gastrointestinal tract, and a cleft of the hard palate.

Staining of whole stage-22 foetuses by the combined alcian blue/Alizarin red S method demonstrated the full syndrome of malformations (Fig. 2). All affected cartilage rudiments, when examined histologically, demonstrated normal organization (not shown).

Morphology of stage-17 and -18 limb buds

Litters were examined during the embryonic period characterized by early outgrowth of the undifferentiated limb bud (stage 16) to stage 22. The earliest gross phenotypic abnormalities occurred in limbs during stage 17 or 18. A precise age that mutant siblings were first distinguished varied between litters, but generally occurred around the 38-somite age if the carrier parent was severely affected. By the 44-somite age, all litters examined yielded the expected proportion of mutant phenotypes. One distinguishing feature was an excess of tissue located preaxially on the autopod giving mutant limb buds an altered contour (Fig. 3). In no litter was this change found before the 38-somite age, and no embryo had abnormally shaped forelimbs unless it also had affected hindlimbs. The microscopic features associated with altered contour were investigated in serial sections of hindlimb buds from 40-somite embryos. Abnormalities were found in the apical ectodermal ridge (AER) and underlying subridge mesenchyme, and were localized to the pre-axial portion of the autopod.

In normal hindlimbs, the AER was a stratified epithelium generally comprising four nuclear layers: an outer periderm of squamosal cells, an innermost basal layer of columnar cells oriented perpendicular to the basement membrane, and two intermediate layers. Within the AER capping the pre-axial half of the normal limb bud, a band of cells appeared to be undergoing degeneration (Fig. 3). Ultrastructurally, many of the basal cells contained large, osmiophilic inclusions indicative of autophagic vacuoles within their cytoplasm; however, these cells did not exhibit signs of nuclear disorganization, cytoplasmic condensation, organelle disruption, or cell fragmentation (Fig. 4). The pre-axial mesenchyme underlying this segment of the AER was composed of cytologically homogeneous, undifferentiated cells which extended filopodial processes along the basal lamina. Some filamentous extracellular material was present in association with the mesodermal face of the basal lamina (Fig. 4). Additionally, a zone of physiologic cell death occurred within the mesenchyme deep to the marginal vein (Fig. 3).

Several abnormalities were found at the pre-axial ectodermal–mesenchymal region of comparably aged mutant hindlimbs. The AER was thicker than normal by one nuclear layer, and there was no sign of cellular degeneration in

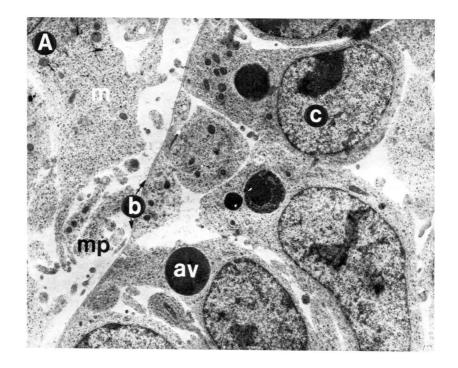

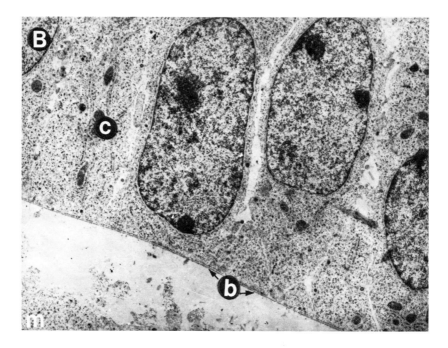

the basal layer. The pre-axial necrotic zone deep to the marginal vein was absent or greatly reduced, and an abundance of mitotic figures occurred at the comparable location (Fig. 3). Ultrastructurally, the sublaminer mesenchymal cells did not extend filopodial processes along the basal lamina, and there was less filamentous extracellular material associated with the mesodermal face of the basal lamina than in normal limbs (Fig. 4).

Hindlimb buds from several embryos were examined, at the 36-somite age, prior to overt manifestation of the mutant phenotype. No cellular degeneration was found within the pre-axial AER or mesenchyme in any embryo, either by light microscopy or by supravital staining with Nile Blue A.

Morphology of stage-19 and -20 limb buds

Hindlimb buds were examined on stages 19 and 20 to identify the early features associated with tibial agenesis and the emergence of supernumerary digits. Mesenchymal cell condensations for the stylopod (femur) and zeugopod (tibia, fibula) occurred during stages 18 and 19. A central block of cells within the zeugopodal condensation of all limb buds had become necrotic by stage 19. Resorption of these cells resulted in separation of tibial (preaxial) and fibular (postaxial) precartilage rudiments by stage 20. Concurrent with overt expression of this cell death in the mutant, a large percentage of presumptive tibial chondroblasts as well as cells within the adjoining myogenic region became necrotic (Fig. 5). On stage 20, the distal portion of the pre-cartilaginous tibia was absent in the mutant.

Continued limb outgrowth led to the formation of a definitive protrusion on the mutant autopod (Fig. 2). Within the AER capping this protrusion, the basal layer in comparison to normal was thicker, devoid of cellular degeneration, and exhibited a higher incorporation of [^{3}H]thymidine (assessed by autoradiography following *in vitro* exposure). The distribution of [^{3}H]thymidine incorporated into the mesenchyme of this protrusion did not differ from normal (Fig. 6), and no cellular degeneration occurred here either. Our regional analysis of cellular degeneration is summarized in Fig. 7 for normal and mutant hindlimb buds.

Fig. 4. Electron micrographs of the pre-axial ectodermal–mesenchymal interface from stage 18 (40-somite) +/+ (A) and Hm^x/+ (B) hindlimb buds. 7200×. (A) Basal cells (*c*) of the AER exhibited osmiophilic vacuoles indicative of autophagic vacuoles (*av*). Mesenchymal processes (*mp*) extended along the basal lamina (*b*), the mesodermal (*m*) face of which was coated with fuzzy, filamentous extracellular material. (B) Basal cells did not exhibit any overt signs of autophagy. Mesenchymal processes did not extend along the basal lamina, and filamentous extracellular material was rarefied.

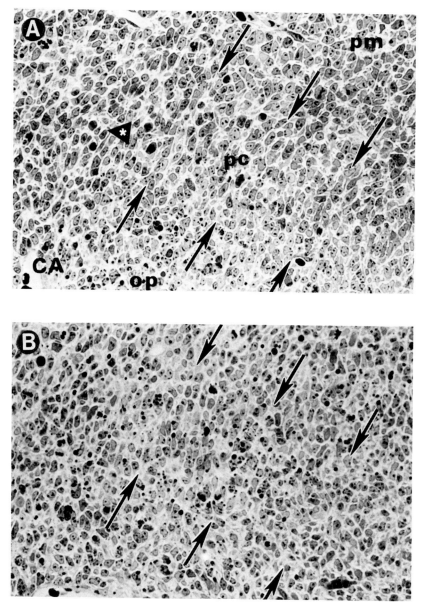

Fig. 5. Longitudinal sections (3 μm) of the pre-cartilaginous tibia (within the arrows) from stage-19 (46-somite) + / + (A) and $Hm^x/$ + (B) embryos. Azure II stain, 660 × . (A) Prechondrogenic (pc) zones were distinguished from adjacent premyogenic (pm) regions by their less rounded nuclei and orientation perpendicular to the long axis of the rudiment. The 'opaque patch' (op) was a zone of cellular degeneration normally occurring between tibial and fibular rudiments. Several mitotic figures (asterisk) were present at the distal limit of the rudiment. Central artery, CA. (B) Cellular degeneration of the 'opaque patch' extended throughout the tibia, as well as the adjacent premyogenic region.

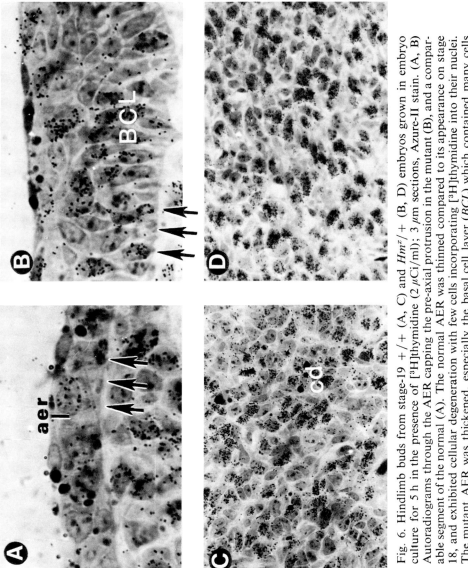

Fig. 6. Hindlimb buds from stage-19 +/+ (A, C) and *Hm^x/+* (B, D) embryos grown in embryo culture for 5 h in the presence of [³H]thymidine (2 μCi/ml); 3 μm sections, Azure-II stain. (A, B) Autoradiograms through the AER capping the pre-axial protrusion in the mutant (B), and a comparable segment of the normal (A). The normal AER was thinned compared to its appearance on stage 18, and exhibited cellular degeneration with few cells incorporating [³H]thymidine into their nuclei. The mutant AER was thickened, especially the basal cell layer (*BCL*) which contained many cells incorporating [³H]thymidine. Sub-ridge mesenchyme was more closely associated with the basement membrane in normal limbs (arrows). 2400×. (C, D) Autoradiograms through the pre-axial sub-ridge mesenchyme exhibited similar radiogranule distribution in mutant (D) and normal (C) limbs, but cellular degeneration (*cd*) was found in normal limbs only. 960×.

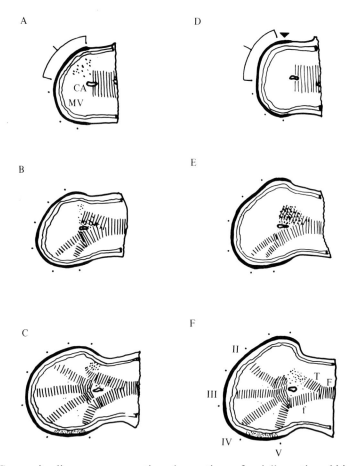

Fig. 7. Composite diagrams representing observations of serially sectioned hindlimb buds from +/+ (A, B, C) and Hm^x/+ (D, E, F) embryos. Cell condensations are represented as transverse lines; mesodermal degeneration as dots; AER degeneration is marked by squares or brackets. (A, D) Stage-18 (40-somite) of gestation. The most pre-axial segment of the AER in the mutant was thickened and free of degeneration (broad arrowhead). Marginal vein, *mv*; central artery, *ca*. (B, E) Stage 19 of gestation. (C, F) Stage 20 of gestation. Femur, *F*; tibia, *T*; fibula, *f*.

DISCUSSION

An autosomal dominant mutation in the mouse, *Hemimelia-extra toe* (Hm^x), induces congenital limb malformations in heterozygotes. The homozygous condition, not investigated for this report, appears to be embryolethal during the eighth to ninth days of gestation (Dr K. S. Brown, personal communication). A 1:1 ratio of normal:mutant offspring was expected from matings between a normal and a mutant parent, but the actual ratio (1:0·82) deviated in favour of normal. Based upon statistical analysis, this deviation was not significant and could be ascribed to random factors. However, these same matings produced an abnormally high number of embryonic resorptions (independent of whether it was the mother or the father which carried the mutant allele), so the decrement in heterozygote number could also be accounted for by an increased sporadic lethality.

All limb malformations were localized to pre-axial elements of the zeugopod and autopod. Variation in gene penetration occurred perhaps because the mouse strain was not inbred. There was always axial shortening of proximal skeletal elements associated with the presence of distal supernumerary elements. Typical expression included tibial and radial 'hemimelia', supernumerary metatarsals, and polydactyly. In the mildest form the zeugopod was unaffected, but there was an absence of some pre-axial metatarsals, with polydactyly or ectrodactyly. Apparently, the regions along the longitudinal axis were affected depending upon the penetrance of the mutant gene, with more proximal elements affected in association with greater penetration. Since, during early limb outgrowth, the longitudinal axis is specified in proximodistal fashion, increased penetration implied that overt expression of the mutation occurred at an earlier developmental age.

The earliest characteristics identified with the mutant phenotype appeared on the eleventh to twelfth days of gestation. The precise age varied between different litters, but generally occurred nearer the 38-somite age (stage 17) when the gene was more penetrant in the heterozygous parent. The full syndrome of malformations was apparent by the middle of the 13th gestational day (stage 20) when the pre-cartilaginous limb pattern was complete. Because histodifferentiation of affected elements proceeded normally, overt effects of the mutation may have been confined to embryonic stages 17–20. As the shapes, sizes and scheduled occurrence of mesenchymal condensations were initially unaltered, it is not likely that the actual mechanisms underlying skeletal pattern formation were disturbed. Rather, abnormal pattern was probably the result of disturbances in other morphogenetic forces that normally shape the limb bud. One prominent disturbance concerned the pattern of physiologic cellular degeneration (PCD) programmed to occur at certain times in three locations within the zeugopod and autopod: (1) the 'opaque patch' (a mesodermal necrotic zone occurring during stages 18 and 19 between tibial and fibular pre-cartilage anlagen (Fell & Canti, 1934; Dawd & Hinchliffe, 1971)); (2) the 'foyer primaire préaxial' (fpp) (a mesenchymal necrotic zone occurring during stages 17 to 20 within the autopod, deep to the most pre-axial segment of the AER (Milaire, 1976)); and (3) cellular degeneration within the most pre-axial segment of the AER itself (Scott *et al.* 1977, 1980).

In mutant limbs, expression of PCD in the opaque patch was accompanied by necrosis within the tibial pre-cartilage. These two necrotic regions were continuous, so spatial overextension of the opaque patch is a possible cause for resorption of the tibia. There are other instances of abnormal skeletogenesis in which genetic or teratogenic alterations in the extent of normal PCD zones directly affect the size of pre-cartilage rudiments. Zwilling (1942, 1959) postulated an overextension in two systems: in the opaque patch as a cause of insulin-induced micromelia in chick embryos (Zwilling, 1959); and in the tail as a basis of dominant *Rumplessness* in chicks (Zwilling, 1942). Furthermore, an absence of cell death in the opaque patch has been associated with fusion of tibial and fibular rudiments in the *talpid*[3] chick mutant (Hinchliffe & Ede, 1967).

Preceding expression of PCD in the opaque patch, distinct abnormalities localized to the pre-axial epithelium and mesenchyme were evident in the autopod of mutant limb buds. The pre-axial AER remained in a thickened state and did not undergo scheduled degression during stages 18 to 20. This was associated with two abnormalities in its basal cell layer: (1) persistence in a rapidly proliferative state at least through stage 19 (assayed by autoradiography following a 5 h incorporation of [³H]thymidine); and (2) failure to undergo cellular degeneration. The former may be related to the latter since withdrawal from the proliferative population commonly precedes morphogenetic cell death, as exemplified by epithelial autolysis during morphogenesis of the secondary palate in rats (Hudson & Shapiro, 1973; Pratt & Martin, 1975). The earliest signs of basal cell degeneration in normal limbs may have been indicative of either phagocytosis, autophagic remodelling of cellular organelles, or an early stage of lysosome-mediated autolysis. There was no evidence of the classic nuclear or cytoplasmic changes found during cell death (Kerr *et al.* 1972; Schweichel & Merker, 1973) at this time. Hence, a lysosome-mediated (type II) autophagy may provide the normal mechanism for programmed degression of the AER, but may or may not lead to autolysis in more advanced stages.

Delayed degression of the preaxial AER in mutant limbs would be expected to prolong its 'mesenchyme inductive' potential (Saunders, 1948; Zwilling, 1956*a*, *b*, *c*), and possibly induce outgrowth of the pre-axial protrusion. The AER has been shown to exert a trophic influence on chick limb mesenchyme *in vitro* maintaining it in a viable (Cairns, 1975), undifferentiated state of which rapid outgrowth is characteristic (Stark & Searls, 1973; Globus & Vethamany-Globus, 1976; Kosher, Savage & Chan, 1979). Although mesenchymal proliferation is recognized as the primary motive force for outgrowth (Ede, Flint & Teague, 1975; Summerbell & Lewis, 1975), the proliferative profile of the mesenchyme within the protrusion was normal. Hence, outgrowth of the protrusion was not associated with an increased rate of proliferation. More likely, failure of PCD to occur at the fpp resulted in a larger population of proliferating cells. A large number of mitotic figures was present during stage 18 just prior to formation of the protrusion, at the expected location of the fpp. By stage 19 the mitotic profile was normal, indicating that a transient mitotic synchrony occurred within the region normally expected to undergo PCD. A delay in the onset of PCD pre-axially in the AER and mesenchyme has been described in the autopods of several other genetic mutants of the chick (*talpid*[2] (Cairns, 1977) and *talpid^3* (Ede, 1971)) and mouse (*Dominant Hemimelia* (Rooze, 1977)), and in teratogenesis induced in rats by the anti-proliferative agents cytosine arabinoside, 5-fluorodeoxyuridine, and 6-mercaptopurine riboside (Scott *et al.* 1977, 1980). In these instances, a high incidence of pre-axial polydactyly was characteristic.

Other abnormalities occurred in the pre-axial mesenchyme of Hm^x/+ limbs, but their relationship to the altered PCD pattern is not clear. The paucity of mesenchymal cell filopodial processes extended along the basal lamina was similar to what has been described in the *talpid^3* chick mutant (Ede, 1971), in which aberrant cellular adhesive properties have been detected in limb mesenchyme (Ede, Flint, Wilby & Colquhoun, 1977). Also in Hm^x/+ limbs, the filamentous extracellular material associated with the mesodermal face of the basal lamina was less abundant than normal. This material, which is probably synthesized by the mesenchyme, may impart cellular adhesive properties to the basal lamina. Regional differences in basement membrane ultrastructure between digital and interdigital zones have been reported by Kelley (1973) for the early limb bud of man. In that study, a collagen-like fibrillar matrix was associated with the basement membrane at interdigital zones (where PCD occurred in the AER and mesenchyme) but was absent at the non-necrotic digital zones. In light of the proximity imposed between apical mesenchyme and the basal layer of the AER, there is a potential role for this matrix during induction of epithelial PCD. Direct correlation between the presence of collagenous stroma and morphogenetic cell death was reported by Ojeda & Hurle (1975) for the fusion of endocardial tubes during chick cardiogenesis, but a causal relationship was not resolved.

This work was supported by a grant from USPHS, HD-10935-03. Special thanks are due to Mrs Debra Lundgren for typing the manuscript.

REFERENCES

CAIRNS, J. M. (1975). The function of the apical ectodermal ridge and distinctive characteristics of adjacent mesoderm in the avian wing bud. *J. Embryol. exp. Morph.* **34**, 155–169.
CAIRNS, J. M. (1977). Growth of normal and talpid^2 chick wing buds: an experimental analysis. In *Vertebrate Limb and Somite Morphogenesis* (ed. D. A. Ede, J. R. Hinchliffe & M. Balls), pp. 123–138. New York: Cambridge University Press.
DAWD, D. S. & HINCHLIFFE, J. R. (1971). Cell death in the 'opaque patch' in the central mesenchyme of the developing chick limb: a cytological, cytochemical, and electron microscopic analysis. *J. Embryol. exp. Morph.* **26**, 401–424.
EDE, D. A. (1971). Control of form and pattern in the vertebrate limb. *Symp. Soc. Exp. Biol.* **25**, 235–257.
EDE, D. A., FLINT, O. P. & TEAGUE, P. (1975). Cell proliferation in the developing wing bud of normal and *talpid^3* mutant chick embryos. *J. Embryol. exp. Morph.* **34**, 589–607.
EDE, D. A., FLINT, O. P., WILBY, O. K. & COLQUHOUN, P. (1977). The development of pre-cartilage condensations in limb bud mesenchyme *in vivo* and *in vitro*. In *Vertebrate Limb and Somite Morphogenesis* (ed. D. A. Ede, J. R. Hinchliffe & M. Balls), pp. 161–179. New York: Cambridge University Press.

FELL, H. B. & CANTI, G. (1934). Experiments on the development in vitro of the avian knee joint. *Proc. Roy. Soc. Lond.* B **116**, 316–351.

FREUND, E. (1936). Congenital defects of the femur, fibula, and tibia. *Arch. Surg.* **33**, 349–391.

GLOBUS, M. & VETHAMANY-GLOBUS, S. (1976). An *in vitro* analogue of early chick limb outgrowth. *Differentiation* **6**, 91–96.

HINCHLIFFE, J. R. & EDE, D. A. (1967). Limb development in the polydactylous talpid[3] mutant of the fowl. *J. Embryol. exp. Morph.* **17**, 385–404.

HUDSON, C. D. & SHAPIRO, B. L. (1973). A radioautographic study of DNA synthesis in embryonic rat palatal shelf epithelium with reference to the concept of programmed cell death. *Archs Oral Biol.* **18**, 77–84.

INOUYE, M. (1976). Differential staining of cartilage and bone in fetal mouse skeleton by alcian blue and Alizarin red S. *Cong. Anom.* **16**, 171–173.

JOHNSON, D. R. (1969). Polysyndactyly, a new mutant gene in the mouse. *J. Embryol. exp. Morph.* **21**, 285–294.

KALTER, H. (1980). A compendium of the genetically induced congenital malformations of the house mouse. *Teratology* **21**, 397–429.

KARNOVSKY, M. J. (1965). A formaldehyde-glutaraldehyde fixative of high osmolarity for use in electron microscopy. *J. Cell Biol.* **27**, 137A.

KELLEY, R. O. (1973). Fine structure of the apical rim–mesenchyme complex during limb morphogenesis in man. *J. Embryol. exp. Morph.* **29**, 117–131.

KERR, J. F. R., WYLLIE, A. H. & CURRIE, A. R. (1972). Apoptosis: a basic biological phenomenon with wide-ranging implications in tissue kinetics. *Br. J. Canc.* **26**, 239–257.

KNUDSEN, T. B. & KOCHHAR, D. M. (1981). Limb development in mouse embryos. III. Cellular events underlying the determination of altered skeletal patterns following treatment with 5'-fluorodeoxyuridine. *Teratology* **23**, 241–251

KOCHHAR, D. M. (1973). Limb development in mouse embryos. I. Analysis of teratogenic effects of retinoic acid. *Teratology* **7**, 289–298.

KOCHHAR, D. M. (1975). The use of in vitro procedures in teratology. *Teratology* **21**, 397–429.

KOCHHAR, D. M. & AGNISH, N. D. (1977). 'Chemical Surgery' as an approach to study morphogenetic events in embryonic mouse limb. *Devl Biol.* **61**, 388–394.

KOCHHAR, D. M., PENNER, J. D. & McDAY, J. A. (1978). Limb development in mouse embryos. II. Reduction defects, cytotoxicity, and inhibition of DNA synthesis produced by cytosine arabinoside. *Teratology* **18**, 71–92.

KOSHER, R. A., SAVAGE, M. P. & CHAN, S-C. (1979). In vitro studies on the morphogenesis and differentiation of the mesoderm subjacent to the apical ectodermal ridge of the embryonic chick limb. *J. Embryol. exp. Morph.* **50**, 75–97.

MILAIRE, J. (1976). Rudimentation digitale au cours du développement normal de l'autopode chez les mammifères. In '*Mécanismes de la rudimentation des organes chez les embryons de vértébrés*' (ed. A. Laynaud). Paris: Editions du CNRS. (Cited by Scott, Ritter & Wilson, 1977.)

OJEDA, J. C. & HURLE, J. M. (1975). Cell death during the formation of tubular heart of the chick embryo. *J. Embryol. exp. Morph.* **33**, 523–534.

PRATT, R. M. & MARTIN, G. R. (1975). Epithelial cell death and cyclic AMP increase during palatal development. *Proc. Natn. Acad. Sci., U.S.A.* **72**, 874–877.

ROOZE, M. A. (1977). The effects of the Dh gene on limb morphogenesis in the mouse. In *Morphogenesis and Malformation of the Limb* (ed. D. Bergsma & W. Lenz), pp. 69–95. New York: Alan R. Liss.

SAUNDERS, J. W. (1948). The proximodistal sequence of origin of the parts of the chick wing and the role of the ectoderm. *J. exp. Zool.* **108**, 363–403.

SAUNDERS, J. W., JR, GASSELING, M. T. & SAUNDERS, L. C. (1962). Cellular death in morphogenesis of the avian wing. *Devl Biol.* **5**, 147–178.

SCHWEICHEL, J-U. & MERKER, H-J. (1973). The morphology of various types of cell death in prenatal tissues. *Teratology* **7**, 253–266.

SCOTT, W. J., RITTER, E. J. & WILSON, J. G. (1977). Delayed appearance of ectodermal cell death as a mechanism of polydactyly induction. *J. Embryol. exp. Morph.* **42**, 93–104.

SCOTT, W. J., RITTER, E. J. & WILSON, J. G. (1980). Ectodermal and mesodermal cell death patterns in 6-mercaptopurine riboside-induced digital deformities. *Teratology* **21**, 271–279.

STARK, R. J. & SEARLS, R. L. (1973). A description of chick wing bud development and a model of limb morphogenesis. *Devl Biol.* **33**, 138–153.

SUMMERBELL, D. & LEWIS, J. H. (1975). Time, place and positional value in the chick wing bud. *J. Embryol. exp. Morph.* **33**, 621–643

THEILER, K. (1972). *The House Mouse.* New York: Springer-Verlag.

WOLPERT, L., TICKLE, C. & SAMPFORD, M. (1979). The effect of cell killing by X-irradiation on pattern formation in the chick limb. *J. Embryol. exp. Morph.* **50**, 175–198.

ZWILLING, E. (1942). The development of dominant Rumplessness in chick embryos. *Genetics* **27**, 641–656.

ZWILLING, E. (1959). Micromelia as a direct effect of insulin – evidence from in vitro and in vivo experimentation. *J. Morph.* **104**, 159–180.

ZWILLING, E. (1956a). Interaction between limb bud ectoderm and mesoderm in the chick embryo. II. Experimental limb duplication. *J. exp. Zool.* **132**, 173–187.

ZWILLING, E. (1956b). Interaction between limb bud ectoderm and mesoderm in the chick embryo. IV. Experiments with a wingless mutant. *J. exp. Zool.* **132**, 241–254.

ZWILLING, E. & HANSBOROUGH, L. A. (1956c). Interactions between limb bud ectoderm and mesoderm in the chick embryo. III. Experiments with polydactylous limbs. *J. exp. Zool.* **132**, 219–240.

J. Embryol. exp. Morph. Vol. 65 (Supplement), pp. 309–325, 1981
Printed in Great Britain © Company of Biologists Limited 1981

The effect of removing posterior apical ectodermal ridge of the chick wing and leg on pattern formation

By DONENE A. ROWE AND JOHN F. FALLON[1]

From the Department of Anatomy, The University of Wisconsin

This paper is dedicated to the memory of our friend and colleague
Professor William A. McGibbon

SUMMARY

Recent experiments, in which barriers were inserted between anterior and posterior tissues of the chick wing bud, resulted in deletion of structures anterior to the barrier (Summerbell, 1979). From these data it was concluded that blockage of morphogen from the polarizing zone by the barrier resulted in the observed failure of specification of anterior structures. We suggest an alternative interpretation, viz. the interruption of the apical ridge by the barrier caused the deletions. This hypothesis was tested by removal of increasing lengths of ridge. This was done beginning at either the anterior or posterior junction of the wing bud with the body wall and proceeding posteriorly or anteriorly, respectively, to each half-somite level between 16/17 and 19/20. With removal of progressively greater lengths of anterior ridge, more anterior limb elements failed to develop. These data were used to construct a map of the ridge responsible for each digit. To test our hypothesis we removed posterior sections of apical ridge, as described above. Removal of posterior ridge to a level which was expected to allow outgrowth of digits anterior to the level of removal resulted in wings without digits in the majority of cases. An exception occurred when ridge posterior to the mid-19 somite level was removed. In almost half of these cases digits 2 and 3 did develop. In most cases the retention of only a half-somite piece of ridge with all other ridge removed, also resulted in deletion of all digits. Again the exception occurred when ridge posterior to somite level mid-19 and anterior to level 18/19 was removed, leaving only that ridge between somite level 18/19 and mid-19. In many of these cases digit 3 did develop. We conclude from these data that, in the wing bud, ridge anterior to the mid-19 somite level must be connected to more posterior ridge to function.

The leg ridge does not exhibit the asymmetrical, low anterior, high posterior configuration, which appears in the wing. Because the leg ridge is symmetrically high anteriorly and posteriorly, we questioned whether or not leg would also require a continuity between anterior and posterior ridge for anterior ridge to function. It did not. When posterior ridge was removed, structures developed under remaining anterior ridge and the elements which developed were complementary to those which developed after anterior ridge removal to the same somite level. Those leg elements, which failed to develop, were truncated at the appropriate proximo-distal levels as indicated by the fate map we have constructed for the leg.

[1] *Author's address* (for reprints): Department of Anatomy, The University of Wisconsin, Madison, WI 53706, U.S.A.

The data reported here do not rule out a role for the polarizing zone in specification of anterior structures. It is apparent that posterior ridge removal in the wing results in loss of structures anterior to the removal. However, this is not true for the leg.

INTRODUCTION

The zone of polarizing activity, or polarizing zone, has been operationally defined as the mesoderm of the limb bud which when grafted to an anterior or mid-distal position of a host limb bud, induces formation of supernumerary limb parts from host anterior tissues (Saunders & Gasseling, 1968). The supernumerary limb elements are polarized so that the most posterior structures form next to the graft. The polarizing zone is limited to the posterior border of the presumptive and developing limb bud during stages 15–28 in chicks (MacCabe, Gasseling & Saunders, 1973; Hornbruch, unpublished) and all amniote limb buds tested to date (Tickle, Shellswell, Crawley & Wolpert, 1976; MacCabe & Parker, 1976a; Fallon & Crosby, 1977).

Clearly, under experimental conditions the polarizing zone can be shown to have morphogenetic activity. However, the role of this region in normal limb development has not been established. The argument that it plays no role in normal development has arisen (Saunders, 1977; Iten & Murphy, 1980a, b). We have suggested an early role in the stabilization and polarization of the limb field before limb-bud outgrowth (Fallon & Crosby, 1975, 1977; Thoms & Fallon, 1980). Yet another hypothesis has proposed that the polarizing zone is acting to polarize the limb elements at limb-bud stages when those elements are determined. The strongest evidence in support of this position has come from experiments by Summerbell (1979). In those experiments permeable and impermeable barriers were inserted through the dorsoventral extent of the limb bud, perpendicular to the body wall and at varying anteroposterior levels, using the somites as a reference. In this way the distal anterior and posterior limb-bud tissues were separated. When impermeable barriers were used at somite level 17/18, 19 of 23 limbs developed without a radius and 14 of 23 developed without digit **2**. After insertion of a permeable barrier at the same level, a radius failed to develop in 14 of 20 cases. If the impermeable barrier was inserted only through the mesoderm, leaving the apical ectoderm intact, a normal limb developed. It was concluded from these data that the action of the polarizing zone upon limb structures can be blocked by a barrier, thus resulting in deletions of limb elements anterior to the barrier. Specifically, it was proposed that a diffusable morphogen from the polarizing zone was prevented from reaching anterior mesoderm and that resulted in the failure of anterior structures to develop. We offer another interpretation: that the interruption of the apical ectoderm by the implanted barrier caused the deletions of limb elements. To test this hypothesis, we have removed pieces of ridge from the anterior and posterior borders of the wing and leg buds and examined whether or not structures, expected to grow out

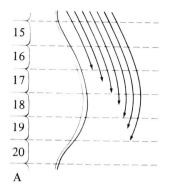 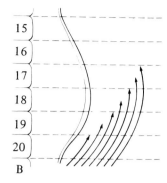

Fig. 1. (A) Diagram of the anterior apical ridge removals in the stage-19 wing bud. The removals begin at a level anterior to somite 15 and proceed posteriorly to one of seven positions as indicated by the arrows. (B) Diagram of the experiments complementary to those in (A). Removal of ridge begins posterior to the somite-20 level and extends anteriorly to one of the seven positions indicated by the arrows.

under the remaining ridge, were deleted. With this approach we determined that structures under remaining anterior ridge fail to develop when posterior ridge is removed from the wing. This is not true for the leg. In the leg, structures which are expected to develop under remaining anterior ridge do develop when posterior ridge is removed.

MATERIALS AND METHODS

Fertilized White Leghorn eggs were incubated for 3 days at 38 °C, candled and fenestrated according to the technique of Zwilling (1959). Embryos of Hamburger & Hamilton (1951) stages 18–20 were used. A portion of the apical ectodermal ridge was removed from the right wing or leg bud with a glass needle. The eggs of operated embryos were sealed with tape and returned to the incubator for 7–9 days. The 10- to 12-day chicks were then fixed in 10 % formalin, stained with Victoria blue and cleared.

Two groups of operations were performed on both wing and leg buds. In the first group, apical ridge was removed beginning at the junction of the limb-bud anterior border (somite level 14/15 for the wing and somite level 27/28 for the leg) with the body wall and proceeding back to a specific level of the bud determined by its relation to the somites. Removals beginning at the 14/15 level were performed to each half-somite level between somites 16/17 and 19/20 for the wing (Fig. 1A) and beginning at the 27/28 level to each half somite level between 28/29 and 31/32 for the leg. The second group of removals was complementary to the first group. In this group, apical ectodermal ridge was removed beginning at the junction of the limb-bud posterior border with the body wall and proceeding anteriorly to each half-somite level between somites 19/20 and

Table 1.

Level of removal	Number of cases	Normal	Missing elements		
			Digit **2**	Digit **3**	Digit **4**
16/17	5	5			
mid-17	10	10			
17/18	11	11			
mid-18	15	0	15 (100 %)		
18/19	12	0	12 (100 %)	11 (92 %)†	
mid-19	10	0	10 (100 %)	10 (100 %)	5 (50 %)*
19/20	7	0	7 (100 %)	7 (100 %)	7 (100 %)

* The distal phalangeal element and the distal end of the proximal element are missing.
† 42 % missing the distal phalangeal element and the distal end of the middle phalangeal element.

16/17 in the wing (Fig. 1 B) and between 31/32 and 28/29 in the leg. Two additional groups of experiments were performed only on the wing bud. These were also complementary types of operations. In one group a small piece of apical ridge, one-half somite in length, was removed. The operations in the other group consisted of removing the entire apical ectoderm with the exception of a piece one-half somite in length.

The numbers which survived postoperatively in each group are given in the results.

RESULTS

Removal of anterior apical ectodermal ridge from the wing bud

Anterior pieces of apical ectodermal ridge were removed to determine which level of ridge is responsible for each digit. The results of these removals at stages 18–19 are given in Table 1 and typical examples of the resulting limbs appear in Fig. 2. At these stages, the amounts of radius and ulna which develop following complete ridge removal vary (Summerbell, 1974). Because of this variability, we will concentrate on the appearance or deletion of the digits, making only minor reference to the radius and ulna. As illustrated in Fig. 2, more posterior digits were deleted after removal of greater lengths of anterior apical ridge.

Removal of anterior ridge back to the level of somite 17/18 and to intervening levels (16/17 and mid 17, Fig. 2A) resulted in development of a normal limb with respect to the digits. When done at stage 18 and early 19, about 50 % of the removals to the 17/18 level resulted in deletion of the radius (Fig. 2B), 100 % of removals to the mid-18 somite level resulted in wings missing digit **2**,

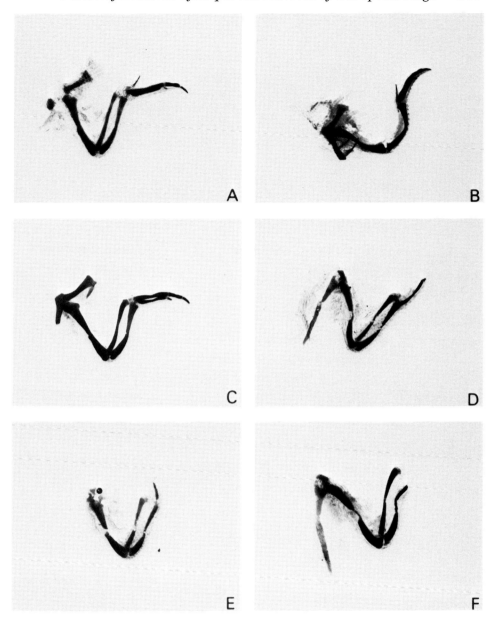

Fig. 2. Photographs of the typical results of anterior apical ridge removal described in Fig. 1A. (A) Somite level mid-17 removal – normal. (B) Somite level 17/18 removal – digits **2**, **3**, **4** and ulna. (C) Somite level mid-18 removal – digits **3**, **4**. (D) Somite level 18/19 removal – digit **4**. (E) Somite level mid-19–proximal digit **4**. (F) Somite level 19/20 removal – no digits.

(Fig. 2C). All 12 cases of removals to the 18/19 somite level resulted in deletion of digit **2** and 92 % in deletion of the proximal part (42 %) or all (58 %) of digit **3**, (Fig. 2D). After removal of ridge to the mid-19 somite level, limbs developed without digits **2** and **3** in all 10 embryos and without the distal portion of digit **4** in 5 of these (Fig. 2E). Removal to the 19/20 level always resulted in deletion

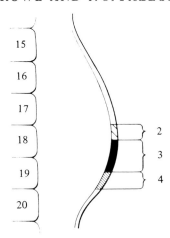

Fig. 3. Diagram of the stage-19 wing bud with a map of the ridge responsible for outgrowth of each digit: digit **2** – between somite levels 17/18 and mid-18, digit **3** – between somite levels mid-18 and mid-19, digit **4** – between somite levels mid-19 and 19/20.

Table 2.

Level of removal	Number of cases	Results			
		Normal	Missing elements		
			Digit **2**	Digit **3**	Digit **4**
16/17	5		5 (100%)	5 (100%)	5 (100%)
mid-17	8		8 (100%)	8 (100%)	8 (100%)
17/18	8		8 (100%)	8 (100%)	8 (100%)
mid-18	11		11 (100%)	11 (100%)	11 (100%)
18/19	10		10 (100%)	10 (100%)	10 (100%)
mid-19	18		10 (56%)	10 (56%)	18 (100%)
19/20	8	8 (100%)			

of all digits (Fig. 2F). The proximal portion of digit **4** was not deleted until the anterior removal of ridge extended to the 19/20 level. However, as noted, the distal portion of digit **4** could be deleted by anterior ridge removal to only the mid-19 level. At stage 18 and early 19, removals to the 19/20 level resulted in the development of limbs missing some or all of the ulna in 67% of cases.

We conclude the following from these data on deletions after anterior ridge removal: that the ridge at the level between somite 17/18 and mid-18 is responsible for the outgrowth of mesoderm which will form digit **2**; the ridge between mid-18 and mid-19 is responsible for digit **3**; the ridge at the level of mid-19 and immediately posterior to that is responsible for digit **4**. These data are summarized in a map of the apical ectodermal ridge in Fig. 3.

Removal of posterior apical ectodermal ridge from the wing bud

Table 2 presents the results of removal of posterior apical ridge from the wing bud. As expected from data in which anterior ridge removal to 16/17, mid-17, and 17/18 somite levels resulted in normal development of digits, removal of the complementary posterior ridge to those levels resulted in deletion of all digits in all cases. However, in most cases, removal of reciprocal pieces of anterior and posterior apical ridge at other levels did not result in formation of complementary digital elements. Recall that removal of anterior ridge to the mid-18 somite level resulted in deletion of digit **2** (Fig. 4A, B). Therefore, formation of digit **2** was expected after removal of the complementary posterior ridge to that level. However, all 11 cases showed deletion of this digit following that operation (Fig. 4C, D). After removal of posterior apical ridge to the level of 18/19 or mid-19, we expected development of structures missing after anterior removal, i.e. digits **2** and **3** (Fig. 4E, F), and digits **2, 3** (Fig. 4I, J), and perhaps distal **4**, respectively. However, in all 10 cases of ridge removal to the 18/19 level, wings developed without digits (Fig. 4G, H). Removal to the mid-19 level resulted in limbs without digits in 56 % of cases (Fig. 4K, L) and limbs missing only digit **4** in 44 % of cases. Only in these 44 % of cases did we find that, after posterior ridge removal, the wing elements developed which were missing after anterior removal. Posterior ridge removal resulted in deletion of structures expected to grow out under remaining anterior apical ridge in a majority of cases.

Removal of half-somite length pieces of apical ectodermal ridge from the wing bud

In order to examine further the precision of the correlation between the level of apical ridge removal and deleted elements, we removed pieces of apical ridge one-half somite in length from the wing bud (Table 3). After removal of ridge between somite levels 17/18 and mid-18, digit **2** did not develop in 67 % of operated wings. However, digit **2** was also deleted after mid-18 to 18/19 (67 %) and 18/19 to mid-19 level removals (50 %). Digit **3** was absent after mid-18 to 18/19 level removals in 89 % of the cases and also in 90 % of the cases after removal of apical ectoderm between the 18/19 and mid-19 somite levels. Finally, with removal at the level of mid-19 to 19/20, digit **4** was deleted in 71 % of cases.

These data are consistent with Fig. 3, in that outgrowth of digit **4** can be attributed to apical ridge between somite levels mid-19 and 19/20 and outgrowth of digit **3** to that between somite levels mid-18 and mid-19. Outgrowth for digit **2** can be attributed to ridge between levels 17/18 and mid-18. However, that digit also was deleted after removal of ridge posterior to the 17/18 to mid-18 level.

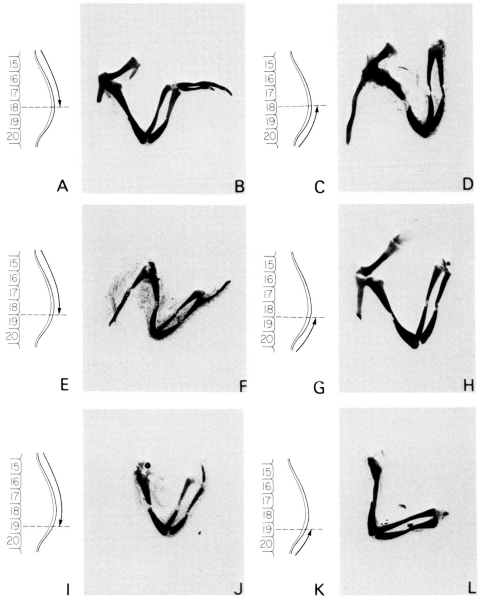

Fig. 4. Diagrams of the level of anterior and posterior ridge removals and photographs of typical resulting wings arranged so they can be compared directly. (A) Drawing of anterior ridge removal to somite level mid-18. (B) Resulting wing with only digits 3 and 4. (C) Drawing of posterior ridge removal to somite level mid-18. (D) Resulting wing with no digits. (E) Drawing of anterior ridge removal to somite level 18/19. (F) Resulting wing with only digit 4. (G) Drawing of posterior ridge removal to somite level 18/19. (H) Resulting wing with no digits. (I) Drawing of anterior ridge removal to somite level mid-19. (J) Resulting wing with only proximal part of digit 4. (K) Drawing of posterior ridge removal to somite level mid-19. (L) Resulting wing with no digits.

Table 3.

Level remaining	Number of cases	Normal	Missing elements		
			Digit **2**	Digit **3**	Digit **4**
mid-17–17/18	11	8 (73 %)	3 (27 %)		
17/18–mid-18	6	1 (17 %)	4 (67 %)	1 (17 %)	
mid-18–18/19	9	1 (11 %)	6 (67 %)	8 (89 %)	
18/19–mid-19	10		5 (50 %)	9 (90 %)	4 (40 %)
mid-19–19/20	7	2 (29 %)		2 (29 %)	5 (71 %)

Table 4.

Level remaining	Number of cases	Normal	Missing elements		
			Digit **2**	Digit **3**	Digit **4**
17/18–mid-18	12		11 (92 %)	12 (100 %)	12 (100 %)
mid-18–18/19	7		5 (71 %)	7 (100 %)	6 (86 %)
18/19–mid-19	7		7 (100 %)	2 (29 %)	7 (100 %)

Retention of half-somite length pieces of apical ectodermal ridge in the wing bud

Next, a complementary set of experiments was performed in which only a piece of apical ridge one-half somite in length was left on the limb bud. The results of these experiments are summarized in Table 4. With ridge remaining between the 17/18 and mid-18 level, digits **3** and **4** should not grow out, according to Fig. 3, but digit **2** should develop. In 92 % of cases, digit **2** was deleted. According to the map in Fig. 3, removal of all ridge except that between mid-18 and 18/19 or 18/19 and mid-19 should result in outgrowth of part of digit **3** and deletion of digits **2** and **4**. But with the mid-18 to 18/19 piece remaining, digit **3** was absent in all seven embryos. However, the expected outgrowth of digit **3** did occur after operations in which ridge between 18/19 and mid-19 was left intact. In these operations, ridge anterior to 18/19 (responsible for digit **2** and part of digit **3** outgrowth) and that posterior to mid-19 (responsible for digit **4** outgrowth) was removed and limbs developed complementary to those from which only ridge between 18/19 and mid-19 was removed. Thus, only in cases where ridge at the level of 18/19 to mid-19 was removed or left intact were complementary wings obtained. In these cases, when ridge posterior to

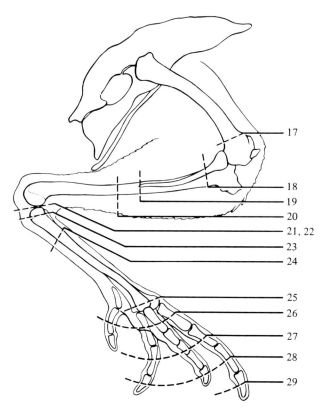

Fig. 5. Drawing of the right chick leg from a right lateral perspective. The stages at which the total ridge removals were done are listed on the right. The lines intersecting the limb represent the levels of truncation of the leg after ridge removal at the indicated stages.

the mid-19 level was removed, digit **3** developed in 71% of embryos. Thus, at levels other than that between 18/19 and mid-19, the digit deleted by removal of a half-somite piece of apical ridge could not be obtained by leaving only that piece of ridge.

Anterior and posterior apical ridge removals from the leg

The ridge of the avian wing bud is asymmetrically distributed on the limb-bud tip. The anterior ridge is low pseudostratified columnar epithelium, while posterior ridge has a high pseudostratified columnar epithelial configuration. The apical ridge of the leg bud, unlike that of the wing, but similar to that of other amniote limb buds, is symmetrically distributed at the limb-bud tip. Therefore, it seemed reasonable to question whether or not the results of posterior ridge removal in the leg would mimic those we found for the wing. In order to determine which structures in the leg are susceptible to deletion with stage-19 to-20 ridge removal, we constructed a proximodistal fate map for the

Table 5.

Level of removal	Number of cases	Results				
		Normal	Missing elements			
			Digit **1**	Digit **2**	Digit **3**	Digit **4**
		Anterior				
28/29	5	5 (100%)				
mid-29	9	9 (100%)				
29/30	10	2 (20%)	8 (80%)			
mid-30	7		7 (100%)	5 (71%)		
30/31	11		11 (100%)	11 (100%)	5 (45%)	
mid-31	8		8 (100%)	8 (100%)	8 (100%)	
31/32	5		5 (100%)	5 (100%)	5 (100%)	5 (100%)
		Posterior				
31/32	8	8 (100%)				
mid-31	12					12 (100%)
30/31	13				13 (100%)	13 (100%)
mid-30	7		1 (20%)	5 (71%)	7 (100%)	7 (100%)
29/30	8			8 (100%)	8 (100%)	8 (100%)
mid-29	5		5 (100%)	5 (100%)	5 (100%)	5 (100%)

leg, employing the technique used by Saunders (1948) and Summerbell (1974) for the wing. The results of these operations are summarized in Fig. 5.

Data from experiments in which anterior or posterior apical ridge were removed from the leg are summarized in Table 5 and typical results shown in Fig. 6, 7. As can be seen in the figures, the removals of posterior ridge from the leg did not result in deletions of structures under remaining anterior ridge, as was the case for the wing. For example, when anterior ridge was removed to the level of 30/31 (Fig. 7A), in all cases digits **1**, **2** and **3** were deleted. With reciprocal removal of posterior ridge to the level of 30/31, legs developed which were missing digit **4** (Fig. 7B). Thus, in contrast to the wing, removal of reciprocal pieces of apical ridge from the leg does result in development of complementary limb elements.

DISCUSSION

The results of experiments reported in this paper show that the apical ecto-dermal ridge of the chick wing bud anterior to the mid-19 somite level must be connected to more posterior ridge to allow outgrowth of anterior mesoderm. If posterior ridge was removed extending anteriorly to at least the mid-19 somite level, limb elements that should develop beneath remaining ridge, failed to do so. The results of half-somite-piece apical ridge removal or retention in the wing bud suggest that the ridge specifically required for anterior ridge

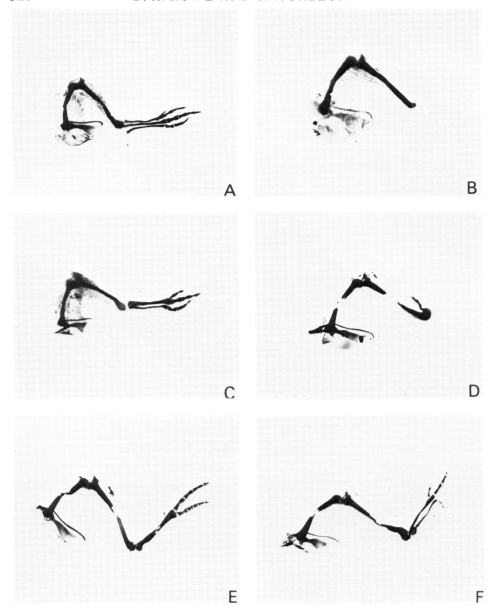

Fig. 6. Photographs of the chick legs resulting from anterior (left) and posterior (right) ridge removals from stage-20 leg buds. (A) Anterior ridge removal to somite level mid-29 – normal leg. (B) Posterior ridge removal to somite level mid-29 – no digits or metatarsals. (C) Anterior ridge removal to somite level 29/30 – digits **2**, **3** and **4** and metatarsals 2, 3 and 4. (D) Posterior ridge removal to somite level 29/30 – digit **1**. (E) Anterior ridge removal to somite level mid-30 – digits **3** and **4** and metatarsals 3 and 4. (F) Posterior ridge removal to somite level mid-30 – digits **1**, **2** and proximal **3** and metatarsals 2 and 3.

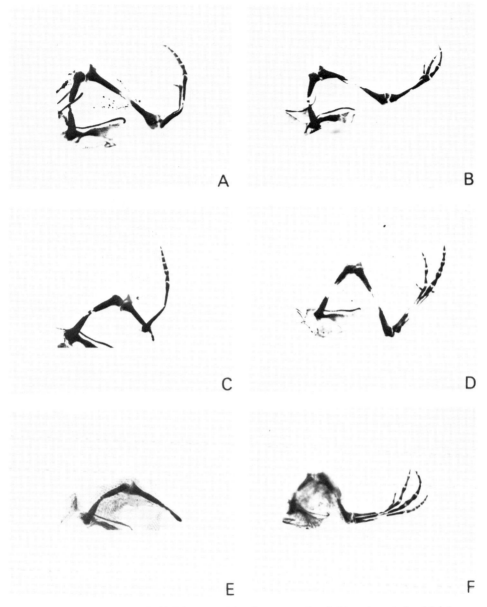

Fig. 7. Photographs of chick legs resulting from anterior (left) and posterior (right) apical ridge removals from stage 20 leg buds. (A) Anterior ridge removal to somite level 30/31 – digit **4** and metatarsals 3 and 4. (B) Posterior ridge removal to somite level 30/31 – digits **1, 2** and proximal **3** and metatarsals 2 and 3. (C) Anterior ridge removal to somite level mid-31 – digit **4** and metatarsal 4. (D) Posterior ridge removal to somite level mid-31 – digits **1, 2, 3** and metatarsals 2, 3 and 4. (E) Anterior ridge removal to somite level 31/32 – no digits or metatarsals. (F) Posterior ridge removal to somite level 31/32 – normal leg.

function may be that at the level of mid-19, the highest ridge epithelium in the wing at the stages used in these experiments. We infer that anterior ridge must be connected to posterior ridge for normal wing development.

The removal of progressively greater amounts of anterior ridge in a posterior direction allows determination of the anteroposterior section of ridge responsible for outgrowth of each limb element. Summerbell (1979) has estimated a fate map for the anteroposterior axis which correlates with that of Stark & Searls (1973). In both of these papers, determinations of which anteroposterior level of mesenchyme would become a particular limb element were made. In contrast, our experiments determine which anteroposterior level of the ridge at stages 18–19 allows outgrowth of digital elements. It is noteworthy that we find that all of the phalangeal elements of a single digit do not seem to develop from mesenchyme in the same horizontal plane with respect to the somites, unless that mesenchyme changes position with respect to overlying ridge. For example, ridge responsible for the outgrowth of the distal part of digit **4** seems to be anterior to that required for outgrowth of the more proximal part of that digit.

Unlike the wing, the leg does not seem to require the presence of the posterior ridge for the normal development of anterior structures. Anterior apical ectodermal ridge functions without the presence of the posterior ridge in the leg. Our original reason for examining whether or not the leg is the same in its requirement for posterior ridge stemmed from the fact that the leg apical ridge is more symmetrical than that of the wing bud. The ridge of the wing bud is higher posteriorly, while that of the leg is high both anteriorly and posteriorly. Whether or not the difference in ridge symmetry can be correlated with the difference in requirement for posterior ridge between wing and leg buds, cannot be determined from the present experiments. Although more anterior ridge is in the high configuration in the leg, as compared to the wing, it seems to contribute little to the formation of the anterior cartilage elements of the leg. At stage 20, ridge can be removed from the anterior junction with the body wall to the midpoint of the leg bud, before digit **1** is deleted. Determination of the factors responsible for the difference in posterior ridge requirement between wing and leg buds will require further examination.

Upon removal of pieces of apical ridge from the wing bud, we obtained results similar to those of Summerbell (1979) after barrier insertion. According to the map in Fig. 3, digits which were expected to develop beneath ridge remaining anterior to the level of removal failed to develop, as they did when a barrier was inserted at the same anteroposterior level. Our data do not permit a conclusion as to whether or not the polarizing zone or its morphogen are involved. Experiments in which dorsal ectoderm immediately proximal to the ridge was removed (unpublished observations) and experiments in which the wing bud was denuded of all ectoderm except the ridge (Zwilling, 1956; Gasseling & Saunders, 1961) give evidence that the deletions occurring after posterior ridge removal are not the result of leakage of a diffusible morphogen through the wound.

The simplest explanation for the results of the experiments reported in this paper is that in order for anterior apical ridge to function, it must be connected to posterior ridge. In support of this, we point out that numerous, very large gap junctions are characteristic of the apical ridge (Fallon & Kelley, 1977). Further, it has been demonstrated that ridge cells transport Lucifer yellow, which is evidence for coupling among ridge cells (Fallon & Sheridan, in preparation). Thus, it may be that ionic or metabolic coupling of posterior to anterior ridge cells is required for anterior ridge function. Aside from this possibility of intercellular coupling, it is also possible that something essential to anterior ridge function emanates from posterior ridge. This reasoning is complicated by the possibility that polarizing zone or its diffusible morphogen may act through the apical ridge (Tickle, 1980). However, there is no direct evidence for this proposal and we do not favour it. Rather, the experiments reported in this paper may demonstrate a property of the ridge of the wing bud which is not related to the polarizing zone directly.

It is pertinent that after removal of the entire posterior half of the wing bud, investigators (Warren, 1934; Amprino & Camosso, 1955) have shown that the anterior half of the wing bud fails to develop. This result is comparable to that obtained with barrier insertion and posterior ridge removal. Further, Hinchliffe & Gumpel-Pinot (1981) have shown that removal of the entire posterior half of the wing bud results in extensive cell death of the remaining anterior mesoderm. This is also true for the leg bud (Fallon & Rowe, unpublished). In the light of chick wing-bud fate maps, analysis of Warren's data leads to the conclusion that anterior structures already determined at the time of removal of the posterior half of the wing bud are absent (e.g., Warren's fig. 22). This is consistent with the observation of massive cell death of anterior mesoderm after posterior wing-bud removal. In contrast, we do not find mesodermal cell death under remaining anterior ridge following removal of posterior ridge (Rowe, Cairns & Fallon, 1980 and in preparation). It is likely than that anterior deletions seen after partial ridge removal and after posterior-half wing-bud removal demonstrate at least two different phenomena. In the former experiment, anterior ridge does not appear to function without continuity with posterior ridge. This is also likely to be true in the latter case, but cell death is an added effect of the removal of posterior mesoderm.

This would seem to implicate the polarizing zone. Simple removal of the polarizing zone can result in normal limb development (MacCabe *et al.* 1973; Fallon & Crosby, 1975). It is probable, however, that the polarizing zone morphogen remains in more anterior tissue for a period after removal of its source, the posterior border mesoderm. In fact, MaCabe & Parker (1976*b*) have shown a gradient of morphogenetic activity with an area of activity at the 'anteroposterior center' of the wing bud, intermediate to the high posterior and null anterior activities. This middle area would remain after removal of polarizing zone alone, but all activity (that from source and middle morphogen containing area) are removed from influencing the anterior mesoderm when the entire posterior half of the bud is excised. This removal from posterior mesodern influence would also occur when a piece of anterior tissue is isolated from the rest of the wing bud, as in MacCabe & Parker's (1975) *in vitro* system. Cell death does occur when anterior mesoderm with its apical ectoderm is grown in culture. However, if polarizing zone is put in the same culture with isolated anterior wing-bud mesoderm, cell death does not occur. Thus, in experimental systems where anterior mesoderm is removed from posterior mesoderm, extensive cell death is observed in anterior wing-bud tissues. It would seem that the posterior-half limb-bud removal and the posterior ridge removal experiments are different. Posterior ridge removal does not result in cell death in anterior wing-bud mesoderm. In contrast, removal of the entire posterior half of the wing bud results in extensive cell death in anterior mesoderm. It is not known whether the barrier experiments done by Summerbell (1979) were accompanied by anterior mesodermal cell death.

We still feel that there is no compelling evidence for a role for polarizing zone morphogen in the specification of the anteroposterior polarity of limb elements during the limb-bud stages of normal development. Consistent with this, we have proposed that the production of morphogen is residual during limb-bud stages (Fallon & Crosby, 1975, 1977). However, considering work in this laboratory and data from the laboratories of MacCabe and Hinchliffe, it is reasonable to assume that factors from the polarizing zone are required for the survival of anterior mesoderm cells during the limb-bud stages of normal development.

This investigation was supported by NSF Grant No. PCM7903980 and NIH Grant No. T32HD7118. We are grateful to Drs Allen W. Clark and David B. Slautterback, and Ms Eugenie Boutin for their constructive criticism of this manuscript. Speicial thanks are due to Ms B. Kay Simandl for technical assistance. We thank Ms Lucy Taylor for making the drawings and Ms Julie Michels and Ms Sue Leonard for typing the manuscript.

REFERENCES

AMPRINO, R. E. & CAMOSSO, M. (1955). Ricerche sperimentali sulla morfogenesi degli arti nel polo. *J. exp. Zool.* **129**, 453–493.

FALLON, J. F. & CROSBY, G. M. (1975). Normal development of the chick wing following removal of the polarizing zone. *J. exp. Zool.* **193**, 449–455.

FALLON, J. F. & CROSBY, G. M. (1977). Polarizing zone activity in the limb buds of amniotes. *Vertebrate Limb and Somite Morphogenesis* (ed. D. A. Ede, J. R. Hinchliffe & M. Balls), pp. 55–71. Cambridge: Cambridge University Press.

FALLON, J. F. & KELLEY, R. O. (1977). Ultrastructural analysis of the apical ectodermal ridge during vertebrate limb morphogenesis. II. Gap junctions as distinctive ridge structures common to birds and mammals. *J. Embryol. exp. Morph.* **41**, 223–232.

GASSELING, M. T. & SAUNDERS, J. W., JR. (1961). Effects of the apical ectodermal ridge on growth of the versene-stripped chick limb bud. *Devl Biol.* **3**, 1–25.

HAMBURGER, V. & HAMILTON, H. (1951). A series of normal stages in the development of the chick embryo. *J. Morph.* **88**, 49–92.

HINCHLIFFE, J. R. & GUMPEL-PINOT, M. (1981). Control of maintenance and anteroposterior skeletal differentiation of the anterior mesenchyme of the chick wing bud by its posterior margin (the ZPA). *J. Embryol. exp. Morph.* **62**, 63–82.

ITEN, L. E. & MURPHY, D. J. (1980*a*). Supernumerary limb structures with regenerated posterior chick wing bud tissue. *J. exp. Zool.* **213**, 327–335.

ITEN, L. E. & MURPHY, D. J. (1980*b*). Pattern regulation in the embryonic chick limb: supernumerary limb formation with anterior (non-ZPA) limb bud tissue. *Devl Biol.* **75**, 373–385.

MACCABE, A. B., GASSELING, M. T. & SAUNDERS, J. W., JR. (1973). Spatiotemporal distribution of mechanisms that control outgrowth and anteroposterior polarization of the limb bud in the chick embryo. *Mech. Aging Develop.* **2**, 1–12.

MACCABE, J. A. & PARKER, B. W. (1975). The *in vitro* maintenance of the apical ectodermal ridge of the chick embryo wing bud: an assay for polarizing activity. *Devl Biol.* **45**, 349–357.

MACCABE, J. A. & PARKER, B. W. (1976*a*). Polarizing activity in the developing limb of the Syrian hamster. *J. exp. Zool.* **195**, 311–317.

MACCABE, J. A. & PARKER, B. W. (1976*b*). Evidence for a gradient of a morphogenetic substance in the developing limb. *Devl Biol.* **54**, 297–303.

ROWE, D. A., CAIRNS, J. M. & FALLON, J. F. (1981). Spatial and temporal patterns of cell death in the limb bud mesoderm after apical ridge removal. *Anat. Rec.* **199**, 218A.

SAUNDERS, J. W., JR. (1948). The proximodistal sequence of origin of the parts of the chick wing and the role of the ectoderm. *J. exp. Zool.* **108**, 363–404.

SAUNDERS, J. W., JR. & GASSELING, M. T. (1968). Ectodermal-mesodermal interactions in the origin of limb symmetry. *Epithelial-Mesenchymal Interactions* (ed. R. Fleischmajer & R. Billingham), pp. 78–97. Baltimore: Williams & Wilkins.

SAUNDERS, J. W., JR. (1977). The experimental analysis of chick limb bud development. *Vertebrate Limb and Somite Morphogenesis* (ed. D. A. Ede, J. R. Hinchliffe & M. Balls), pp. 1–24. Cambridge: Cambridge University Press.

STARK, R. J. & SEARLS, R. L. (1973). A description of chick wing bud development and a model of limb morphogenesis. *Devl Biol.* **33**, 138–153.

SUMMERBELL, D. (1974). A quantitative analysis of the effect of excision of the AER from the chick limb bud. *J. Embryol. exp. Morph.* **32**, 651–660.

SUMMERBELL, D. (1979). The zone of polarizing activity: evidence for a role in normal chick limb morphogenesis. *J. Embryol. exp. Morph.* **50**, 217–233.

THOMS, S. D. & FALLON, J. F. (1980). Pattern regulation and the origin of extra parts following axial misalignments in the urodele limb bud. *J. Embryol. exp. Morph.* **60**, 33–55.

TICKLE, C., SHELLSWELL, G., CRAWLEY, A. & WOLPERT, L. (1976). Positional signalling by mouse limb polarizing region in the chick wing bud. *Nature, Lond.* **259**, 396–397.

TICKLE, C. (1980). The polarizing region and limb development. In *Development in Mammals* (ed. M. H. Johnson) pp. 101–136. Amsterdam: Elsevier.

WARREN, A. E. (1934). Experimental studies on the development of the wing in the embryo of *Gallus domesticus*. *Am. J. Anat.* **54**, 449–485.

ZWILLING, E. (1956). Interaction between limb and ectoderm mesoderm in the chick embryo. II. Experimental limb duplication. *J. exp. Zool.* **132**, 173–187.

ZWILLING, E. (1959). A modified chorioallantoic grafting procedure. *Transplant. Bull.* **6**, 238–247.

Index of Authors

Date Due